Dam Safety. Overtopping and Geostructural Risks

Dam Safety. Overtopping and Geostructural Risks

Editors

Miguel Á. Toledo
Rafael Morán

MDPI • Basel • Beijing • Wuhan • Barcelona • Belgrade • Manchester • Tokyo • Cluj • Tianjin

Editors
Miguel Á. Toledo
Universidad Politécnica de Madrid
Spain

Rafael Morán
Universidad Politécnica de Madrid
Spain

Editorial Office
MDPI
St. Alban-Anlage 66
4052 Basel, Switzerland

This is a reprint of articles from the Special Issue published online in the open access journal *Water* (ISSN 2073-4441) (available at: https://www.mdpi.com/journal/water/special_issues/dam_safety).

For citation purposes, cite each article independently as indicated on the article page online and as indicated below:

LastName, A.A.; LastName, B.B.; LastName, C.C. Article Title. *Journal Name* **Year**, *Volume Number*, Page Range.

ISBN 978-3-0365-5655-0 (Hbk)
ISBN 978-3-0365-5656-7 (PDF)

Cover image courtesy of Prof. Dr. Miguel Á. Toledo and Prof. Dr. Rafael Morán

Contents

About the Editors . . . vii

Preface to "Dam Safety. Overtopping and Geostructural Risks" . . . ix

Miguel Á. Toledo and Rafael Moran
Dam Safety-Overtopping and Geostructural Risks
Reprinted from: *Water* **2022**, *14*, 2826, doi:10.3390/w14182826 . . . 1

Ricardo Monteiro-Alves, Miguel Á. Toledo, Rafael Moran and Luis Balairón
Failure of the Downstream Shoulder of Rockfill Dams Due to Overtopping or Throughflow
Reprinted from: *Water* **2022**, *14*, 1624, doi:10.3390/w14101624 . . . 5

Juan Carlos López, Miguel Ángel Toledo and Rafael Moran
A Unified View of Nonlinear Resistance Formulas for Seepage Flow in Coarse Granular Media
Reprinted from: *Water* **2021**, *13*, 1967, doi:10.3390/w13141967 . . . 33

Nils Solheim Smith, Ganesh H. R. Ravindra and Fjóla Guðrún Sigtryggsdóttir
Numerical Modeling of the Effects of Toe Configuration on Throughflow in Rockfill Dams
Reprinted from: *Water* **2021**, *13*, 1726, doi:10.3390/w13131726 . . . 57

Francisco Javier Caballero, Miguel Ángel Toledo, Rafael Moran and Javier San Mauro
Hydrodynamic Performance and Design Evolution of Wedge-Shaped Blocks for Dam Protection against Overtopping
Reprinted from: *Water* **2021**, *13*, 1665, doi:10.3390/w13121665 . . . 77

Rafael Moran, Miguel Ángel Toledo, Javier Peraita and Raffaella Pellegrino
Energy Dissipation in Stilling Basins with Side Jets from Highly Convergent Chutes
Reprinted from: *Water* **2021**, *13*, 1343, doi:10.3390/w13101343 . . . 109

Patricia Alocén, Miguel Á. Fernández-Centeno and Miguel Á. Toledo
Prediction of Concrete Dam Deformation through the Combination of Machine Learning Models
Reprinted from: *Water* **2022**, *14*, 1133, doi:10.3390/w14071133 . . . 127

Chen Wang, Hanyun Zhang, Yunjuan Zhang, Lina Guo, Yingjie Wang and Thiri Thon Thira Htun
Influences on the Seismic Response of a Gravity Dam with Different Foundation and Reservoir Modeling Assumptions
Reprinted from: *Water* **2021**, *13*, 3072, doi:10.3390/w13213072 . . . 147

Raffaella Pellegrino, Miguel Ángel Toledo and Víctor Aragoncillo
Discharge Flow Rate for the Initiation of Jet Flow in Sky-Jump Spillways
Reprinted from: *Water* **2020**, *12*, 1814, doi:10.3390/w12061814 . . . 169

Vera M. van Bergeijk, Vincent A. Verdonk, Jord J. Warmink and Suzanne J. M. H. Hulscher
The Cross-Dike Failure Probability by Wave Overtopping over Grass-Covered and Damaged Dikes
Reprinted from: *Water* **2021**, *13*, 690, doi:10.3390/w13050690 . . . 195

Nicola Rossi, Mario Bačić, Meho Saša Kovačević and Lovorka Librić
Fragility Curves for Slope Stability of Geogrid Reinforced River Levees
Reprinted from: *Water* **2021**, *13*, 2615, doi:10.3390/w13192615 . . . 217

Nicola Rossi, Mario Bačić, Meho Saša Kovačević and Lovorka Librić
Development of Fragility Curves for Piping and Slope Stability of River Levees
Reprinted from: *Water* **2021**, *13*, 738, doi:10.3390/w13050738 . . . . **239**

About the Editors

Miguel Á. Toledo

Prof. Dr. Miguel Á. Toledo coordinates the Teaching Unit "Dams" and the Research Group "SERPA" at the Department of Civil Engineering: Hydraulics, Energy and Environment of the Universidad Politécnica de Madrid. He has been involved in many dam projects and studies. His research interests are focused on two main research lines related to dam safety: understanding and evaluation of the danger and protection of dams against overtopping, and the assessment of dam safety through the analysis of monitoring data by means of machine learning techniques. He is currently involved in research projects on the automatic detection of arch dam cracking and the failure process of embankment dams made up of widely graded materials.

Rafael Morán

Prof. Dr. Rafael Morán has taught "Dams" and "Hydraulic Infrastructures" at UPM since 2009. Previously, he worked for 9 years as a professional engineer and specialized in the design and construction of dams. During the last 15 years, his research work has focused on dam safety in the fields of spillway hydraulics and protections against the overtopping of dams. He is currently a member of the "European Working Group on Overtopping Erosion" of the European Club of the International Committee of Large Dams and co-leads the subgroup of Dam Protections against Overtopping. Prof. Dr. Morán is the co-founder of "ACIS innovación + ingeniería", a spin-off company of UPM.

Preface to "Dam Safety. Overtopping and Geostructural Risks"

Today, the main concerns regarding dam safety and recent improvements are focused on the threat of overtopping and the need to develop tools for early detection and prevention of geostructural risks. The purpose of this book is to promote and gather recent advances in those two key areas of dam safety.

The scientific approach to the aforementioned problems is mandatory for solving them in a reliable and efficient way that makes informed decision-making possible. This text is intended to support scientists working in the area of dam safety, but also dam owners and practitioners who deal with the dam safety problems of real dams that must be efficiently solved with always limited economic and personal resources.

The authors contributing to this book are experts that come from quite varying knowledge areas, such as hydraulics of granular materials, dam overtopping, machine learning, and risk analysis, among others, to give us a wide and rich view of the dam safety topics touched on here. Our acknowledgements go to all of them for their high-quality contributions, and also to all the funders, institutions, and collaborators that made the work of the authors possible.

Miguel Á. Toledo and Rafael Morán
Editors

water

Editorial

Dam Safety-Overtopping and Geostructural Risks

Miguel Á. Toledo [1,*] and Rafael Moran [1,2]

[1] Department of Civil Engineering: Hydraulics, Energy and Environment, Universidad Politécnica de Madrid, 28040 Madrid, Spain
[2] Centre Internacional de Metodes Numerics en Enginyeria, Campus Norte UPC, Gran Capitán s/n, 08034 Barcelona, Spain
* Correspondence: miguelangel.toledo@upm.es

There is a growing concern about the safety of dams and dikes in modern society. The new regulations demand an increasing level of safety. Therefore, the technical community related to dams is making an effort to understand the failure mechanisms that threaten dam safety, prioritize actions with informed criteria, and develop more efficient solutions to heighten the safety of new and existing dams with the limited resources available.

The safety of dams and dikes involves multiple issues such as the understanding of the behavior, even in extreme conditions near failure; proper design, construction, and exploitation; and also logical and efficient management and assignment of available economical and personal resources.

The overtopping of embankment dams has been the main cause of dam failure in the last decades. Consequently, the main challenge related to dam safety is to find efficient solutions to quantify risk and avoid the failure of new and existing dams due to overtopping. Hydrological safety is one of the most active areas in dam engineering research, involving a considerable technical community all over the world.

The rest of the failure mechanisms (internal erosion, sliding, concrete cracking, etc.), related to dam behavior and geostructural safety usually affect parameters such as seepage, movements, or interstitial pressures that can be controlled by means of visual inspection and the analysis of the data provided by the monitoring system.

Predictive models are developed with the purpose of detecting anomalous dam behavior that could potentially be a symptom of the onset of an incident or dam failure. Physically based models, such as finite element models, or data models are useful for that. Both types of models have experienced enormous development in the last years and today a huge effort is being made to enhance their prediction accuracy.

Data models were traditionally based on multiple linear regression (HST model and a long list of models derived from that). Machine learning and artificial intelligence techniques are now being investigated to develop models that more closely adapt to the complexity of the dam-foundation system. Better accuracy and more profound understanding is being achieved by these methods, previously developed in different complex areas of knowledge such as sociology or the Internet.

In the field of dam safety management, risk analysis has also experienced a rapid dissemination. A lot of effort is concentrated on developing fragility curves for the different types of failure mechanisms, which are essential for a rigorous application of risk analysis.

A deep understanding of the physical processes involved in dam failure mechanisms is essential for modeling the behavior of dams and dikes in extreme situations, close to catastrophic failure, and the definition of reliable fragility curves.

The Special Issue *Dam Safety. Overtopping and Geostructural Risks* covers recent advances in the understanding and improvement of hydrological and geostructural dam safety related to the abovementioned subjects. It includes eleven papers (10 research papers and 1 review paper; 2 of them are Feature papers and 1 is Editor's Choice).

Citation: Toledo, M.Á.; Moran, R. Dam Safety-Overtopping and Geostructural Risks. *Water* **2022**, *14*, 2826. https://doi.org/10.3390/w14182826

Received: 7 September 2022
Accepted: 8 September 2022
Published: 11 September 2022

The reader will find five papers related to overtopping; two of them are about understanding of dam failure and related actions, and three papers are about the protection of embankment and concrete dams. Three papers deal with geostructural safety, two of them about the use of data and physically based models, and one more related to sky jumps. Estimation of failure probability and development of fragility curves correspondent to overtopping and different geostructural risks is the subject of another three papers.

Understanding the failure process is the necessary first step for the assessment of the safety of dams against overtopping. This process strongly depends on the type of dam and its particular characteristics. Most of the research effort has been dedicated in the past to understanding of the failure of clayey homogeneous embankment dams, and several codes with empirical basis were developed to model the failure process of this type of dam.

Based on extensive experimental work, Monteiro-Alves et al. [1] analyzed the failure process of rockfill dams in the case that the rockfill of the shoulders is highly permeable. They provided a description of the failure process and a formula for the estimation of the unit overtopping discharge that causes the complete failure of the shoulder, which can be expected to be quite similar to the one that leads to the catastrophic dam failure.

In this type of rockfill dam, the failure process and the critical unit discharge of the overtopping highly depend on the permeability of the rockfill. Therefore, the correct assessment of the permeability is relevant for modeling the failure. In coarse granular materials, the relation between seepage velocity and hydraulic gradient, the resistance formula of the flow, is not linear, as is usual in fine materials, but parabolic. López et al. [2] collected a wide variety of resistance formulas available and compiled them under a unified view in such a way that every formula is a particular case of the general formulation.

Three contributions to this special issue refer to different solutions to protect dams against overtopping [3–5]. A rockfill toe can be added to the downstream shoulder of a rockfill dam to increase the resistance in the event of overflow. Several geometrical configurations are possible for that toe, from completely external to completely internal to the dam shoulder. Smith et al. [3] analyzed, by a combination of physical and numerical modeling, the effect of the rockfill toe configuration on the throughflow. It is quite relevant for the safety of the dam, considering that the failure strongly depends on the position of the water surface inside the dam. They conclude that "the internal and combined toe configurations are effective in lowering the phreatic line within the dam, for enhanced slope stability compared to the cases without a toe or an external toe" [3] (p. 18).

Different techniques are available to protect the downstream surface of an earth or rockfill dam to avoid erosion or unraveling in an overtopping scenario. Wedge-Shaped Blocks (WSBs) have a long history of development in different countries. Using a testing facility specifically designed for this purpose, Caballero et al. [4] presented a new enhanced WSB and provide a considerable quantity of data on the pressures on the faces of the blocks, also the base and the riser, the hydrodynamic forces and percolation through the block contacts. Remarkable conclusions are obtained about the hydrodynamic performance of the new WSB, like the suctions registered at the base that attract the block towards the dam body, or the position of the highest negative pressures in the upper part of the riser, where the aeration holes are located.

Concrete dams are far less vulnerable to overtopping than embankment dams. However, the overflow can cause significant and dangerous erosion in the rock mass in the area of contact with the dam. The challenge is to evacuate the flood in a safe way, and an effective solution for that is to use Highly Convergent Chutes that gather water flowing over the dam in the lateral areas, outside of the spillway. This type of solution was investigated by Moran et al. [5] by means of a singular testing facility that allows modeling spillways with Highly Converging Chutes of different slope and position. The research is focused on the behavior of the stilling basin and its ability to dissipate the water energy. Especially relevant is the conclusion that an existing hydraulic jump stilling basin may serve to accommodate a flood greater than the one considered during the design phase if a part of the extra flow is introduced through highly converging chutes. This makes this type of solution very attractive for solving the improvement of safety against overtopping of gravity dams and even arch dams.

Research work related to geostructural safety involves a wide variety of techniques and failure mechanisms. The application of machine learning and artificial intelligence to monitoring data for the assessment of geostructural dam safety is one of the most active research areas today. Different algorithms like Artificial Neural Networks, Support Vector Machines or Random Forest, among others, have successfully been applied for detecting behavior anomalies. It is well known that the most appropriate algorithm depends on the particular features of the data set. Alocén et al. [6] improved the precision and robustness of the data models by using several machine learning techniques (experts). They compared the Stacking and Blending strategies for combining algorithms and conclude that Stacking provides more accurate predictions than Blending. A complete methodology is proposed to combine experts.

Many dams are being built in locations with high seismic hazards. Although dams are structures with considerable resistance to seismic action, compared with other types of structures, such as buildings or bridges, it is necessary to confidently assess the risk of failure in such cases. Due to the difficulty in performing physical models that reproduce the effect of earthquakes, numerical models are used to evaluate dam behavior. These models are quite complex, and a considerable number of assumptions must be made. Based on extensive numerical experimentation, Wang et al. [7] analyzed the influence of some of the most significant assumptions. They conclude that the dam-foundation interaction has a significant effect in the dam response, and also that the method used to model the water-dam interaction significantly influences dam stresses, although the effect on displacements is negligible. They proposed a reservoir length three times the dam height for the numerical model.

The risk associated with rock scour downstream of ski jump spillways has drawn attention during the last years. Assessment of the impact area is required for the later evaluation of that risk. The impact area corresponding to high discharge flow rates is easy to determine because it is clear that the water jumps in the direction determined by the angle of the ski jump. Pellegrino et al. [8] presented an experimentally verified formula and a complete method to estimate the minimum discharge that causes the jump to occur. For a discharge under that minimum value, water falls at the ski jump toe, which must be protected. With the calculation of that minimum value and the maximum corresponding to the design flood, the impact area can be assessed and proceed to estimate the risk associated with the rock scour using one of the methodologies available.

Three papers are directly related to the development of fragility curves that are necessary for the risk assessment [9–11]. The fragility curves represent the probability of failure, or a certain level of damage, depending on a parameter that defines the intensity of the main action that might cause the failure. Fragility curves corresponding to the different potential failure mechanisms are needed to quantify the global risk of a dam.

Van Bergeijk et al. [9] developed fragility curves for dikes covered with grass subject to wave overtopping. The dike is considered failed when erosion reaches a depth of 20 cm, which is approximately the length of grass roots. They observed a very significant influence of grass cover, as compared with a clayey surface, and also its quality, on the probability of failure. They included the effect of transitions and damages.

Fragility curves for river levees reinforced with geogrid, related to slope stability, were developed by Rossi et al. [10]. They concluded that the global uncertainty is much more conditioned by the uncertainty of the friction angle value of the dam material than that associated with the geogrid material. However, the contribution of the geogrid layer located at the highest level is greater than that of the rest layers. Based on a probabilistic characterization of soil properties and the use of numerical modeling, the same authors [11] developed fragility curves for piping and slope stability of river levees and applied them to the river Drava levee. Probability of failure depends on the water level in the river, including overflow scenarios.

An idea is common to the three papers dealing with fragility curves: conservative assumptions were made to cover the lack of knowledge that led to a conservative estimation of failure probability. Therefore, it is expected that the increase of basic knowledge in relation to the different failure mechanisms will permit us in the future to develop more accurate fragility curves. It will have the consequence of lowering the cost of corrective measures adopted to improve the safety level when needed.

As highlighted here, research is very active to understand overtopping and develop efficient solutions for protecting dams and levees against overtopping, and also to understand and evaluate the geostructural risks associated with different failure mechanisms. This research will support informed decisions to efficiently assess and guarantee the demanded safety of our dams.

Funding: This research received no external funding.

Acknowledgments: As Guest Editors, we want to thank the authors for their contribution to this Special Issue. Our thanks also go to the editors and referees, who contributed to improve the quality of the eleven published papers.

Conflicts of Interest: The authors declare no conflict of interest.

References

1. Monteiro-Alves, R.; Toledo, M.Á.; Moran, R.; Balairón, L. Failure of the downstream shoulder of Rockfill Dams Due to Overtopping or throughflow. *Water* **2022**, *14*, 1624. [CrossRef]
2. López, J.C.; Toledo, M.Á.; Moran, R. A Unified View of Nonlinear Resistance Formulas for Seepage Flow in Coarse Granular Media. *Water* **2021**, *13*, 1967. [CrossRef]
3. Smith, N.S.; Ravindra, G.H.R.; Sigtryggsdóttir, F.G. Numerical Modeling of the Effects of Toe Configuration on Throughflow in Rockfill Dams. *Water* **2021**, *13*, 1726. [CrossRef]
4. Caballero, F.J.; Toledo, M.Á.; Moran, R.; San Mauro, J. Hydrodynamic Performance and Design Evolution of Wedge-Shaped Blocks for Dam Protection against Overtopping. *Water* **2021**, *13*, 1665. [CrossRef]
5. Moran, R.; Toledo, M.Á.; Peraita, J.; Pellegrino, R. Energy dissipation in Stilling Basins with Side Jets from highly convergent cites. *Water* **2021**, *13*, 1343. [CrossRef]
6. Alocén, P.; Fernández-Centeno, M.; Toledo, M.Á. Prediction of concrete dam deformation through the Combination of machine learning models. *Water* **2022**, *14*, 1133. [CrossRef]
7. Wang, C.; Zhang, H.; Zhang, Y.; Guo, L.; Wang, Y.; Htun, T.T.T. Influences on the Seismic Response of a Gravity Dam with Different Foundation and Reservoir Modeling Assumptions. *Water* **2021**, *13*, 3072. [CrossRef]
8. Pellegrino, R.; Toledo, M.; Aragoncillo, V. Discharge flow rate for the Initiation of Jet Flow in Sky-Jump Spillways. *Water* **2020**, *12*, 1814. [CrossRef]
9. van Bergeijk, V.M.; Verdonk, V.A.; Warmink, J.J.; Hulscher, S.J.M.H. The Probability of Cross-Dike Failure by Wave Overtopping over Grass-Covered and Damaged Dikes. *Water* **2021**, *13*, 690. [CrossRef]
10. Rossi, N.; Bai, M.; Kovačević, M.S.; Librić, L. Fragility Curves for Slope Stability of Geogrid Reinforced River Levees. *Water* **2021**, *13*, 2615. [CrossRef]
11. Rossi, N.; Bačić, M.; Kovačević, M.S.; Librić, L. Development of Fragility Curves for Piping and Slope Stability of River Levees. *Water* **2021**, *13*, 738. [CrossRef]

water

MDPI

Article

Failure of the Downstream Shoulder of Rockfill Dams Due to Overtopping or Throughflow

Ricardo Monteiro-Alves [1,*], Miguel Á. Toledo [1], Rafael Moran [1,2,*] and Luis Balairón [3]

1 Civil Engineering Department: Hydraulics, Energy, and Environment, E.T.S. de Ingenieros de Caminos, Canales y Puertos, Universidad Politécnica de Madrid, 28040 Madrid, Spain; miguelangel.toledo@upm.es
2 International Centre for Numerical Methods in Engineering, Universitat Politècnica de Catalunya, Campus Norte, 08034 Barcelona, Spain
3 Hydraulics Laboratory, Centro de Estudios Hidrográficos, Centro de Estudios y Experimentación de Obras Públicas (CEDEX), 28005 Madrid, Spain; luis.balairon@cedex.es
* Correspondence: ricardo.monteiro@upm.es (R.M.-A.); r.moran@upm.es (R.M.)

Abstract: This paper presents the results of an extensive laboratory set of tests aimed to study the failure of the downstream shoulder of highly permeable rockfill subjected to overflow. The experimental research comprised testing 114 physical models by varying the following elements: (i) the median size of the uniform gravels (7 to 45 mm); (ii) the configuration of the dam, i.e., upstream and downstream shoulders and crest or just the downstream shoulder; (iii) the dam height (from 0.2 to 1 m), (iv) the crest length (from 0.4 to 2.5 m), (v) the downstream slope (from 1 to 3.5 H:V), (vi) the type of impervious element (i.e., central core, upstream face, and no impervious element). The tests allowed us to identify two failure mechanisms, slumping and particle dragging. In addition, the downstream slope was observed to be one of the most important variables in this parametric study, as it influenced the pore water pressures inside the dam, the failure discharge, and the occurrence of one or the other mechanism of failure.

Keywords: rockfill dam; overtopping; dam failure; overflow; dam safety; floods; dam breach

Citation: Monteiro-Alves, R.; Toledo, M.Á.; Moran, R.; Balairón, L. Failure of the Downstream Shoulder of Rockfill Dams Due to Overtopping or Throughflow. *Water* **2022**, *14*, 1624. https://doi.org/10.3390/w14101624

Academic Editor: Giuseppe Pezzinga

Received: 15 April 2022
Accepted: 12 May 2022
Published: 18 May 2022

1. Introduction

Rockfills may be formed by natural processes or as a direct result of human action in civil engineering structures. Examples of natural rockfills are moraine dams [1] and landslide or avalanche dams [2–4]. On the other hand, constructed rockfill structures include levees, dikes, and dams built to fulfill different human needs, and embankment-like deposits of homogeneous coarse rockfill, usually produced by mining activities, also referred to as rock drains [5].

The two main causes of failure of large rockfill dams registered up to 1986, excluding dams constructed in Japan pre-1930 and in China, are overtopping (55.6% of cases corresponding to 5 failures) and piping (11.1% of cases corresponding to 1 failure) [6]. Due to the high permeability of clean rockfill, both overtopping and piping lead to the formation of a seepage profile at the base of the downstream shoulder [7–18], that finally exits the dam at the toe [5,9,19–22]. In the toe, delimited upstream by the first emergence point [10,20], the hydraulic gradients and seepage forces are maximum and, besides that, point outward of the dam [22], making this area prone to erosion and a zone of primary engineering concern [5,19]. As a consequence, failure initiates at the toe for a discharge that must overcome a given threshold [3,23–26] and may occur by slumping, internal migration of particles, or surface unraveling erosion resulting in concentrated flow paths [22,26–30].

In natural rockfill dams, formed without any impervious element, the prediction of the final breach geometry and dimensions is crucial for the estimation of the peak outflow. Diverse studies have been performed in this area for both cohesive and non-cohesive materials [31–38], as well as literature reviews [39–41]. In dams constructed with an impervious element, prediction of the breach geometry and dimensions is also important but, in this case, the stability of this element (internal core or upstream face), which becomes unprotected after the removal of the downstream shoulder, must also be analyzed. In these cases, the failure of the impervious element controls the breach hydrograph [42–47]. Either way, knowing the discharge that completes the failure of the downstream rockfill shoulder, i.e., the 'failure discharge' (Q_f) whereby damages to the downstream shoulder reach the crest, is relevant to understanding how far the dam is from a catastrophic failure.

Failure progression, patterns, and mechanisms are affected by the dam's geometric characteristics and material gradings [3]. Based on an extensive laboratory test campaign, this paper provides a parametric analysis to understand how the failure discharge is affected by some characteristics of the dam body. Based on the results of the parametric analysis, an empirical formulation is calibrated to estimate the 'failure discharge'.

2. Methodology

2.1. Test Overview

A total of 114 physical models (PM), all tested in horizontal flumes with rectangular sections, are summarized in Table 1. They were tested by varying the following elements: (i) the size of the uniform gravel, characterized by its D50; (ii) the configuration of the cross-section, using partial (PPM) or complete physical models (CPM) (complete configurations are trapezoidal and include both upstream (USS) and downstream shoulders (DSS) and crest, while partial configurations are triangular and include only the downstream shoulder); (iii) geometrical parameters as the height (H) of the physical models, the width of the flume (W), the width of the crest (l_c), the downstream and upstream slopes (Z_{dss} and Z_{uss}, respectively); (iv) the type of impervious element (IE) (central core (CC), upstream face (UF), and no impervious element (NIE)).

Table 1. Summary of the 114 physical models tested to study the failure of the rockfill downstream shoulder (values in parentheses represent the number of models tested for that particular configuration). NA means 'not available'.

H (m)	W (m)	Z_{dss} (H:V)	Z_{uss} (H:V)	l_c (m)	IE	Material
0.229 (4 models)	0.4 (4/4)	1.5 (1/4) 2.5 (1/4) 3.5(2/4)	1.5 (4/4)	0.057 (4/4)	UF (4/4)	M3 (4/4)
0.5 (44 models)	0.4 (2/44) 0.6 (25/44) 1.32 (11/44) 2.46 (6/44)	1.0 (2/44); 1.1 (1/44) 1.3 (1/44); 1.4 (1/44) 1.5 (5/44); 1.6 (2/44) 1.75 (3/44); 1.9 (1/44) 1.95 (1/44); 2.0 (4/44) 2.1 (2/44); 2.2 (3/44) 2.25 (1/44); 2.3 (1/44) 2.4 (1/44); 2.5 (1/44) 2.6 (2/44); 2.7 (2/44) 2.75 (1/44); 2.8 (2/44) 2.9 (2/44); 3.0 (3/44) 3.1 (1/44); 3.3 (1/44)	NA (42/44) 1.5 (2/44)	NA (42/44) 0.1 (2/44)	NIE (38/44) CC (6/44)	M4 (22/44) M6 (2/44) M7 (20/44)
0.6 (4 models)	2.5 (4/4)	1.5 (1/4) 1.75 (1/4) 2.5 (1/4) 2.7 (1/4)	1.5 (4/4)	0.2 (4/4)	NIE (4/4)	M2 (4/4)
0.8 (3 models)	1.0 (3/3)	1.5 (1/3) 2.5 (1/3) 3.5 (1/3)	1.5 (3/3)	0.2 (3/3)	UF (3/3)	M7 (3/3)

Table 1. *Cont.*

H (m)	W (m)	Z_{dss} (H:V)	Z_{uss} (H:V)	l_c (m)	IE	Material
1 (61 models)	1.0 (37/61) 1.32 (2/61) 1.5 (2/61) 2.46 (17/61) 2.5 (3/61)	1.5 (18/61) 1.6 (1/61) 1.9 (1/61) 2.2 (20/61) 2.5 (2/61) 3.0 (19/61)	NA (2/61) 1.5 (59/61)	NA (2/61) 0.2 (59/61)	NIE (34/61) CC (12/61) UF (15/61)	M1 (18/61) M4 (13/61) M5 (3/61) M6 (13/61) M7 (12/61) M8 (2/61)

All of these variables combined can be grouped into four types of physical models as shown in Figure 1. These groups are (i) complete configuration with upstream face (CPM/UF), (ii) complete configuration without impervious element (CPM/NIE), (iii) partial or complete configuration with a central core (PPM/CC), and (iv) partial configuration without impervious element (PPM/NIE).

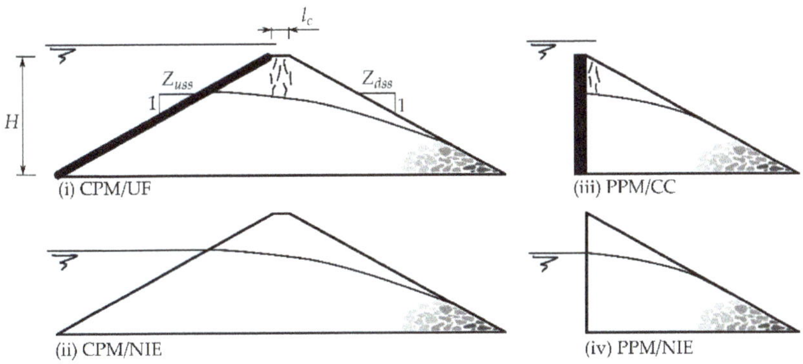

Figure 1. Schematic drawing of the different types of physical models tested at the laboratory: (**i**) Complete Physical Models with Upstream Face, (**ii**) Complete Physical Models with No Impervious Element, (**iii**) Partial Physical Models with Central Core and (**iv**) Partial Physical Models with No Impervious Element.

2.2. Specific Tests

Within the main campaign, we performed specific tests to evaluate the variability of the results and the scale effect. Regarding variability, five groups of tests were performed (from A to E), all CPM/NIE, consisting of repeating the same physical model a given number of times to assess if the test procedure could substantially affect the 'unit failure discharge' (q_f) or any other factors, such as the hydraulic pressures inside the downstream shoulder or the 'failure path', i.e., the evolution of the failure progress with the throughflow discharge [25,48]. It must be noted, though, that the discharge steps were not (in general) the same throughout the tests within the same group. Tests in the same group all had the same geometry and dimensions as well as the same granular material:

- [Group A]: $W = 2.50$ m, $H = 1$ m, $Z_{dss} = 2.2$, $Z_{uss} = 1.5$, Gravel M5 (tests 5, 6, 7).
- [Group B]: $W = 2.46$ m, $H = 1$ m, $Z_{dss} = 3.0$, $Z_{uss} = 1.5$, Gravel M7 (tests 8, 9).
- [Group C]: $W = 1.00$ m, $H = 1$ m, $Z_{dss} = 2.2$, $Z_{uss} = 1.5$, Gravel M6 (tests 81, 82, 85, 87, 88).
- [Group D]: $W = 1.00$ m, $H = 1$ m, $Z_{dss} = 1.5$, $Z_{uss} = 1.5$, Gravel M6 (tests 89, 90, 91, 92).
- [Group E]: $W = 1.00$ m, $H = 1$ m, $Z_{dss} = 3.0$, $Z_{uss} = 1.5$, Gravel M6 (tests 94, 95).

Regarding the scale effect, the aim was to analyze if the Froude similitude could be applied to scale q_f. So, for a scale factor $s_L = 1 : 3.5$, we tested 0.23 m high 'small-scale' physical models (tests nº 109, 111, and 130) and 0.8 m high 'prototypes' (tests nº 108, 110, and 112). This scale factor was applied to all lengths (except for the flume width) including the gravels' D50. So, gravels M3 and M7 were respectively used in the 'small-scale' model and 'prototype'. In total, we tested three Z_{dss} = 1.5, 2.5, and 3.5, and for each of these slopes, we tested one 'small-scale' physical model and a 'prototype'.

2.3. Test Setup and Procedure

The physical models were constructed by pouring and extending the granular material without compaction. Nevertheless, some unintentional compaction resulted from walking over the models during construction, mainly those constructed in the larger flumes involving the placement of tons of material. To obtain the final geometry, the physical model surfaces were evened with an aluminum straight guide.

Tests were based on a stepwise flow increment methodology until total failure of the downstream shoulder occurred. By total failure, we mean that the damages inflicted to the downstream shoulder reached the crest of the dam. Although the location of the upstream impervious face or the internal clay core was slightly different, this criterion permitted a homogeneous analysis of the results arising from tests with different types of impervious elements. The failure forefront (Figure 2) is the border that separates intact areas of the slope from those damaged by failure. The maximum advance of failure (B_f), i.e., the most upward point of the failure forefront, was used to define the complete failure of the shoulder. So, the physical models were defined as completely failed when B_f reached the downstream edge of the crest. For practical purposes, the discharge that produced complete failure (Q_f) was defined as being the average value between the highest discharge in which failure did not reach the crest, and the lowest in which failure surpassed it.

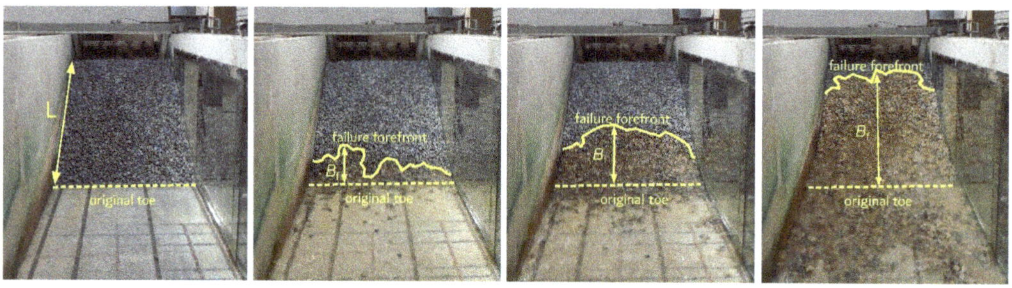

Figure 2. Scheme of the failure process sequence for discharges of 0, 0.0188, 0.0256, and 0.0423 m^3 s^{-1} (test nº 133, H = 1 m, W = 1.32 m, Z_{dss} = 1.6, gravel M7).

Each discharge was kept constant until steady-state conditions were reached, i.e., until no additional damage was observed to the shoulder or any change in the water elevation and pressures. Several long preliminary tests (more than 1 h per step) showed that a step duration of 30 min was long enough for reaching the stationary state. The number of steps varied from test to test and, following the initiation of failure (first damage observed on the shoulder), five steps were performed, on average. The minimum and the maximum number of flow steps were three and ten, respectively. Once the stationary state condition of every discharge step was reached, all measurements were performed.

2.4. Materials

Tests were performed with eight uniform limestone gravels (M1 to M8) of different sizes, ranging D50 from 0.00736 to 0.04509 m, and with a coefficient of uniformity (C_u = D60/D10) ranging from 1.46 to 2.28. Their main characteristics are summarized in Table 2 and Figure 3. These eight gravels were obtained by sieving four raw gravels with size ranges 4–12, 12–20, 20–40, and 40–80 mm. Materials M1 to M3 resulted from gravel 4–12 mm, M4 to M6 from gravel 12–20 mm, M7 from gravel 20–40 mm, and M8 from gravel 40–80 mm.

Table 2. Summary of the main geotechnical characteristics of the gravels used in the construction of the physical models. NA means 'not available'.

Raw Materials				
Variable	4–12 mm	12–20 mm	20–40 mm	40–80 mm
D10 particle size [mm]	5.27	10.18	20.05	NA
D50 particle size [mm]	8.5	15.3	26.5	NA
Coefficient of uniformity (C_u)	1.75	1.60	1.40	NA
Fine percentage (%)	0.80	0.10	0.95	NA
Specific gravity (G)	2.70	2.70	2.70	NA
Dry unit weight (γ_d) [kN·m^{-3}]	14.7	14.5	15.0	NA
Saturated unit weight (γ_d) [kN·m^{-3}]	18.9	18.5	19.0	NA
Porosity (n) [%]	42.3	41.1	41.0	NA
Void ratio (e) [%]	73.5	69.5	66.0	NA
Coefficient of permeability (k) [m·s^{-1}] †	0.0008	0.0016	0.0051	NA
Internal friction angle (φ') [degrees]	43.98	48.85	53.86	NA

Sieved Materials								
Variable	M1	M2	M3	M4	M5	M6	M7	M8
D10 particle size [mm]	4.89	5.97	5.72	8.36	11.11	10.62	23.68	21.58
D50 particle size [mm]	7.36	8.20	9.98	12.64	16.49	17.33	35.04	45.09
Coefficient of uniformity (C_u)	1.63	1.46	1.87	1.54	1.58	1.80	1.56	2.28
Specific gravity (G)	NA	NA	2.60	NA	NA	NA	2.50	2.60
Dry unit weight (γ_d) [kN·m^{-3}]	NA	NA	14.7	16.1	NA	15.7	14.5	15.0
Saturated unit weight (γ_d) [kN·m^{-3}]	NA	NA	18.9	19.9	NA	19.7	18.5	19.0
Porosity (n) [%]	NA	NA	42.6	39.3	NA	40.8	41.2	41.0
Void ratio (e) [%]	NA	NA	73.5	64.7	NA	68.9	69.5	66.0
Angle of repose (φ_{repose}) [degrees]	NA	NA	NA	36.9	NA	NA	40.4	42.8
Resistance law term a [s·m^{-1}]	NA	NA	1.44	2.71	NA	1.53	0.82	0.65
Resistance law term b [s^2·m^{-2}]	NA	NA	144.77	65.35	NA	84.66	52.82	16.96

† These values were obtained for hydraulic gradients of 0.54 (4–12 mm), 0.20 (12–20 mm), and 0.06 (20–40 mm).

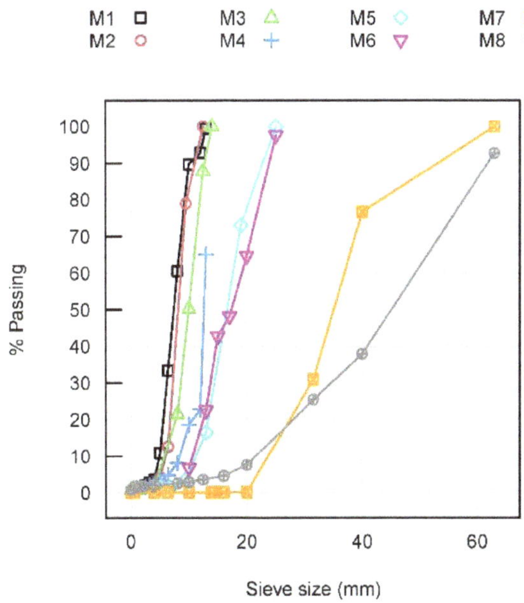

Figure 3. Particle size distribution of gravels from M1 to M8.

Some characterization tests were consigned to an external laboratory, the Geotechnics Laboratory of CEDEX, to obtain the particle size distribution (UNE EN 933-1), the specific gravity (G) of the soil solids (ASTM D5550-06), and the soil density (UNE 103301:1994) of gravels M3, M7, and M8. With these two last standards, we obtained the void ratio (e) with Equation (1), porosity (n) with Equation (2), and the saturated specific weight (γ_{sat}) with Equation (3). The particle size distribution was obtained for the rest of the materials following the same standards. The porosity of gravels M4 and M6 was obtained by filling a bucket full of gravel with water. The quadratic resistance law of flow through coarse granular materials expressed by Equation (4) relating the hydraulic gradient (i) to the flow velocity (v) was obtained for gravels M4, M7, and M8 following a methodology described in the state of the art [10,49], and for gravels M3 and M6 with a horizontal permeameter of large dimensions [50].

$$\gamma_d = \frac{G \cdot \gamma_w}{1+e} \tag{1}$$

$$n = \frac{e}{1+e} \tag{2}$$

$$\gamma_{sat} = \gamma_d + n \cdot \gamma_w \tag{3}$$

$$i = av + bv^2 \tag{4}$$

The angle of repose was obtained for gravels M4, M7, and M8 by scanning the surface of the mounds of these gravels using a 2D laser (LMS200-30106 by SICK^{TM}, scanning range for objects up to 10 m; angular range up to 180° with a maximum angular resolution of 0.25°; a systematic error of ±0.015 m; a statistical error of ±0.005 m). The repose angles were obtained by fitting a linear regression to the external surface of the mounds [45]. The Geotechnics Laboratory of CEDEX was also consigned to perform the characterization of the first three raw gravels to obtain the particle distribution (UNE 103101:1995), soil density (UNE 103301:1994), permeability (UNE 103403:1999), friction angles (UNE 103401:1998), and specific gravity of soil solids using a gas pycnometer (ASTM D5550-06).

2.5. Facilities and Instrumentation

Tests were conducted in four U-shaped flumes (rectangular section) located in two different laboratories: one flume at the Hydraulics Laboratory of the *E.T.S.I. de Caminos, Canales y Puertos* of the *Universidad Politécnica de Madrid* (UPM), and three at the Hydraulics Laboratory of the *Centro de Estudios Hidrográficos* of the *Centro de Estudios y Experimentación de Obras Públicas* (CEDEX), both laboratories located in Madrid (Spain).

The UPM flume, straight with horizontal bottom, is 13.7 m long, 2.5 m wide, and 1.3 m high (inner dimensions) with an inspection window 4.6 m long and 1.1 m high placed in the left-side wall (Figure 4a). In this flume, we tested physical models with different widths ranging from 0.6 to 2.5 m; hence, when a smaller width had to be tested, it was necessary to build a longitudinal central wall. Figure 4b,c show images of a test performed at UPM.

This flume was supplied using a constant water level tank (in which the water level was kept constant employing a pump with a variable-frequency drive) connected to the flume through a pipe 0.3 m in diameter with a manual/automated valve. This system could supply approximately 0.080 m^3s^{-1} with this valve fully opened. An extra hydraulic pump, connecting directly the underground main tank with the flume through a different pipe, 0.5 m in diameter, could supply a constant inflow of up to approximately 0.120 m^3s^{-1}. Flows were measured downstream of the physical models using a sharp rectangular weir with lateral contraction (crest length and height were 0.502 and 0.28 m, respectively) located in the 0.8 m wide flume that returned water to the underground main tank (270 m^3 capacity through an area of 180 m^2). The water level upstream of the weir was measured with a P8000 ultrasonic sensor with a digital display (Dr. D. Wehrhahn, Hannover, Germany) measuring between 0.07 and 2 m with an accuracy of ±0.0001 m), located 0.69 m from the weir. The records of the water level were obtained visually by registering the displayed values. The hydraulic pressures were measured with a set of 84 piezometers spread over seven transversal rows and twelve longitudinal lines (Figure 4a). Measurements were obtained by visual inspection using a millimetric ruler.

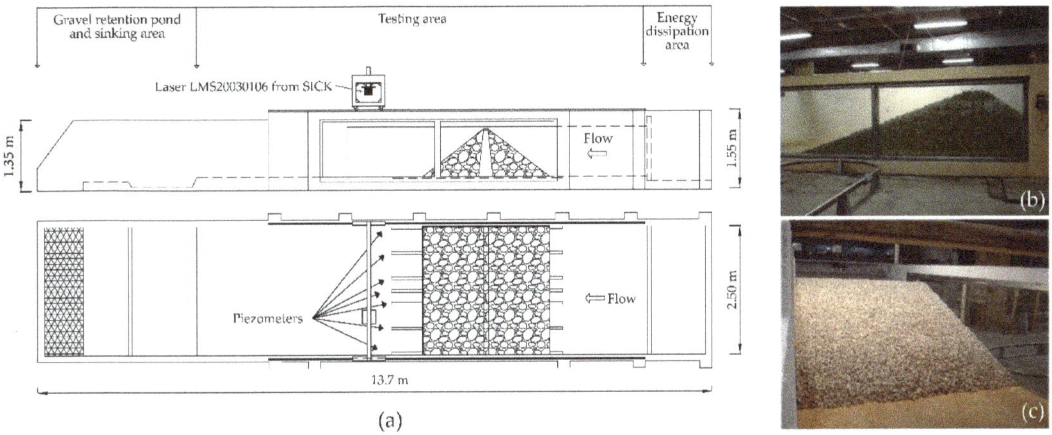

Figure 4. The UPM flume. (**a**) Scheme of the flume dimensions and areas, (**b**) image of a test taken from the outside of the flume through the inspection window located on the left-side wall, and (**c**) taken from inside the flume.

At the CEDEX laboratory, three flumes were used (Figure 5). The smaller one was a tilting metallic flume 12 m long, 0.4 m wide, and 0.6 m high. Although the slope of the flume could be controlled, it was kept horizontal throughout the tests. This flume was supplied by a constant water level tank where discharges were measured upstream of the physical model using a thin-plate rectangular weir 0.487 m long with no lateral contraction. This constant level tank was supplied with water by pumping from the main tank located below the laboratory floor (3000 m^3) with a Worthington hydraulic pump (three-phase motor GEAL 220/380 V, 4.4 kW, 6 hp) capable of pumping up to 0.06 $m^3\ s^{-1}$ with 5 mwc. The medium-size concrete flume was 12 m long, 1.0 m wide, and 1.1 m high. Supplied by a constant water level tank, discharges were also measured upstream of the physical model with a 90° thin-plate triangular weir. The water level upstream of the weirs was measured with P8000 ultrasonic level sensors with digital displays (Dr. D. Wehrhahn, Hannover, Germany). This constant water level tank was also supplied with water from the main tank with a Worthington hydraulic pump (three-phase motor Alcanza 220/380 V, 45 kW, 6 hp) capable of pumping up to 0.2 $m^3\ s^{-1}$ with 14 mwc. In this flume, the hydraulic pressures were measured using the intelligent pressure instrumental system Scanivalve, placed at the base of the flume. This system was composed of 44 measuring points distributed in ten transversal rows along a distance of 2.9 m. Finally, the bigger metal-glazed flume, 100 m long, 1.5 m wide, and 1.5 m high, was supplied directly from the main tank using two Jeumont-Schneider hydraulic pumps, one capable of pumping up to 1.7 $m^3\ s^{-1}$ with 4.4 mwc (DC motor 107 kW and 440 V), and the other capable of pumping up to 0.8 $m^3\ s^{-1}$ with 4.25 mwc (DC motor 55 kW and 440 V).

Figure 5. Some views of the three CEDEX flumes: (**a**) small, (**b**) medium, and (**c**) large.

2.6. Dimensionless Variables

Using the height of the physical models (H) and the acceleration of gravity (g) as the basic variables, and by applying the *Buckingham* Π theorem we can obtain the dimensionless unit discharge (q^*) expressed by Equation (5) and the dimensionless equivalent Darcy's coefficient of permeability (k^*_{eq}) expressed by Equation (6).

$$q^* = \frac{q}{\sqrt{g \cdot H^3}} \quad (5)$$

$$k^*_{eq} = \frac{k_{eq}}{\sqrt{g \cdot H}} \quad (6)$$

Even though Darcy's law is not applicable in coarse materials such as those used in this study, the equipotential lines at the toe of a rockfill shoulder with linear and nonlinear models are nearly vertical [51]. Assuming the maximum hydraulic gradient at the toe of the rockfill dam as being $i_{max} = 1/Z_{dss}$, then parameters a and b of the nonlinear resistance law (Equation (4)) can be converted into a single equivalent Darcy's coefficient of permeability (k_{eq}) using Equation (7) [10]. The velocity v_{max} is that occurring for the maximum gradient i_{max} at the toe of the dam.

$$k_{eq} = \frac{v_{max}}{i_{max}} = \frac{Z_{dss} \cdot \left(-a + \sqrt{a^2 + 4b/Z_{dss}}\right)}{2b} \quad (7)$$

To compare physical models with different geometries and dimensions, the horizontal lengths were also converted to non-dimensional (x^*) using Equation (8). This dimensionless variable ranges from zero to one, from the downstream edge of the crest to the toe of the dam.

$$x^* = \frac{x}{Z_{dss} \cdot H} \quad (8)$$

2.7. Statistical Analyses

The statistical analyses presented in this paper were all performed with the statistical software R (version 4.0.3). The regression models were obtained with the 'lm' function (R 'stats' Package), the hypotheses contrasts were performed with the 't.test' function (R 'stats' Package), and the power analyses with the 'cohen.d' function (Package 'effsize') to calculate the effect size statistics and the 'pwr.2p2n.test' function (Package 'pwr') to compute the power of the test. The contrast hypotheses between the two groups of samples were performed assuming the two variances as equal.

3. Results

3.1. Failure Initiation and Progress

The failure progress was observed to be the same for all tests. These observations were in line with those obtained by other authors [3,19,23,28] for highly permeable materials.

Given the test procedure and regardless of the type of flow (overtopping or throughflow), material gradings, and dam geometry, failure always initiated at the toe of the dam for a unit discharge that must overcome a given threshold (q_{fi}). Once this threshold was overcome, failure progressed upwards until a new equilibrium state was achieved for a given constant inflow discharge. The failure progress or 'failure paths' obtained during the specific campaign for analyzing the variability of the results are shown in Figure 6.

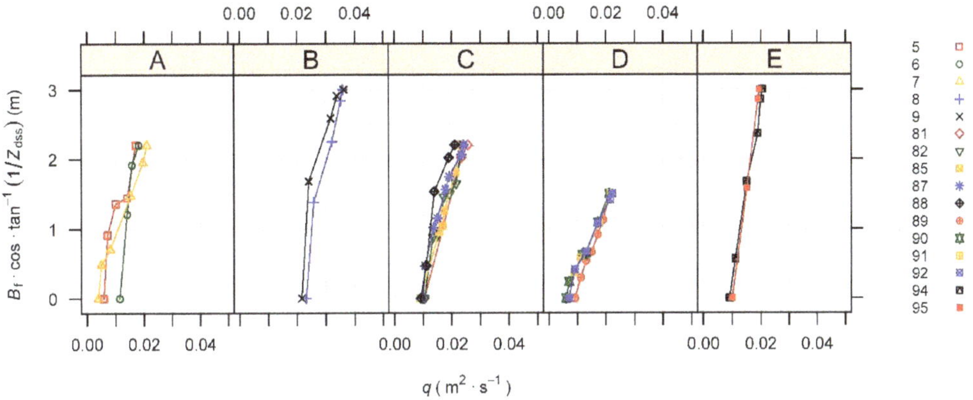

Figure 6. Failure paths obtained with the set of tests performed to assess the variability of the results. Each series represents a physical model with several discharge steps.

3.2. Failure Discharge

The failure discharge was assumed to be that by which damages inflicted to the downstream shoulder reached the crests of the physical models. For practical reasons, in general, it was defined as the average value between the last discharge step in which damages did not reach the crest ($Q_{f,pre}$) and the first in which damages surpassed it ($Q_{f,pos}$). Table A1 in Appendix A summarizes the average failure discharges ($Q_f = Q_{f,ave}$) obtained for every physical model. Although difficult to compare with other studies from the state of the art (different materials, criteria, etc.), these results are roughly in line with the results of other authors. Test nº 84 ($H = 0.5$ m, $W = 0.4$ m, $Z_{dss} = 1.5$, D50 = 17.3 mm) resulted in a $q_f = 0.0168$ m^2 s^{-1}, in the same order of magnitude as similar tests performed by Franca and Almeida [28] ($H = 0.5$ m, $W = 2$ m, $Z_{dss} = 1.5$, D50 = 18.9 mm), which obtained a value of $q_f = 0.0138 \pm 0.009$ m^2 s^{-1}.

3.3. Hydraulic Pressures

The hydraulic pressures at the base of the dam were measured for the majority of the tests and all discharge steps. Nonetheless, for simplicity, here, only those pressures relevant for the analysis of the results are presented. Table A2 in Appendix A summarizes the average hydraulic pressures at the bases of the physical models nº 85, 87, 90, 94, and 95 for a given discharge step in the early stages of each test, and Table A3 the average hydraulic pressures from tests nº 87, 94, and 95 for a transversal section roughly located at $x^* = 0.3$ (i.e., 30% of the base length from the crest). The average values were obtained using the records of each piezometer in the same transversal row.

3.4. Failure Mechanisms

The laboratory tests allowed the identification of two dominant failure mechanisms: particle dragging (PD) and mass sliding or slumping (MS). Slumping occurs predominantly in embankments with steep slopes. In these cases, failure of the downstream slope affected the entire width of the physical model (Figure 7a). Particle dragging is the predominant failure mechanism in embankments with gentle slopes. In these cases, we observed the formation of one or more erosion channels whose final width was smaller than the total width of the physical model (Figure 7b). Figure 7 shows digital elevation models (DEM) resulting from the difference between the original undeformed and the failed embankment for different discharge steps (from top to bottom, the discharge is increasing). Light colors represent eroded areas, while dark colors represent areas of deposition.

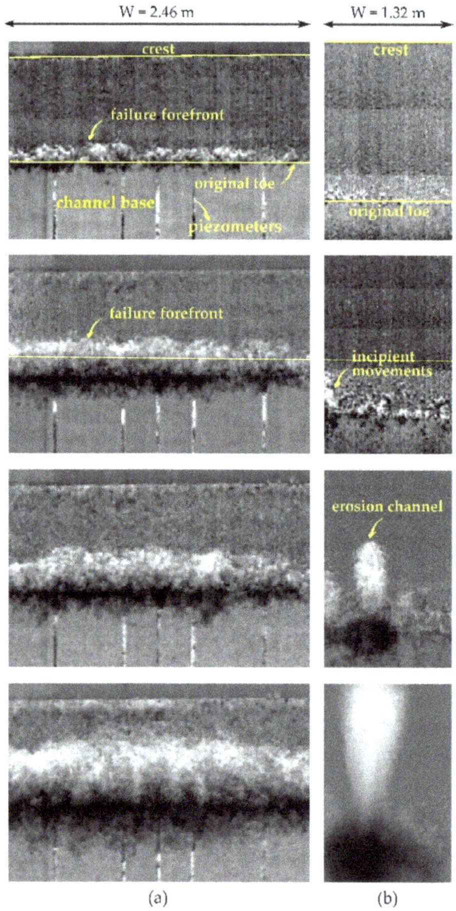

Figure 7. Two dominant mechanisms of failure: (**a**) mass sliding and (**b**) particle dragging.

4. Discussion

4.1. The Mechanics of Failure

Particle dragging is all about the individual stability of each particle when subjected to throughflow forces and gradients as well as skimming flow over the shoulder surface. Once a given discharge threshold is overcome, the motion of a single particle (not a group of particles) is observed, changing the stability conditions of the adjacent particles. If this is a 'key' particle, a type of chain reaction will be triggered, leading the adjacent particles to also move downstream, forming an erosion channel. If the particle is not categorized as key, the adjacent particles will remain stable and in place. In the early stages of failure, this mechanism could lead to the formation of several incipient erosion channels along the entire toe of the dam, but in a given moment, only a few will prevail and grow upstream. Here, the seepage conditions could change significantly, leading to the concentration of flow in the prevailing channels and forcing failure to progress through them. Eventually, only one will prevail, completing the failure process of the downstream shoulder. These erosion channels are hourglass-shaped (Figure 7b) with steeper walls than the original slope and widths smaller than the total flume width. If key particles are displaced, these and the corresponding adjacent particles will fall radially into the erosion channel, making it progress upwards as well as laterally. Nevertheless, if the displaced particles are not key, this will deepen the erosion channel. The deepening of the channel also occurs naturally as the erosion channel progresses upstream. This phenomenon could eventually lead to slumping. In this case, slumping is not the main failure mechanism but a consequence of particle dragging.

Mass sliding or slumping is related to a problem of global instability affecting a certain mass of material and associated with pore water pressures inside the dam. The sliding mechanism is difficult to detect because the sliding surfaces are usually shallow and quasi-parallel to the slope. A way of detecting sliding is by observing the simultaneous movement (not consecutive) of a group of particles. The physical models where this failure mechanism is dominant are not immune to particle dragging. In these cases, if key particles are displaced as a consequence of sliding, this would also trigger a chain reaction like that described previously, leading the failure forefront, which in these cases usually covers the entire width of the model (Figure 7a), to progress upwards.

4.2. The Scale Effect

To compare both 'small-scale' physical models and 'prototypes', the Froude similitude was applied to q_f. From this similitude theory, it follows that the unit flow discharge scale factor is $s_q = s_L^{3/2}$. So, comparing both scaled and prototype results of q_f (Table 3), we obtain errors of 2.47%, 2.43%, and 2.75% for models with $Z_{dss} = 1.5$, 2.5, and 3.5, respectively. These errors represent the difference between both results relative to the 'prototype' value. The mean value for the errors is then 2.55% (mean) ±0.14% (standard deviation).

Table 3. Summary of the specific campaign to assess the scale effect.

'Prototype' Tests			'Small-Scale' Models			
Test	Z_{dss}	q_f (m^2s^{-1})	Test	Z_{dss}	q_f ($m^2\ s^{-1}$)	$q_{f,scaled}$ ($m^2\ s^{-1}$)
108	1.5	0.032749	109	1.5	0.005125	0.033558
110	2.5	0.042304	111	2.5	0.006618	0.043334
112	3.5	0.042505	130	3.5	0.006313	0.041337

Even though it is a simplistic approach, the Froude similitude to scale between 'physical model' and 'prototype' by scaling D50 seems to give good results for this scale factor. Given that it is the phreatic surface elevation flowing through the downstream shoulder and the first emergence point (intersection between the phreatic surface and the downstream slope) that govern failure and define how far it will progress, then, an alternative approach could be scaling the 'unit discharge–permeability' ratio, as this would more accurately scale the water table elevation. By scaling D50, we are indirectly scaling the permeability, albeit by a factor that we do not know and that should be, in theory, $s_L^{1/2}$, the scale factor for a velocity. Given the small errors between the prototype and scaled q_f, we can conclude that for these uniform gravels, the 'unit discharge–permeability' ratio is being somehow properly scaled by scaling D50 with the limits of size existing in this study.

Another problem associated with scaling D50 is that we are managing gravels with different repose angles in the 'prototypes' and 'small-scale' physical models. A failed shoulder profile presents three different slopes, (i) the original slope not yet affected by the failure, (ii) a zone over the phreatic surface with the dry repose angle, and finally (iii) a zone below the phreatic surface and flow with a submerged repose angle. Because it is the slope over the phreatic surface with the dry repose angle that is defines whether failure reaches the crest of the dam or not, then complete failure of the downstream shoulder depends greatly on the gravel that is being tested. Because the repose angle flattens by reducing the size of the granular materials, that could imply that physical models tested with smaller materials could reach complete failure earlier than if they were tested with coarser materials.

4.3. Repeatability

The set of tests dedicated to analyzing repeatability was used to quantify the variability of the results. First, the hydraulic pressures were compared between tests of the same group. Here, we preferred to compare pressures for an early stage of failure, preferably before any major damage was observed to the downstream shoulder because, as failure progresses, the flow net also changes, especially for those cases where particle dragging is the dominant failure mechanism. In these cases, characterized by the formation of one or more erosion channels acting as boreholes or wells, the flow net could change significantly along the width of the physical models suffering a dropdown where the erosion channels are located. As can be observed in Figure 8, the differences between tests nº 85 and 87 (Group C) and nº 94 and 95 (Group E) are negligible. This figure presents the hydraulic pressures measured for roughly the same discharge, 0.0051 and 0.0067 $m^2\ s^{-1}$ in tests 85 and 87, respectively, and of 0.0071 and 0.0072 $m^2\ s^{-1}$ in tests 94 and 95. A maximum difference of 0.021 m was observed between tests 85 and 87 for the most upstream measuring section.

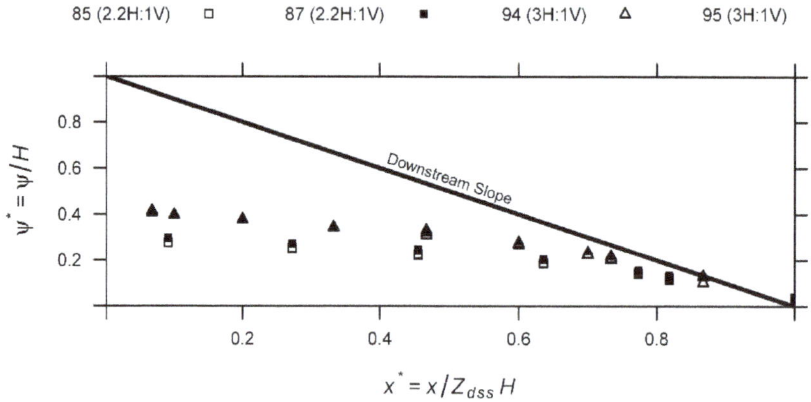

Figure 8. Hydraulic pressures measured for roughly the same discharge in tests nº 85 and 87 (Group C) and nº 94 and 95 (Group E). These pressures correspond to discharges of 0.0051 $m^3\ s^{-1}$ and 0.0067 $m^3\ s^{-1}$ in tests 85 and 87, respectively, and of 0.0071 $m^3\ s^{-1}$ and 0.0072 $m^3\ s^{-1}$ in tests 94 and 95.

By observing the failure paths plotted in Figure 6, it can also be concluded that, in general, tests performed under the same group of tests presented the same trajectories. When failure was complete, groups A, B, C, D, and E resulted, respectively, in unit failure discharges around 0.0187 $m^2\ s^{-1}$ (mean) $\pm$ 0.0020 $m^2\ s^{-1}$ (one standard deviation), 0.0357 $\pm$ 0.0005, 0.0234 $\pm$ 0.0016, 0.0216 $\pm$ 0.0005, and 0.02 $\pm$ 0.0007. The ratios of the standard deviation to the mean, i.e., the coefficients of variation (CV), were, also respectively, 10.9%, 1.4%, 6.9%, 2.1%, and 3.7%, being the mean value of CV = 5.0% $\pm$ 3.9%. It can be stated that the results varied within reasonable ranges given the great amount of uncertainty associated with this kind of test.

4.4. The Effect of the Downstream Slope

The downstream slope was observed to greatly affect the failure mechanism. The chart presented in Figure 9 was plotted using only the results for the physical models wider than 1 m to avoid having the flume walls affecting the correct development of the erosion channels if that was the case. Also, we only used those models where one of the two mechanisms was dominant at the final stages of failure. In summary, a total of forty-one physical models were used. The transition between mechanisms occurs for a range of Z_{dss} varying roughly between 2.0 and 2.5. If we focus on the tests with $Z_{dss} = 2.2$, it can be observed that eight out of the nine models failed by particle dragging and resulted in the formation of erosion channels. So, for practical purposes, steep slopes were defined to be those steeper than 2.2 and gentle slopes as those smoother than this value. This could be a simplistic way of categorizing between slopes given that the repose angle may most certainly affect it. Nonetheless, the small variation in these angles within the rockfill materials must be taken into account.

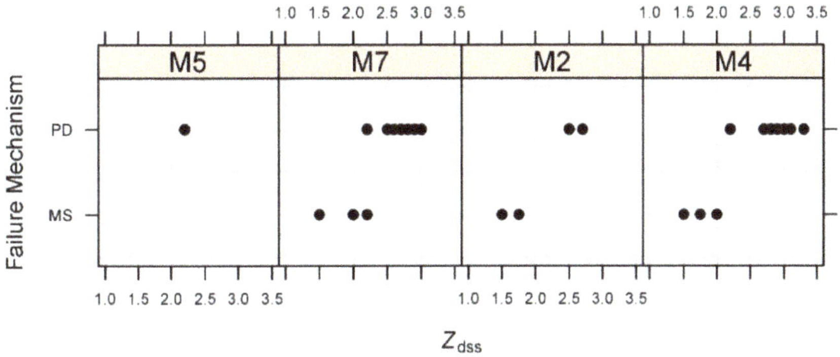

Figure 9. Relationship between the failure mechanisms and the downstream slope (Z_{dss}). The acronyms PD and MS refer to 'particle dragging' and 'mass sliding', respectively.

The effect of the slope was also noticed in the hydraulic pressures, which increased as the slope was gentler for the same unit flow. This fact is shown in Figure 10, which presents the hydraulic pressures measured for the first discharge step in tests nº 87, 90, 94, and 95. All of these tests were of the type CPM/NIE, i.e., complete configurations without impervious element and, thus, throughflow. For relative distances of $x^* = 0.2$, 0.5, and 0.7, an embankment with a slope $Z_{dss} = 3.0$ resulted, respectively, in hydraulic pressures 47.3%, 59.6%, and 78.9% higher than those resulting from an embankment with a slope $Z_{dss} = 1.5$.

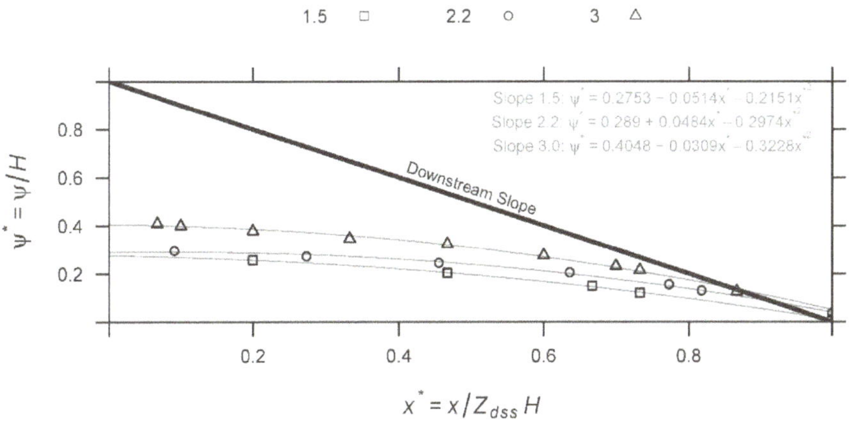

Figure 10. Hydraulic pressures measured for roughly the same discharge (0.007 $m^3\ s^{-1}$) in tests nº 87 [$Z_{dss} = 2.2$], nº 90 [$Z_{dss} = 1.5$], and nº 94 and 95 [$Z_{dss} = 3.0$]. The plotted pressures for $Z_{dss} = 3.0$ were averaged between both tests (Figure 8). Although pressures measured in test nº 85 [$Z_{dss} = 2.2$] were similar to that measured in test nº 87 (Figure 8), in this figure it was decided not to average them because in test nº 85 they correspond to a discharge of 0.005 $m^3\ s^{-1}$.

Figure 11 presents the evolution of the hydraulic pressure measured during tests nº 87, 94, and 95 in a single section of the physical models, roughly located at $x^* = 0.3$ (30% of the base length from the crest).

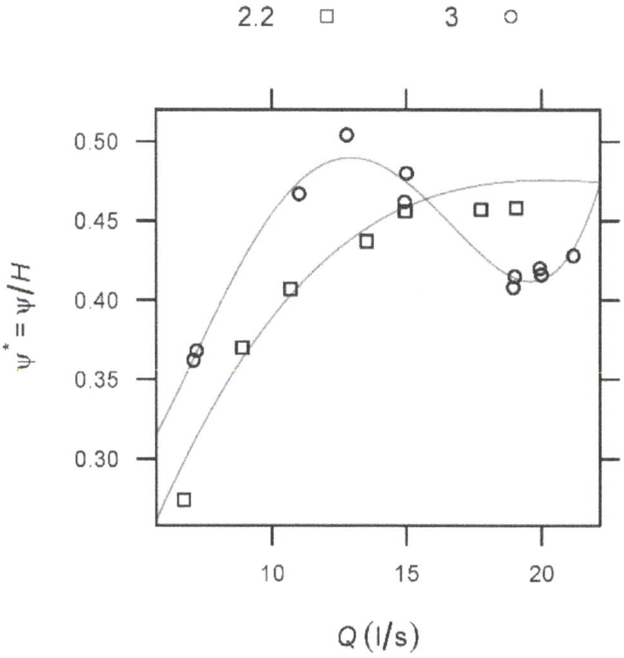

Figure 11. Hydraulic pressures measured in a single section of the physical models located at $x^* = 0.3$ (30% of the base length from the crest) in tests nº 87 [$Z_{dss} = 2.2$] and nº 94 and 95 [$Z_{dss} = 3.0$]. The plotted pressures for $Z_{dss} = 3.0$ were averaged between both tests (Figure 8).

Here, it can also be seen that, for a given unit flow, the gentler the slope, the higher the hydraulic pressures, at least in the early stages of the tests. In a given moment, i.e., for a given discharge step, pressures measured in the physical models with $Z_{dss} = 3.0$ suffered a sudden decrease that matched the formation of an erosion channel, as can be observed in Figure 12. It must be noted that this pressure dropdown was not observed in test nº 87 ($Z_{dss} = 2.2$) even though the failure mechanism of this test was the same as that of tests nº 94 and 95 ($Z_{dss} = 3.0$), as can be observed in Figure 13. This observation denotes that the failure mechanism is not enough to explain the pressure dropdown and that other variable/s should also be considered—for example, the geometry and dimensions of the erosion channel.

The repeatability campaign groups C ($Z_{dss} = 2.2$), D ($Z_{dss} = 1.5$), and E ($Z_{dss} = 3.0$) were compared with each other to assess the possible effect of this variable on q_f. From a physical point of view, and considering only the range of slopes for which the dominant failure mechanism is mass sliding or slumping, it could be expected that the steeper slopes within this range resist higher flow discharges than gentler slopes as a result of the higher hydraulic gradients that lead to lower phreatic surfaces and lower pressures. This hypothesis was confirmed in Figure 10. If we now expand the range of slopes and compare those equal to 1.5 and 3.0 through a one-sided test, we obtain that steep slopes completely fail for higher flow discharges than gentle slopes for a p-value $= 0.086$ and a power of 64.9%. We could accept the alternative hypothesis (steep slopes resist higher flow discharges) for a 0.1 significance level (α), thus having a 10% and 35.1% chance of committing Type I and II errors, respectively. When contrasting the slopes 2.2 and 3.0

also through a one-sided test, we obtain a p-value $= 0.035$ and a power of 74.0%. We could accept the alternative hypothesis even for a lower significance level, $\alpha = 0.05$, thus having a 5% and 26.0% chance of committing Type I and II errors, respectively. Therefore, steeper slopes seem to resist higher unit discharges than gentler slopes, a trend that is also observed in Figure 14. Nonetheless, it must be noted that this tendency was not observed, for example, during the scale effect campaign, summarized in Table 3, nor when comparing slopes 1.5 and 2.2 also through a one-sided test, where we obtained a p-value $= 0.955$ and a power of 0.06% (for $\alpha = 0.1$). These are contradictory results that should be examined more deeply in future investigations. They could be related more to testing variability (tests performed throughout different R&D projects, different laboratories, measurement techniques, discharge steps, etc.) rather than the variability of the results that we already saw to be small when repeating the same test in the same conditions.

Figure 12. Images of test nº 95 [$Z_{dss} = 3.0$] before and during the pressure dropdown (Figure 11).

Figure 13. Images of test nº 87 [$Z_{dss} = 2.2$].

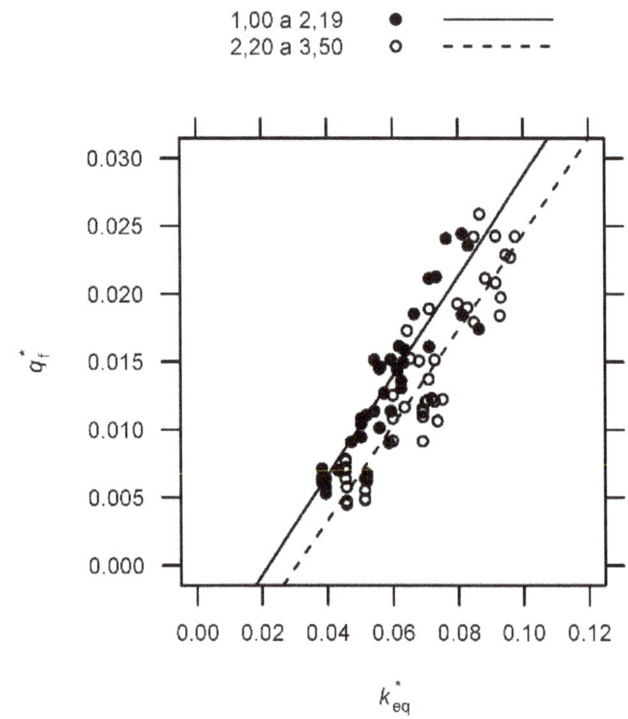

Figure 14. Variation of q_f^* with k_{eq}^* differentiating by the downstream slope steepness.

One last observation must be taken into consideration. Here, we would like to emphasize that we were dealing with slopes that were unstable in throughflow conditions, i.e., we were dealing with physical models that failed. So, if a gentle slope, stable to slumping, is not subjected to flow discharges capable of dragging its particles, then this embankment will remain stable. On the other hand, the same dam constructed with a steeper slope unstable to slumping would fail, so in this case, gentler slopes would be more resistant.

4.5. Other Effects

There are clear differences between the physical processes related to overtopping and throughflow. Nonetheless, characteristics of throughflow such as pressures or the phreatic surface elevation (including the first emergence point) are not significantly different between both types of flow [10]. So, because it is the water level inside the rockfill dam and the position of the first emergence point that governs the failure of the downstream shoulder by determining how far it will progress, and since these do not change significantly when changing the type of flow, we would expect to obtain a similar failure discharge q_f in each case, for example, CPM/UF vs. CPM/NIE or PPM/CC vs. PPM/NIE (Figure 1). Results of the statistical analysis performed for tests nº 140, 141, 142 (CPM/NIE) and nº 143 and 146 (CPM/UF) support this idea. The p-value $= 0.027$, resulting from a two-tailed test, allows us to reject the alternative hypothesis for $\alpha = 0.01$ and claim that the type of flow does not affect q_f. The statistical power of this analysis is about 93%, so we have a small probability of also committing the Type II statistical error. In other words, we have a 7% chance of rejecting the alternative hypothesis when it should be accepted. Besides this comparison, Figure 15 shows the relation between k_{eq}^* and q_f^* differentiating between tests performed with overtopping and throughflow. Differences between both scatter plots are negligible, especially in the lower part of the chart.

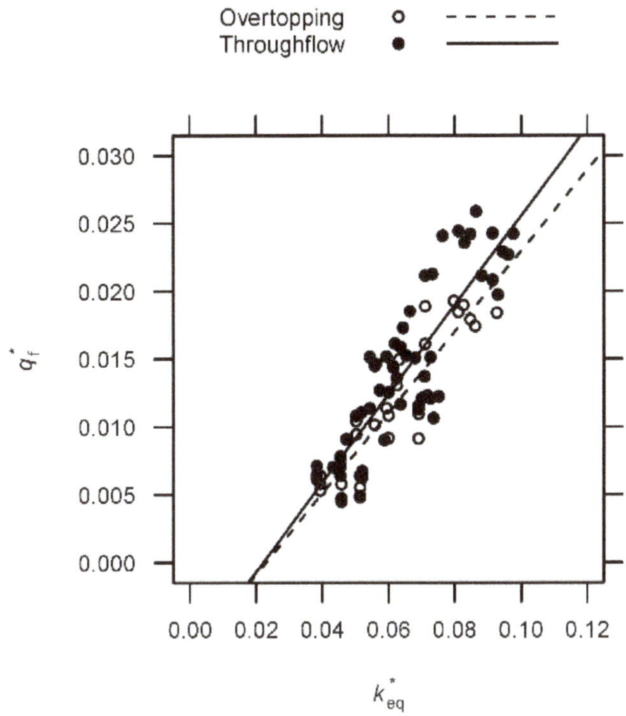

Figure 15. Variation of q_f^* with k_{eq}^* differentiating by the type of flow.

Those physical models where the differentiating variable was the type of impervious element (central core or upstream face) were also compared. In this case, we could somehow expect to observe some differences in the value of q_f. The process through which water infiltrates was the same, overtopping, but the flow paths inside the body of the dam were longer for dams with an upstream face, which did not necessarily imply a difference in the position of the first emergence point. Results from the statistical analysis performed on tests nº 22 and 123 (PPM/CC) and nº 124 and 128 (CPM/UF) indicate that the type of impervious element does not have a significant impact on the value of q_f. The p-value = 0.124 allows us to reject the alternative hypothesis for an $\alpha = 0.05$. The statistical power is about 73%, so here we have a 27% chance of rejecting the alternative hypothesis when it should be accepted. Crossing these results with Figure 16 makes us confident to state, for now, that the type of impervious element does not affect q_f. Figure 17 presents the images of both tests and discharge steps used in the construction of the Figure 16 plot. Besides this comparison, Figure 18 shows the relation between k_{eq}^* and q_f^* differentiating between tests performed with upstream face (UF) and central core (CC). Here, we also cannot see any clear separation between both scatter plots.

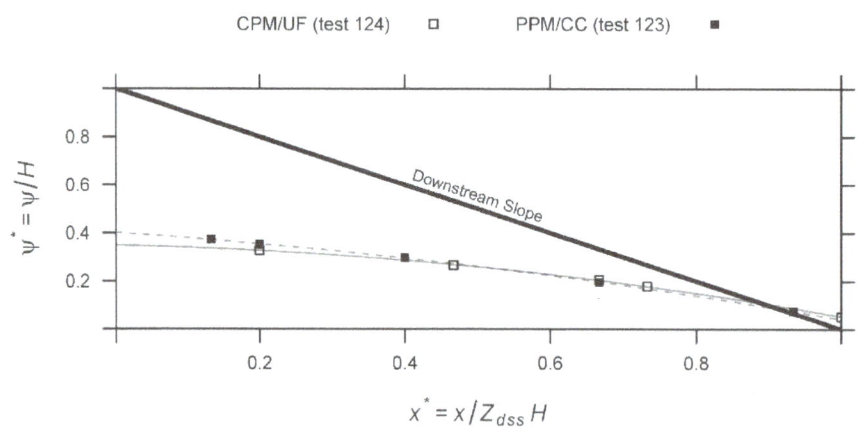

Figure 16. Hydraulic pressures measured for a discharge of 0.0091 and 0.0090 $m^3\ s^{-1}$ in tests nº 123 and 124, respectively.

Figure 17. Failure progress in tests nº 123 and 124 for a discharge of roughly 0.0090 $m^3\ s^{-1}$.

4.6. Research Scope and Limitations

The inferential analysis seeks patterns from a population through samples of it. In this kind of analysis, the aphorism 'more is better' referring to the sample size is true because samples with a higher number of observations imply smaller confidence intervals and more reliable conclusions. But the truth is that researchers are most of the time far from this ideal goal working even with extremely small samples. In these cases, statistical power analysis assumes an important role by including the probability of committing Type II errors besides the traditional assessment of the Type I error [52]. Alongside size sampling, this study contains a series of other limitations, including, for example, the size of the physical models that could lead to scale effects, the use of uniform limestone materials, preventing a thorough analysis of the effect of other material parameters, and the small number of tests repeated in the same conditions.

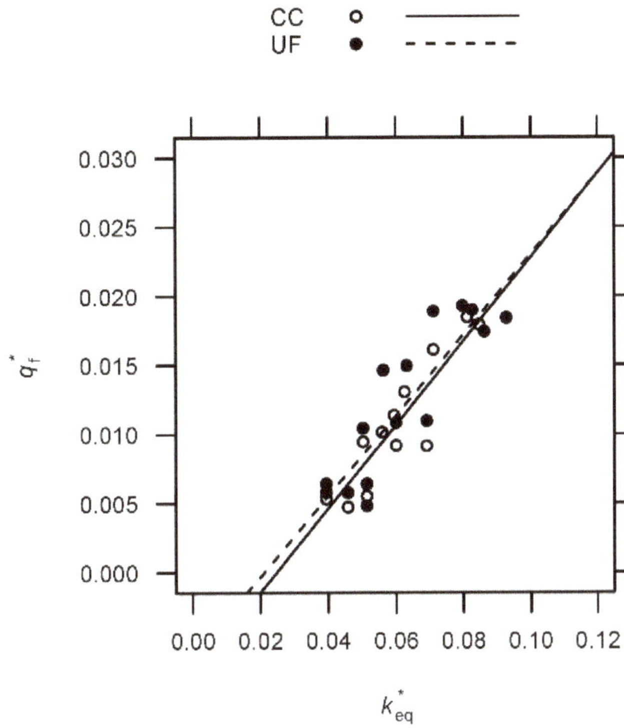

Figure 18. Variation of q_f^* with k_{eq}^* differentiating by the impervious element (Central Core and Upstream Face).

4.7. A Regression Model for the Failure Discharge

Taking into account the discussion of the results presented in the previous sections and the scope and limitations of the research, we propose a regression model that can be used to estimate the failure discharge of a rockfill downstream shoulder (q_f), i.e., that overtopping or throughflow discharge with a failure degree that reaches the crest of the dam. This model depends on the equivalent Darcy's coefficient of permeability (k_{eq}), expressed by Equation (7), on the downstream slope (Z_{dss}) and height (H) of the dam, and on the acceleration of gravity (g). To be precise, two regression models were calibrated, one for 'steep' and one for 'gentle' slopes, with the critical slope $Z_{dss} = 2.2$ for the granular materials used in this research and an angle of repose around 41°. Both models pass through the origin since this is a non-tested data point, i.e., in the limit, a dam with $k_{eq}^* = 0$ should need no overtopping to fail. Equations (9) and (10) should be used for gentle and steep slopes, respectively. If no distinction is desired between these two categories, then Equation (11) should be used. The first two equations both have a coefficient of determination $R^2 = 0.97$, while the third uses 0.95. It should be noted that tests 114 and 129 were excluded from Equation (9) because these were outliers with Cook's distances of roughly 1 and 0.5, respectively, and that these calibrations included only those tests where the permeability of the materials was obtained. Equations (9) and (10) can be observed fitted to data in Figure 19.

$$q_{f,gentle} = 2.659 \cdot \sqrt{\frac{H}{g}} \cdot {k_{eq}}^2 \tag{9}$$

$$q_{f,steep} = 3.610 \cdot \sqrt{\frac{H}{g}} \cdot {k_{eq}}^2 \tag{10}$$

$$q_f = 2.937 \cdot \sqrt{\frac{H}{g}} \cdot k_{eq}{}^2 \quad (11)$$

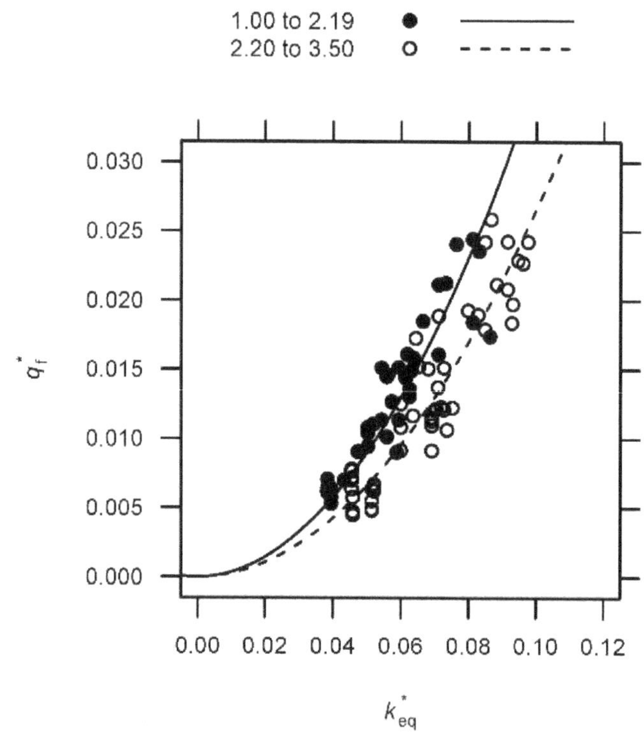

Figure 19. Regression models proposed to estimate the failure discharge (q_f) for dams with steep slopes ($Z_{dss} < 2.2$) and gentle slopes ($Z_{dss} > 2.2$).

5. Conclusions

This study allowed the identification of two different failure mechanisms for highly permeable rockfills subjected to overflowing: slumping and particle dragging. The occurrence of one or the other is heavily dependent on the slope of the embankment and the size of the particles. Slumping is related to a problem of global instability of a certain mass of material and is predominant in embankments with steep slopes ($Z_{dss} < 2.2$). In these cases, failure of the downstream slope affected the entire width of the physical models. On the other hand, when particle dragging is the predominant failure mechanism, it means that we are dealing with slopes that are stable to slumping. In these cases, which are all about the individual stability of each particle when subject to seepage forces and hydraulic gradients, we observed the formation of erosion channels whose width was smaller than the total width of the physical models. The critical slope, i.e., the slope that defines the limit for the occurrence of one mechanism or the other, was identified to be $Z_{dss} = 2.2$ for the characteristics of the materials used in this study.

The downstream slope was also observed to affect both the hydraulic pressures at the base of the physical models and the value of the failure discharge, i.e., the inflow that forces failure to reach the crest of the dam. For a discharge of 0.007 $m^3\ s^{-1}$, the hydraulic pressures measured in a dam with $Z_{dss} = 3.0$ were, on average, 61.9% higher than in a dam with $Z_{dss} = 1.5$. Regarding failure, steep slopes ($Z_{dss} < 2.2$) were prone to fail for higher discharges than gentle slopes ($Z_{dss} > 2.2$) or, in other words, tended to be, on average, more resistant. Given this observation, two regression models were proposed to estimate the unit failure discharge (q_f) of dams with gentle and steep slopes, Equations (9) and (10), respectively, both functions of the equivalent Darcy's coefficient of permeability (k_{eq}), expressed by Equation (7), the downstream slope (Z_{dss}), the height of the embankment (H), and the acceleration of gravity (g). If no distinction is desired to be made between slopes, then q_f should be estimated with Equation (11).

Other factors such as the type of flow (overtopping or throughflow) or the type of impervious element (central core or upstream face) were not observed to affect the hydraulic pressures or the failure discharge.

Author Contributions: Conceptualization, R.M.-A., M.Á.T. and R.M.; methodology, M.Á.T. and R.M.; software, R.M.-A.; validation, R.M.-A., M.Á.T.; formal analysis, R.M.-A. and M.Á.T.; investigation, R.M.-A., R.M. and L.B.; resources, M.Á.T. and L.B.; data curation, R.M.-A.; writing—original draft preparation, R.M.-A.; writing—review and editing, M.Á.T. and R.M.; visualization, R.M.-A.; supervision, M.Á.T. and R.M.; project administration, R.M.-A., M.Á.T., R.M. and L.B.; funding acquisition, M.Á.T., R.M. and L.B. All authors have read and agreed to the published version of the manuscript.

Funding: This research was funded by the Spanish Ministry of Science and Innovation, grant number BIA2010-21350-C03-03 (Project EDAMS—Rotura del elemento impermeable de presas de materiales sueltos en situación de sobrevertido y análisis de protecciones combinando modelación física e inteligencia artificial), and by the Spanish Ministry of Economy and Competitiveness, grant number BIA2007-68120-C03-02 (Project XPRES—Caracterización de la rotura de las presas de escollera por sobrevertido y desarrollo de criterios para evaluar la seguridad del conjunto presa-área afectada durante una avenida).

Acknowledgments: Within the scope of the funding research projects, we would like to thank Ángel Lara, Rafael Cobo, Cristina Lechuga, Isabel Berga, and Hibber Campos for their participation in these projects.

Conflicts of Interest: The authors declare no conflict of interest. The funders had no role in the design of the study; in the collection, analyses, or interpretation of data; in the writing of the manuscript, or in the decision to publish the results.

Nomenclature

The following symbols and acronyms are used in this paper:

a	Seepage resistance coefficient of the quadratic flow equation (fundamental units $T{\cdot}L^{-1}$)
b	Seepage resistance coefficient of the quadratic flow equation (fundamental units $T^2{\cdot}L^{-2}$)
B_f	Maximum advance of failure in relation to the embankment toe, i.e., the distance between the most upward point of the failure forefront and the downstream toe as shown in Figure 2 (fundamental units L)
CC	Central core
CPM	Complete physical model or complete configuration (both upstream and downstream slopes and crest)
CPM/UF	Complete configuration with upstream face
CPM/NIE	Complete configuration without impervious element
C_u	Coefficient of uniformity, the ratio D60/D10 where D60 and D10 are the sieve sizes through which 60 and 10% of the granular material (dimensionless) passes
CV	Coefficient of variation
D50	Sieve size passing 50% of the particles (fundamental units L)
DSS	Downstream shoulder
e	Void ratio (dimensionless)
g	Acceleration of gravity (fundamental units $L{\cdot}T^{-2}$)
H	Height of the dam (fundamental units L)
i	Hydraulic gradient (dimensionless)
k_{eq}	Equivalent Darcy's coefficient of permeability (fundamental units $L{\cdot}T^{-1}$)
k^*_{eq}	Dimensionless equivalent Darcy's coefficient of permeability (dimensionless)
L	Length, fundamental dimension
l_c	Width of the embankment crest (fundamental units L)
M	Mass, fundamental dimension

n	Porosity (dimensionless)
NIE	No impervious element
PM	Physical model
PPM	Partial physical model (partial configurations)
PPM/CC	Partial or complete configuration with central core
PPM/NIE	Partial configuration without impervious element
q	Unit discharge (fundamental units $L^2 \cdot T^{-1}$)
q^*	Unit discharge (dimensionless)
q_f	Unit failure discharge, i.e., the discharge in which failure reaches the crest of the embankment (fundamental units $L^2 \cdot T^{-1}$)
q_{fi}	First unit discharge producing any visible damage to the downstream slope (fundamental units $L^2 \cdot T^{-1}$)
$q_{f,pre}$	Last unit discharge step in which failure does not reach the crest of the embankment (fundamental units $L^2 \cdot T^{-1}$)
$q_{f,pos}$	First unit discharge step in which failure surpasses the crest of the embankment (fundamental units $L^2 \cdot T^{-1}$)
R^2	Coefficient of determination
s_L	Scale factor
sd	Standard deviation
T	Time, fundamental dimension
UF	Upstream face
USS	Upstream shoulder
W	Width of the test flumes (fundamental length L)
x^*	Horizontal lengths (dimensionless)
Z_{dss}	Slope of the downstream rockfill shoulder (dimensionless)
Z_{uss}	Slope of the upstream rockfill shoulder (dimensionless)
φ'	Internal friction angle of gravels (°)
γ	Dry unit weight (fundamental units $M \cdot L^{-3}$)
γ_s	Specific gravity of solid particles (fundamental units $M \cdot L^{-3}$)
γ_{sat}	Saturated unit weight (fundamental units $M \cdot L^{-3}$)
γ_w	Density of water (fundamental units $M \cdot L^{-3}$)
ψ	Pressure head (fundamental units L)
ψ^*	Pressure head (dimensionless)

Appendix A

Table A1. Summary of the failure discharge (Q_f) obtained for every physical model tested at the laboratory. The subscript 'pre' refers to the last discharge in which failure did not reach the crest, and 'pos' refers to the first in which failure surpasses it. The subscript 'ave' refers to the average value of the previous two. NA means 'not available'.

Test	Laboratory	H (m)	W (m)	Z_{dss} (H:V)	Z_{uss} (H:V)	l_c (m)	IE	Gravel Code	$Q_{f,pre}$ ($L \cdot s^{-1}$)	$Q_{f,pos}$ ($L \cdot s^{-1}$)	Q_f ($L \cdot s^{-1}$)
1	UPM	0.6	2.50	2.50	1.50	0.20	NIE	M2	18.000	21.000	19.500
2	UPM	0.6	2.50	1.50	1.50	0.20	NIE	M2	18.400	21.500	19.950
3	UPM	0.6	2.50	1.75	1.50	0.20	NIE	M2	23.400	26.000	24.700
4	UPM	0.6	2.50	2.70	1.50	0.20	NIE	M2	20.200	23.000	21.600
5	UPM	1.0	2.50	2.20	1.50	0.20	NIE	M5	35.000	50.000	42.500
6	UPM	1.0	2.50	2.20	1.50	0.20	NIE	M5	39.240	51.290	45.265
7	UPM	1.0	2.50	2.20	1.50	0.20	NIE	M5	49.000	55.800	52.400
8	UPM	1.0	2.46	3.00	1.50	0.20	NIE	M7	86.009	87.944	86.977
9	UPM	1.0	2.46	3.00	1.50	0.20	NIE	M7	83.075	94.468	88.772
10	UPM	1.0	2.46	2.50	1.50	0.20	NIE	M7	85.206	94.468	89.837
11	UPM	1.0	2.46	2.20	1.50	0.20	NIE	M7	94.468	98.569	96.519
12	UPM	1.0	2.46	1.50	1.50	0.20	NIE	M7	72.204	94.468	83.336
13	UPM	1.0	2.46	2.20	1.50	0.20	UF	M7	75.512	91.380	83.446

Table A1. *Cont.*

Test	Laboratory	H (m)	W (m)	Z_{dss} (H:V)	Z_{uss} (H:V)	l_c (m)	IE	Gravel Code	$Q_{f,pre}$ ($L{\cdot}s^{-1}$)	$Q_{f,pos}$ ($L{\cdot}s^{-1}$)	Q_f ($L{\cdot}s^{-1}$)
14	UPM	1.0	2.46	3.00	1.50	0.20	UF	M7	76.560	92.300	84.430
15	UPM	1.0	2.46	1.50	1.50	0.20	UF	M7	71.040	89.610	80.325
16	UPM	1.0	2.46	3.00	1.50	0.20	CC	M7	63.740	77.070	70.405
17	UPM	1.0	2.46	2.20	1.50	0.20	CC	M7	62.790	78.440	70.615
18	UPM	1.0	2.46	1.50	1.50	0.20	CC	M7	65.180	80.400	72.790
20	UPM	1.0	2.46	3.00	1.50	0.20	CC	M4	39.400	45.430	42.415
21	UPM	1.0	2.46	2.20	1.50	0.20	CC	M4	33.470	38.990	36.230
22	UPM	1.0	2.46	1.50	1.50	0.20	CC	M4	39.560	42.280	40.920
23	UPM	1.0	2.46	2.20	1.50	0.20	NIE	M4	30.630	38.470	34.550
24	UPM	1.0	2.46	3.00	1.50	0.20	NIE	M4	34.830	40.420	37.625
25	UPM	1.0	2.46	3.00	1.50	0.20	UF	M4	36.990	37.070	37.030
34	UPM	0.5	2.46	1.50	NA	NA	CC	M4	26.272	28.969	27.621
35	UPM	0.5	2.46	1.75	NA	NA	CC	M4	30.167	31.706	30.937
36	UPM	0.5	2.46	2.00	NA	NA	CC	M4	33.752	37.335	35.544
38	UPM	0.5	2.46	1.50	NA	NA	CC	M7	41.292	46.392	43.842
40	UPM	0.5	2.46	2.00	NA	NA	CC	M7	46.365	54.097	50.231
41	UPM	0.5	2.46	2.20	NA	NA	CC	M7	46.475	51.055	48.765
42	UPM	0.5	0.60	1.00	NA	NA	NIE	M4	5.054	7.013	6.034
43	UPM	0.5	0.60	1.50	NA	NA	NIE	M4	9.019	10.230	9.625
44	UPM	0.5	0.60	1.75	NA	NA	NIE	M4	8.973	11.171	10.072
45	UPM	0.5	0.60	2.00	NA	NA	NIE	M4	9.025	9.025	9.025
46	UPM	0.5	0.60	2.25	NA	NA	NIE	M4	8.959	11.283	10.121
47	UPM	0.5	0.60	2.50	NA	NA	NIE	M4	9.024	10.981	10.003
48	UPM	0.5	0.60	2.75	NA	NA	NIE	M4	7.035	9.112	8.074
49	UPM	0.5	0.60	3.00	NA	NA	NIE	M4	9.050	11.044	10.047
50	UPM	0.5	0.60	1.00	NA	NA	NIE	M7	4.855	7.130	5.993
51	UPM	0.5	0.60	1.50	NA	NA	NIE	M7	13.037	15.053	14.045
52	UPM	0.5	0.60	1.75	NA	NA	NIE	M7	14.842	17.151	15.997
53	UPM	0.5	0.60	2.00	NA	NA	NIE	M7	15.120	17.337	16.229
54	UPM	0.5	0.60	2.20	NA	NA	NIE	M7	15.014	17.164	16.089
55	UPM	0.5	0.60	2.40	NA	NA	NIE	M7	13.024	15.074	14.049
57	UPM	0.5	0.60	2.60	NA	NA	NIE	M7	15.143	17.101	16.122
58	UPM	0.5	0.60	1.40	NA	NA	NIE	M4	6.994	8.096	7.545
59	UPM	0.5	0.60	1.60	NA	NA	NIE	M4	7.876	8.978	8.427
60	UPM	0.5	0.60	1.90	NA	NA	NIE	M4	9.096	10.139	9.618
61	UPM	0.5	0.60	1.95	NA	NA	NIE	M4	10.056	11.371	10.714
62	UPM	0.5	0.60	2.10	NA	NA	NIE	M4	10.036	11.033	10.535
63	UPM	0.5	0.60	1.10	NA	NA	NIE	M7	9.046	10.049	9.548
64	UPM	0.5	0.60	1.30	NA	NA	NIE	M7	11.296	13.290	12.293
65	UPM	0.5	0.60	1.60	NA	NA	NIE	M7	13.126	15.121	14.124
66	UPM	0.5	0.60	2.10	NA	NA	NIE	M7	15.162	16.159	15.661
67	UPM	0.5	0.60	2.30	NA	NA	NIE	M7	17.191	17.191	17.191
68	UPM	0.5	1.32	2.70	NA	NA	NIE	M4	16.531	18.722	17.627
69	UPM	0.5	1.32	2.80	NA	NA	NIE	M4	18.721	21.357	20.039
70	UPM	0.5	1.32	2.90	NA	NA	NIE	M4	17.536	18.374	17.955
71	UPM	0.5	1.32	3.00	NA	NA	NIE	M4	16.443	18.913	17.678
72	UPM	0.5	1.32	3.10	NA	NA	NIE	M4	14.282	16.815	15.549
73	UPM	0.5	1.32	3.30	NA	NA	NIE	M4	16.612	19.146	17.879
74	UPM	0.5	1.32	2.60	NA	NA	NIE	M7	30.024	30.848	30.436
75	UPM	0.5	1.32	2.70	NA	NA	NIE	M7	27.689	30.017	28.853
76	UPM	0.5	1.32	2.80	NA	NA	NIE	M7	33.028	33.898	33.463
77	UPM	0.5	1.32	2.90	NA	NA	NIE	M7	32.132	34.200	33.166
78	UPM	0.5	1.32	3.00	NA	NA	NIE	M7	34.330	36.508	35.419
81	CEDEX	1.0	1.00	2.20	1.50	0.20	NIE	M6	23.318	25.514	25.514
82	CEDEX	1.0	1.00	2.20	1.50	0.20	NIE	M6	21.520	23.557	23.557
83	CEDEX	0.5	0.40	2.20	1.50	0.10	NIE	M6	7.134	8.173	7.654
84	CEDEX	0.5	0.40	1.50	1.50	0.10	NIE	M6	6.322	7.080	6.701

Table A1. *Cont.*

Test	Laboratory	H (m)	W (m)	Z_{dss} (H:V)	Z_{uss} (H:V)	l_c (m)	IE	Gravel Code	$Q_{f,pre}$ (L·s^{-1})	$Q_{f,pos}$ (L·s^{-1})	Q_f (L·s^{-1})
85	CEDEX	1.0	1.00	2.20	1.50	0.20	NIE	M6	21.260	23.049	23.049
87	CEDEX	1.0	1.00	2.20	1.50	0.20	NIE	M6	23.288	24.773	24.031
88	CEDEX	1.0	1.00	2.20	1.50	0.20	NIE	M6	21.000	21.058	21.058
89	CEDEX	1.0	1.00	1.50	1.50	0.20	NIE	M6	19.011	21.375	21.375
90	CEDEX	1.0	1.00	1.50	1.50	0.20	NIE	M6	17.146	21.144	21.144
91	CEDEX	1.0	1.00	1.50	1.50	0.20	NIE	M6	17.576	21.491	21.491
92	CEDEX	1.0	1.00	1.50	1.50	0.20	NIE	M6	21.491	22.929	22.210
94	CEDEX	1.0	1.00	3.00	1.50	0.20	NIE	M6	19.933	21.144	20.539
95	CEDEX	1.0	1.00	3.00	1.50	0.20	NIE	M6	19.011	19.988	19.500
108	CEDEX	0.8	1.00	1.50	1.50	0.20	UF	M7	30.265	35.233	32.749
109	CEDEX	0.229	0.40	1.50	1.50	0.06	UF	M3	1.964	2.136	2.050
110	CEDEX	0.80	1.00	2.50	1.50	0.20	UF	M7	39.279	45.328	42.304
111	CEDEX	0.229	0.40	2.50	1.50	0.06	UF	M3	1.495	3.798	2.647
112	CEDEX	0.80	1.00	3.50	1.50	0.20	UF	M7	40.126	44.883	42.505
113	CEDEX	1.00	1.50	1.50	1.50	0.26	UF	M8	60.130	103.480	81.805
114	CEDEX	1.00	1.50	2.50	1.50	0.26	UF	M8	70.150	107.800	88.975
123	CEDEX	1.00	1.00	1.50	1.50	0.20	CC	M4	15.971	17.196	16.584
124	CEDEX	1.00	1.00	1.50	1.50	0.20	UF	M4	17.121	19.091	18.106
125	CEDEX	1.00	1.00	2.20	1.50	0.20	UF	M4	17.121	19.011	18.066
126	CEDEX	1.00	1.00	3.00	1.50	0.20	UF	M4	18.878	21.087	19.983
127	CEDEX	1.00	1.00	3.00	1.50	0.20	UF	M4	18.878	21.087	19.983
128	CEDEX	1.00	1.00	1.50	1.50	0.20	UF	M4	18.878	21.144	20.011
129	CEDEX	0.229	0.40	3.50	1.50	0.06	UF	M3	1.465	1.651	1.558
130	CEDEX	0.229	0.40	3.50	1.50	0.06	UF	M3	2.441	2.608	2.525
132	UPM	1.0	1.32	1.90	NA	NA	NIE	M4	27.639	30.279	28.959
133	UPM	1.0	1.32	1.60	NA	NA	NIE	M7	44.460	46.904	45.682
134	CEDEX	1.0	1.00	3.00	1.50	0.20	NIE	M1	NA	NA	9.138
135	CEDEX	1.0	1.00	3.00	1.50	0.20	NIE	M1	NA	NA	10.336
136	CEDEX	1.0	1.00	3.00	1.50	0.20	NIE	M1	NA	NA	10.139
137	CEDEX	1.0	1.00	3.00	1.50	0.20	NIE	M1	NA	NA	9.979
138	CEDEX	1.0	1.00	2.20	1.50	0.20	NIE	M1	NA	NA	8.957
139	CEDEX	1.0	1.00	2.20	1.50	0.20	NIE	M1	NA	NA	7.480
140	CEDEX	1.0	1.00	1.50	1.50	0.20	NIE	M1	NA	NA	9.055
141	CEDEX	1.0	1.00	1.50	1.50	0.20	NIE	M1	NA	NA	8.616
142	CEDEX	1.0	1.00	1.50	1.50	0.20	NIE	M1	NA	NA	9.105
143	CEDEX	1.0	1.00	1.50	1.50	0.20	UF	M1	NA	NA	7.945
144	CEDEX	1.0	1.00	2.20	1.50	0.20	UF	M1	NA	NA	8.409
145	CEDEX	1.0	1.00	3.00	1.50	0.20	UF	M1	NA	NA	8.393
146	CEDEX	1.0	1.00	1.50	1.50	0.20	UF	M1	NA	NA	8.159
147	CEDEX	1.0	1.00	3.00	1.50	0.20	CC	M1	NA	NA	8.206
148	CEDEX	1.0	1.00	3.00	1.50	0.20	CC	M1	NA	NA	8.440
149	CEDEX	1.0	1.00	2.20	1.50	0.20	CC	M1	NA	NA	7.156
150	CEDEX	1.0	1.00	2.20	1.50	0.20	CC	M1	NA	NA	8.123
151	CEDEX	1.0	1.00	1.50	1.50	0.20	CC	M1	NA	NA	7.510

Table A2. Summary of the average hydraulic pressures ψ at the base of the physical models nº 85, 87, 90, 94, and 95 for a discharge step in the early stages of each test.

Test	Z_{dss} (H:V)	Q (L·s^{-1})	x^* to Toe	x^* to Crest	$\psi_{average}$ (m)
85	2.2	5.1	0.000	1.000	0.035
85	2.2	5.1	0.182	0.818	0.120
85	2.2	5.1	0.227	0.773	0.144
85	2.2	5.1	0.364	0.636	0.187
85	2.2	5.1	0.545	0.455	0.224
85	2.2	5.1	0.727	0.273	0.252
85	2.2	5.1	0.909	0.091	0.275

Table A2. *Cont.*

Test	Z_{dss} (H:V)	Q ($L \cdot s^{-1}$)	x^* to Toe	x^* to Crest	$\psi_{average}$ (m)
87	2.2	6.7	0.000	1.000	0.036
87	2.2	6.7	0.182	0.818	0.129
87	2.2	6.7	0.227	0.773	0.155
87	2.2	6.7	0.364	0.636	0.205
87	2.2	6.7	0.545	0.455	0.244
87	2.2	6.7	0.727	0.273	0.273
87	2.2	6.7	0.909	0.091	0.296
90	1.5	7.1	0.000	1.000	0.009
90	1.5	7.1	0.267	0.733	0.120
90	1.5	7.1	0.333	0.667	0.149
90	1.5	7.1	0.533	0.467	0.203
90	1.5	7.1	0.800	0.200	0.257
94	3	7.1	0.133	0.867	0.122
94	3	7.1	0.267	0.733	0.207
94	3	7.1	0.300	0.700	0.224
94	3	7.1	0.400	0.600	0.270
94	3	7.1	0.533	0.467	0.325
94	3	7.1	0.667	0.333	0.343
94	3	7.1	0.800	0.200	0.376
94	3	7.1	0.900	0.100	0.399
94	3	7.1	0.933	0.067	0.408
95	3	7.2	0.133	0.867	0.131
95	3	7.2	0.267	0.733	0.221
95	3	7.2	0.300	0.700	0.239
95	3	7.2	0.400	0.600	0.282
95	3	7.2	0.533	0.467	0.322
95	3	7.2	0.667	0.333	0.349
95	3	7.2	0.800	0.200	0.380
95	3	7.2	0.900	0.100	0.401
95	3	7.2	0.933	0.067	0.410

Table A3. Summary of the average hydraulic pressures ψ at the base of the physical models nº 87, 94, and 95 for a transversal section roughly located at $x^* = 0.3$ (30% of the base length from the crest).

Test	Z_{dss} (H:V)	Q ($L \cdot s^{-1}$)	x^* to Toe	x^* to Crest	$\psi_{average}$ (m)
87	2.2	6.7	0.727	0.273	0.274
87	2.2	8.9	0.727	0.273	0.370
87	2.2	10.7	0.727	0.273	0.407
87	2.2	13.5	0.727	0.273	0.437
87	2.2	15	0.727	0.273	0.456
87	2.2	17.8	0.727	0.273	0.457
87	2.2	19.1	0.727	0.273	0.458
94	3	12.8	0.667	0.333	0.504
94	3	15	0.667	0.333	0.480
94	3	19	0.667	0.333	0.415
94	3	20	0.667	0.333	0.416
94	3	21.2	0.667	0.333	0.428
95	3	5	0.667	0.333	0.178
95	3	7.1	0.667	0.333	0.240
95	3	11.7	0.667	0.333	0.324
95	3	13	0.667	0.333	0.342
95	3	17.1	0.667	0.333	0.403
95	3	21.1	0.667	0.333	0.437

References

1. Costa, J.E.; Schuster, R.L. The formation and failure of natural dams. *Geol. Soc. Am. Bull.* **1988**, *100*, 1054–1068. [CrossRef]
2. Wishart, J.S. Overtopping Breaching of Rock-Avalanche Dams. Master's Thesis, University of Canterbury, Upper Riccarton, Christchurch, New Zealand, 2007.
3. Gregoretti, C.; Maltauro, A.; Lanzoni, S. Laboratory experiments on the failure of coarse homogeneous sediment natural dams on a sloping bed. *J. Hydraul. Eng.* **2010**, *136*, 868–879. [CrossRef]
4. Yan, K.; Zhao, T.; Liu, Y. Experimental Investigations on the Spillway Section Shape of the Breaching Process of Landslide Dams. *Int. J. Geomech.* **2022**, *22*, 04022045. [CrossRef]
5. Hansen, D.; Zhao, W.Z.; Han, S.Y. Hydraulic performance and stability of coarse rockfill deposits. *Proc. Inst. Civ. Eng. Water Manag.* **2005**, *158*, 163–175. [CrossRef]
6. Foster, M.; Fell, R.; Spannagle, M. The statistics of embankment dam failures and accidents. *Can. Geotech. J.* **2000**, *37*, 1000–1024. [CrossRef]
7. Li, B.; Garga, V.K. Theoretical solution for seepage flow in overtopped rockfill. *J. Hydraul. Eng.* **1998**, *124*, 213–217. [CrossRef]
8. Li, B.; Garga, V.K.; Davies, M.H. Relationships for non-Darcy flow in rockfill. *J. Hydraul. Eng.* **1998**, *124*, 206–212. [CrossRef]
9. Stephenson, D.J. *Rockfill in Hydraulic Engineering*; Elsevier: Amsterdam, The Netherlands, 1979.
10. Monteiro-Alves, R.; Toledo, M.Á.; Morán, R. Characterization of the Overtopping Flow through the Downstream Shell of Rockfill Dams. *J. Hydraul. Eng.* **2019**, *145*, 04019015. [CrossRef]
11. Sedghi-Asl, M.; Rahimi, H.; Farhoudi, J.; Hoorfar, A.; Hartmann, S. One-dimensional fully developed turbulent flow through coarse porous medium. *J. Hydrol. Eng.* **2014**, *19*, 1491–1496. [CrossRef]
12. Ferdos, F.; Worman, A.; Ekstrom, I. Hydraulic conductivity of coarse rockfill used in hydraulic structures. *Transp. Porous Media* **2015**, *108*, 367–391. [CrossRef]
13. Ferdos, F.; Yang, J.; Wörman, A. Characterization of hydraulic behaviours of coarse rock materials in a large permeameter. *J. Geosci. Environ. Prot.* **2013**, *1*, 1–6. [CrossRef]
14. Salahi, M.-B.; Sedghi-Asl, M.; Parvizi, M. Nonlinear flow through a packed-column experiment. *J. Hydrol. Eng.* **2015**, *20*, 04015003. [CrossRef]
15. Ergun, S. Fluid flow through packed columns. *Chem. Eng. Prog.* **1952**, *48*, 89–94.
16. Venkataraman, P.; Rama Mohan Rao, P. Darcian, transitional, and turbulent flow through porous media. *J. Hydraul. Eng.* **1998**, *124*, 840–846. [CrossRef]
17. Soualmia, A.; Jouini, M.; Masbernat, L.; Dartus, D. An analytical model for water profile calculations in free surface flows through rockfills. *J. Theor. Appl. Mech.* **2015**, *53*, 209–215. [CrossRef]
18. Morteza, S.; Reza, S.M.; Habibollah, B. Concept of hydraulic porosity and experimental investigation in nonlinear flow analysis through rubble-mound breakwaters. *J. Hydrol.* **2014**, *508*, 266–272. [CrossRef]
19. Hansen, D.; Roshanfekr, A. Assessment of potential for seepage-induced unraveling failure of flow-through rockfill dams. *Int. J. Geomech.* **2012**, *12*, 560–573. [CrossRef]
20. Hansen, D.; Roshanfekr, A. Use of index gradients and default tailwater depth as aids to hydraulic modeling of flow-through rockfill dams. *J. Hydraul. Eng.* **2012**, *138*, 726–735. [CrossRef]
21. Parkin, A.K. *Bulletin No. 7: Rockfill Dams with Inbuilt Spillways—Stability Characteristics*; Water Research Foundation of Australis: Adelaide, Australia, 1963.
22. Toledo, M.Á. Rockfill Dams under Overtopping Scenario. Study of the Throughflow and Stability to Sliding. Ph.D. Thesis, Universidad Politécnica de Madrid, Madrid, Spain, 1997. (In Spanish).
23. Javadi, N.; Mahdi, T.F. Experimental investigation into rockfill dam failure initiation by overtopping. *Nat. Hazards* **2014**, *74*, 623–637. [CrossRef]
24. Larese, A.; Rossi, R.; Oñate, E.; Toledo, M.Á.; Morán, R.; Campos, H. Numerical and experimental study of overtopping and failure of rockfill dams. *Int. J. Geomech.* **2015**, *15*, 1–23. [CrossRef]
25. Alves, R.M.; Toledo, M.Á.; Morán, R. Overflow for the complete failure of the downstream shell of a rockfill dam. In Proceedings of the 2nd International Seminar on Dam Protection against Overtopping, Fort Collins, CO, USA, 7–9 September 2016.
26. Siddiqua, S.; Blatz, J.A.; Privat, N.C. Evaluating the behaviour of instrumented prototype rockfill dams. *Can. Geotech. J.* **2013**, *50*, 298–310. [CrossRef]
27. Toledo, M.Á.; Campos, H.; Lara, Á.; Cobo, R. Failure of the downstream shoulder of rockfill dams in overtopping or accidental leakage scenario. In Proceedings of the Dam Protections against Overtopping and Accidental Leakage; Toledo, M.Á., Morán, R., Oñate, E., Eds.; CRC Press: Madrid, Spain, 2015; pp. 89–99.
28. Franca, M.J.; Almeida, A.B. Experimental tests on rockfill dam breaching process. In Proceedings of the International Symposium on Hydraulic and Hydrological Aspects of Reliability and Safety Assessment of Hydraulic Structures, St. Petersburg, FL, USA, 29 May–2 June 2002.
29. Garga, V.K.; Hansen, D.; Townsend, D.R. Mechanisms of massive failure for flowthrough rockfill embankments. *Can. Geotech. J.* **1995**, *32*, 927–938. [CrossRef]
30. Morán, R.; Toledo, M.Á. Research into protection of rockfill dams from overtopping using rockfill downstream toes. *Can. J. Civ. Eng.* **2011**, *38*, 1314–1326.
31. Xu, Y.; Zhang, L.M. Breaching parameters for earth and rockfill dams. *J. Geotech. Geoenviron. Eng.* **2009**, *135*, 1957–1970. [CrossRef]

32. Feliciano Cestero, J.A.; Imran, J.; Chaudhry, M.H. Experimental investigation of the effects of soil properties on levee breach by overtopping. *J. Hydraul. Eng.* **2015**, *141*, 04014085. [CrossRef]
33. Wahl, T.L. Uncertainty of predictions of embankment dam breach parameters. *J. Hydraul. Eng.* **2004**, *130*, 389–397. [CrossRef]
34. Wahl, T.L. *Hydraulic Laboratory Report HL-2014-02—Evaluation of Erodibility-Based Embankment Dam Breach Equations*; Bureau of Reclamation: Denver, CO, USA, 2014.
35. Coleman, S.E.; Andrews, D.P.; Webby, M.G. Overtopping breaching of noncohesive homogeneous embankments. *J. Hydraul. Eng.* **2002**, *128*, 829–838. [CrossRef]
36. Chiganne, F.; Marche, C.; Mahdi, T.-F. Evaluation of the overflow failure scenario and hydrograph of an embankment dam with a concrete upstream slope protection. *Nat. Hazards* **2014**, *71*, 21–39. [CrossRef]
37. Froehlich, D.C. Embankment dam breach parameters and their uncertainties. *J. Hydraul. Eng. ASCE* **2008**, *134*, 1708–1721. [CrossRef]
38. Jing, X.; Chen, Y.; Williams, D.J.; Serna, M.L.; Zheng, H. Overtopping Failure of a Reinforced Tailings Dam: Laboratory Investigation and Forecasting Model of Dam Failure. *Water* **2019**, *11*, 315. [CrossRef]
39. Wahl, T.L. *Prediction of Embankment Dam Breach Parameters—A Literature Review and Needs Assessment*; U.S. Department of the Interior, Bureau of Reclamation, Dam Safety Office: Denver, CO, USA, 1998.
40. ASCE/EWRI Task Committee on Dam/Levee Breaching. Earthen embankment breaching. *J. Hydraul. Eng. ASCE* **2011**, *137*, 1549–1564. Available online: https://ascelibrary.org/doi/abs/10.1061/%28ASCE%29HY.1943-7900.0000498 (accessed on 14 April 2022). [CrossRef]
41. Zhong, Q.; Wang, L.; Chen, S.; Chen, Z.; Shan, Y.; Zhang, Q.; Ren, Q.; Mei, S.; Jiang, J.; Hu, L.; et al. Breaches of embankment and landslide dams—State of the art review. *Earth-Sci. Rev.* **2021**, *216*, 103597. [CrossRef]
42. Powledge, G.R.; Ralston, D.C.; Miller, P.; Chen, Y.H.; Clopper, P.E.; Temple, D.M. Mechanics of overflow erosion on embankments. II: Hydraulic and design considerations. *J. Hydraul. Eng.* **1989**, *115*, 1056–1075. [CrossRef]
43. Mohamed, M.; Samuels, P. Improving the accuracy of prediction of breach formation through embankment dams and flood embankments. In *Proceedings of the River Flow 2002—International Conference on Fluvial Hydraulics*; Bousmar, D., Zech, Y., Eds.; Balkema: Louvain-la-Neuve, Belgium, 2002; Volume 2002.
44. Ricoy, L.F.; Toledo, M.Á.; Morán, R. A model for the analysis of the structural failure of the clay core in rockfill dams due to overtopping. In Proceedings of the Protections 2016: 2nd International Seminar on Dam Protection Against Overtopping, Fort Collins, CO, USA, 7–9 September 2016.
45. Monteiro-Alves, R. Proceso de Rotura de las Presas de Escollera Por Sobrevertido. Ph.D. Thesis, Universidad Politécnica de Madrid, Madrid, Spain, 2021.
46. Zhong, Q.M.; Chen, S.S.; Deng, Z. A simplified physically-based breach model for a high concrete-faced rockfill dam: A case study. *Water Sci. Eng.* **2018**, *11*, 46–52. [CrossRef]
47. Li, Y.; Qiu, W.; Chen, Z.; Wen, L.; Wang, L. Experimental study on the process of overtopping breach of concrete-faced sand-gravel dam. *Eur. J. Environ. Civ. Eng.* 2021. Available online: https://www.tandfonline.com/doi/abs/10.1080/19648189.2021.1939791?journalCode=tece20 (accessed on 14 April 2022).
48. González Tejada, I.; Monteiro-Alves, R.; Morán, R.; Toledo, M.Á. Cellular automata modeling of rockfill dam failure caused by overtopping or any other extreme throughflow. *Eng. Struct.* **2021**, *245*, 112933. [CrossRef]
49. Morán, R. Mejora de la Seguridad de las Presas de Escollera Frente a Percolación Accidental Mediante Protecciones Tipo Repié. Ph.D. Thesis, Universidad Politécnica de Madrid, Madrid, Spain, 2013.
50. Lara, Á.; Cobo, R.; Viña, M.P.; García, F. Pérdidas de energía del flujo en medios granulares. *Ing. Civ.* **2013**, *172*, 97–108.
51. Parkin, A.K. Field solutions for turbulent seepage flow. *J. Soil Mech. Found. Div.* **1971**, *97*, 209–218. [CrossRef]
52. de Winter, J.C.F. Using the Student's t-test with extremely small sample sizes. *Pract. Assess. Res. Eval.* **2013**, *18*, 10. Available online: https://scholarworks.umass.edu/pare/vol18/iss1/10/ (accessed on 14 April 2022).

water

MDPI

Review

A Unified View of Nonlinear Resistance Formulas for Seepage Flow in Coarse Granular Media

Juan Carlos López [1,*], Miguel Ángel Toledo [2] and Rafael Moran [2,3]

[1] Department of Civil Engineering, Hydraulics, Energy and Environment, E.T.S. de Ingenieros de Caminos, Canales y Puertos, Universidad Politécnica de Madrid (UPM), Profesor Aranguren s/n, 28040 Madrid, Spain

[2] SERPA Dam Safety Research Group, Department of Civil Engineering, Hydraulics, Energy and Environment, E.T.S. de Ingenieros de Caminos, Canales y Puertos, Universidad Politécnica de Madrid (UPM), Profesor Aranguren s/n, 28040 Madrid, Spain; miguelangel.toledo@upm.es (M.Á.T.); r.moran@upm.es (R.M.)

[3] Centre Internacional de Metodes Numerics en Enginyeria (CIMNE), Campus Norte UPC, Gran Capitán s/n, 08034 Barcelona, Spain

* Correspondence: jclopez@grupoitinere.com; Tel.: +34-620-248-574

Abstract: There are many studies on the nonlinear relationship between seepage velocity and hydraulic gradient in coarse granular materials, using different approaches and variables to define the resistance formula applicable to that type of granular media. On the basis of an analysis of the existing formulations developed in different studies, we propose an approach for comparing the results obtained by some of the most important studies on state-of-the-art seepage flow in coarse granular media.

Keywords: porous media; Forchheimer equation; non-Darcy flow; high velocity; crushed rock; rounded materials; hydraulic mean radius; intrinsic permeability; shape of particles; angularity of particles; surface roughness of particles

Citation: López, J.C.; Toledo, M.Á.; Moran, R. A Unified View of Nonlinear Resistance Formulas for Seepage Flow in Coarse Granular Media. *Water* **2021**, *13*, 1967. https://doi.org/10.3390/w13141967

Academic Editor: Anargiros I. Delis

Received: 17 May 2021
Accepted: 6 July 2021
Published: 17 July 2021

1. Introduction and Objectives

It is essential to know the relationship between filtration speed and hydraulic gradient to understand the interactions that occurs between infiltrated water and structures composed of gravel or rockfill, such as dams, levees, drainage structures, and coastal dikes. These porous media, made up of coarse particles, possess special characteristics due to their large pores which, under certain conditions, may give rise to non-laminar seepage flow, invalidating Darcy's Law. Calculating the seepage flow through this type of porous media requires the use of so-called nonlinear resistance formulas. The flow regime in which this is applied is known as non-Darcy flow.

Various nonlinear relationships have been proposed to describe the flow in coarse porous media. They can be grouped into two types of equations as follows:

$$i = a \cdot V^b \quad (1)$$

$$i = r \cdot V + s \cdot V^2 \quad (2)$$

where V is the seepage velocity, defined as the average fluid velocity in the whole transversal section; i is the hydraulic gradient; a and r are parameters depending on the characteristics of the porous medium and of the fluid; s is only a function of the characteristics of the porous medium; and b is a function parameter of the conditions of the flow.

Equation (1) is the so-called exponential equation and Equation (2) is the quadratic equation. To obtain relationships among the parameters a, b, r, and s, different physical parameters have been considered; experimental data with different intervals of size, shape, and particle angularity have been used, as well as a wide range of gradient intervals. However, there is currently no formula that can completely create Equations (1) and (2) and combine the determination criteria of their coefficients based on physical parameters.

In order to have a unified vision of the relationships between gradients and seepage velocity developed in different studies, we conduct an analysis of the existing formulations and identify similarities by mathematically comparing the physical parameters considered in these studies. These relationships make it easier to compare the results and formulations obtained by each of the studies considered in this article.

2. Review of Resistance Formulas in Nonlinear Porous Media

2.1. Conceptual Approach

Various studies have been completed with the aim of developing expressions for parameters r (linear coefficient) and s (quadratic coefficient) of Forchheimer's Law (1901) [1] Equation (2), which defines a macroscopic hydraulic behaviour. The first term of Equation (2) (i.e., the term with coefficient r) represents the loss of energy due to the viscous forces and depends on the properties of both the porous medium and the fluid. The second term of Equation (2) (i.e., the term with coefficient s) considers the loss of energy due to the forces of inertia, and depends only on the properties of the porous medium.

In most of these studies, the development of formulas was based on the analogy of the flow in pipes, through the application of two dimensionless groups that, in this paper, are referred to as the generalised friction factor f of the Darcy–Weisbach Equation (3) and generalised Reynolds number R_e Equation (4) represented as:

$$f = L_c \cdot 2g \cdot \frac{i}{V_p^2} \tag{3}$$

$$R_e = \frac{L_c \cdot V_p}{v} \tag{4}$$

where L_c is the characteristic length adopted in each case, g is the gravitational acceleration, i is the hydraulic gradient, v is the kinematic viscosity, and V_p is the pore velocity, determined by Equation (5):

$$V_p = \frac{V}{n} \tag{5}$$

being n the porosity of the porous medium.

The characteristic lengths, on which most studies have been based, can be grouped into three types:

(a) the representative size of the particle (D)

(b) the square root of the intrinsic permeability (K_0), as a macroscopic property of the porous medium. For the laminar regime, it is determined by Equation (6):

$$K_0 = \frac{v}{g} \cdot \frac{1}{i} \cdot V \tag{6}$$

(c) the hydraulic mean radius R_h that was first defined by Taylor (1948) [2] and determined by Equation (7):

$$R_h = \frac{n}{S_e \cdot (1-n)} \tag{7}$$

where S_e is the average specific surface area of the solid particles that make up the porous medium and depends on the shape, angularity, and surface roughness of the particles (Crawford, C.W. et al., 1986 [3]; Sabin G.C.W. and Hansen D., 1994 [4]) and is defined by the Equation (8):

$$S_e = \frac{SP}{VP} \tag{8}$$

where SP is the average particle surface area and VP is the average particle volume.

Among the first studies are those developed by Blake (1922) [5] and S.P. Burke and W.B. Plumer (1928) [6], who used dimensionless groups defined by Equations (3) and (4), adopting L_c as the characteristic length, and S as the specific surface area of the packing of porous medium. The relationship with the average particle specific surface area S_e determined by Equation (9) as follows:

$$S = S_e \cdot (1 - n) \tag{9}$$

In their tests, they used commercial porous media (glass beads, glass rings, solid glass cylinders, and Raschig rings), and the experimental data corresponding to a porous medium with the same geometry and size. The results were represented a generalised diagram [R_e, f, Equation (10) and the values fitted to a smooth curve:

$$f = \frac{A_1}{R_e} + A_2 \tag{10}$$

where A_1 is referred to as a linear generalised dimensionless coefficient, and A_2 as a quadratic generalised dimensionless coefficient.

Equation (10) is referred to as a generalised equation $[R_e, f]$.

Figure 1 shows a $[R_e, f]$, schematic diagram similar to that used by Bear (1988) [7] with asymptotic values for increased Reynolds numbers R_e.

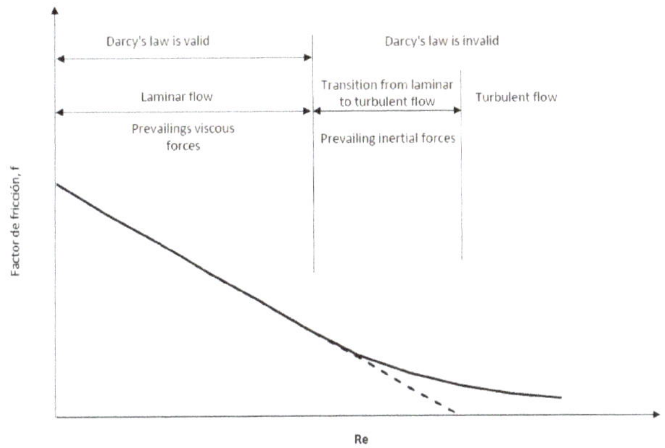

Figure 1. Schematic diagram of $[R_e, f]$ Equation (10). Adapted from Bear (1988) [7].

As indicated by Sabri Ergun and A. A. Orning (1949) [8] "This transition from the dominance of viscous to kinetic effects, for most packed systems, is smooth, indicating that there should be a continuous function relating pressure drop to flow rate."

Ward (1964) [9] stated the same thing when he asserted, "The smooth transition from laminar to turbulent flow in porous media is expected. In the laminar flow region, the flow is laminar in all parts of the porous media. In the laminar transition region, the flow is laminar in most parts of the porous medium, but there are parts where the flow is turbulent. In the turbulent transition region, the flow is turbulent in most parts of the porous medium, but there are still parts where laminar flow conditions persist. Finally, turbulent flow exists in all parts of the porous medium at high values of R_k. Simultaneous existence of laminar and turbulent flow in different parts of a porous medium is possible because of the irregularities and variation in pore size."

Dudgeon (1966) [10] indicated that "the only likely solution to the problem of a generalised $[R_e, f]$ plot is in terms of a set of graphs for each family of geometrically similar porous media."

Other authors (Blake (1922) [5] and S.P. Burke and W.B. Plumer (1928) [6]; Morcon A.R. (1946) [11]; Ergun (1952) [12], Kadlec, H.R., and Knight, L.R. (1966) [13]; Ahmed and Sunada (1969) [14]; Kovacs (1969) [15], Arbhabhirama and Dinoy (1973) [16]; Stephenson (1979) [17]; Li, B et al. (1998) [18];and more recently Sidiropuolou et al. (2007) [19]; Moutsopoulos et al. (2009) [20]; Sedghi-Asl and Rahimi (2013) [21]; and Salahi et al.

(2015) [22]), obtained continuous curves such as those given by Equation (10) through the corresponding adjustment of the experimental data used, always within the range of the Reynolds number R_e on which the tests were developed.

The existence of these continuous curves in porous media contrasts with the turbulent flow in pipes where there are sudden jumps in the Reynolds number interval R_e between 2000 and 4000 (White, F.W. (2003) [23]).

If we substitute the generalised values of f and R_e, given by Equations (3) and (4) in Equation (10) we get Equation (11):

$$L_c \cdot 2 \cdot g \cdot \frac{i}{V_p^2} = \frac{A_1 \cdot v}{L_c \cdot V_p} + A_2 \tag{11}$$

Resolving i in Equation (11) we get what we refer to as a quadratic generalised equation, i.e., Equation (12):

$$i = \frac{v}{2 \cdot g} \cdot A_1 \cdot \frac{1}{L_c^2} \cdot V_p + \frac{1}{2 \cdot g} \cdot A_2 \cdot \frac{1}{L_c} \cdot V_p^2 \tag{12}$$

In accordance with the Equation (12) the parameters r and s of the Forchheimer equation, i.e., Equation (2) determine Equations (13) and (14):

$$r = \frac{v}{2 \cdot g} \cdot A_1 \cdot \frac{1}{L_c^2} \cdot \frac{1}{n} \tag{13}$$

$$s = \frac{1}{2 \cdot g} \cdot A_2 \cdot \frac{1}{L_c} \cdot \frac{1}{n^2} \tag{14}$$

In the case of the fully developed turbulent regime, it may be possible to disregard the linear expression r, in such a way that we could obtain the exponential equation, Equation (2), considering the coefficient a as equal to the quadratic expression s and the exponent b as equal to 2.

Various researchers have worked on the exponential law, Equation (1), in regimes of transition, and therefore with the exponent values b below 2: Wilkins (1956) [24], $b = 1.85$; Dudgeon (1966) [10], $1.2 < b < 1.91$; Parkin (1991) [25], $b = 1.85$; Moutsopoulos K. N. et al. (2009) [20], $1.280 < b < 1.687$; Sedhi-Asl et al. (2013) [21] $1.479 < b < 1.804$.

In this respect, Stephenson (1979) [17] pointed out the fact that not obtaining an adjustment of b equal to 2, with the experimental data, is due to "the tests being carried out on a small scale and as a result of low Reynolds numbers." Ferdos, F. et al. (2015) [26] obtained an exponent b equal to 2 in their tests as a result of using elevated Reynolds numbers R_e. These studies both used the particle Reynolds number R_d determined by Equation (15) as follows:

$$R_d = \frac{D \cdot V_p}{v} \tag{15}$$

They achieved values of R_d = 220,000 for the size interval of 100–160 mm and R_d = 320,000 for the range of 160–240 mm. The values were both very much higher as compared with the value proposed by Stephenson (i.e., R_d = 10,000) in order to reach the fully developed turbulent flow.

It is important to consider that the flow through for highly permeable porous media may contain various flow regimes, as the generalised Reynolds number R_e increases. In general, most of the studies have shown that there are four flow regimes: laminar, nonlinear laminar, turbulent transition, and fully developed turbulent flow (Ward, 1964 [9]; Wright (1968) [27]; Kovacks (1969) [15], Dybbs and Edwards (1975) [28]; R.M. Fand et al. (1987) [29]; H. Huang and S. Ayoub (2007) [30]) and, consequently, there is a debate regarding whether the dimensionless coefficients A_1 and A_2 considered in Equations (10) and (12) for a determined porous medium, are constants throughout a wide range of seepage velocity spanning the four flow regimes, in other words maintaining the relationship determined by Equation (10) without producing sudden jumps as occurred in the flow in tubes.

For that purpose, it is interesting to linearise Equation (10) multiplying both parts of the equation by R_e to obtain Equation (16):

$$\lambda = A_1 + A_2 \cdot R_e \quad (16)$$

where λ, the linearised generalised friction faction, is determined by Equation (17):

$$\lambda = f \cdot R_e \quad (17)$$

Equation (16) is referred to as a linearised generalised equation $[R_e, \lambda]$.

Figure 2 shows a $[R_e, \lambda]$ diagram similar to that used by R.M Fand et al. (1987) [29]. In accordance with Equation (16), each porous material is represented by a line whose curve corresponds with the quadratic generalised dimensionless coefficient A_2; the cut on the ordinate axis corresponds to the linear generalised dimensionless coefficient A_1. The laminar regime is represented by a horizontal line. This type of graph is used to check if, for the same porous material, each non-Darcy flow regime remains determined by their corresponding value pairs $[A_1, A_2]$ or, on the contrary, such values are constant for the three non-Darcy flow regimes.

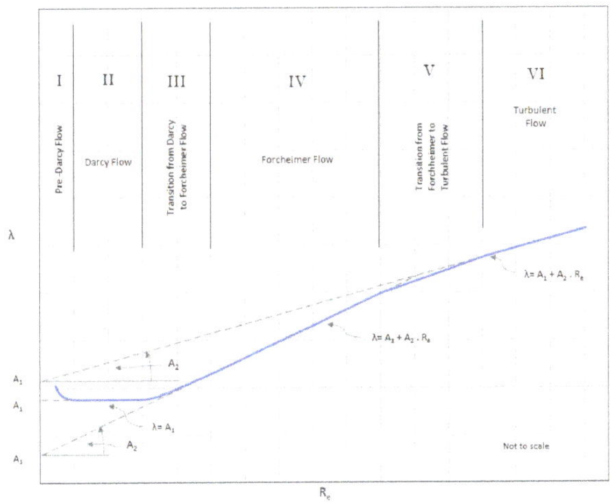

Figure 2. Schematic diagram $[R_e, \lambda]$ Equation (16). Adapted from Fand et al. (1987) [30].

In this debate, the following studies stand out: McCorquodale et al. (1978) [31] and Fand et al. (1987) [29]. The first group used granular materials, of various sizes, shapes, and angularity, whereas the second worked with spheres with a range of sizes from 2.00 mm < D < 4.00 mm. The studies both proposed different coefficient values, A_1 and A_2, for the transition zones that they detected in their tests: nonlinear laminar and turbulent transition.

2.2. *Resistance Formulas*

Next, we describe the main nonlinear resistance formulas based on the adopted characteristic length L_c.

2.2.1. Resistance Formulas Based on the Representative Diameter of the Particles

In 1952, Sabri Ergun [12] carried out a study to develop a formula for general application based on the dimensionless groups:

$$f_E = D_p \cdot g \cdot \frac{n^3}{(1-n)} \cdot \frac{i}{V^2} \tag{18}$$

$$R_E = \frac{D_p \cdot V}{v} \tag{19}$$

where f_E is Ergun's friction factor and R_E is Ergun's Reynolds number. The parameter D_p was adopted as the representative diameter of the particles, that is, the average effective diameter of the granular material and determined by Equation (20):

$$D_p = \frac{6}{S_e} \tag{20}$$

where D_p corresponds with the diameter of a sphere that has the same specific surface area, S_e, as the particle.

The continuous curve, as determined by Equation (10) and defined by the author, we refer to it as the Ergun equation $[R_E, f_E]$ as follows:

$$f_E = 72 \cdot \alpha \cdot \frac{(1-n)}{R_E} + \frac{6}{8} \cdot \beta \tag{21}$$

where α is the linear dimensionless coefficient and β is the quadratic dimensionless coefficient (by Sabri Ergun and A. A. Orning).

Finally, substituting the values of f_E and R_E of Equations (18) and (19) into Equation (21) we get the quadratic equation from Ergun (1952) [12] as Equation (22):

$$i = \frac{v}{g} \cdot 72 \cdot \alpha \cdot \frac{1}{D_p^2} \cdot \frac{(1-n)^2}{n^3} \cdot V + \frac{1}{g} \cdot \frac{6}{8} \cdot \beta \cdot \frac{(1-n)}{n^3} \cdot \frac{1}{D_p} \cdot V^2 \tag{22}$$

This equation proposed by Ergun (1952) [12] had, in fact, been previously developed by Sabri Ergun and A. A. Orning (1949) [8] who adopted as characteristic length L_c, the specific surface area of the particles S_e. The equation is obtained by substituting in Equation (22) the value of the effective diameter D_p defined by Equation (20). These authors developed this equation based on Kozeny's (1927) hypothesis [32], which is based on a capillary model that, as the authors show "the granular bed is equivalent to a group of parallel and equal-sized channels, such that the total internal surface and the free internal volume, are equal to the total packing surface area and the void volume, respectively, of the randomly Packed bed." However, they added that "For a packed bed, the flow path is sinuous, and the stream lines frequently converge and diverge. The kinetic losses, which occur only once for the capillary, occur with a frequency that is statistically related to the number of particles per unit length. For these reasons, a correction factor must be applied to each term. These factors may be designated as α and β." For these reasons, in agreement with the authors, it is essential to consider some correction factors in each linear and quadratic expression of Equation (22) to consider these characteristics of the porous medium. The authors designated them as α and β, respectively.

Equation (22) considers the existence of a function of porosity for each expression: f_L for the linear expression (r) and f_T for the quadratic expression (s) of Equation (2) where:

$$f_L = \frac{(1-n)^2}{n^3} \tag{23}$$

$$f_T = \frac{(1-n)}{n^3} \tag{24}$$

As per Ergun, Leva M., and Grimmer M. (1947) [33] they confirmed these functions of porosity. Ergun himself carried out various tests on a wide range of porosity variation n to check the validity of these functions.

With respect to the shape, angularity, and surface roughness of the particles, these physical parameters seem implicit in the effective diameter, D_p, which relate to the specific surface area S_e through Equation (20).

Finally, with respect to the packed porous material he indicates that "the orientation of the randomly packed beds is not susceptible to exact mathematical formulation", and, consequently, "the effect of the orientation was not included."

Ergun, worked on 640 experiments including his own, made up of crushed porous solids, and those obtained through other authors such as Burke and Plummer (1928) [6], who worked with lead shot (spherical shape) of reduced sizes 1.48, 3.08, and 6.34 mm and Morcon (1949) [11], who worked with capsules, cylinders, nodules, and spheres. Ergun represented the three series of data in the diagram $[R_E, f_E]$ checking that they correctly matched Equation (21) The fluids used in this case were the gases of CO_2, N_2, CH_4, and H_2. As a result of this, he obtained the universal values of α = 2.08 and β = 2.33 applicable to all porous media.

According to Ergun, most dispersions happen with porous materials that include a mixture of sizes (non-uniform materials), and with those in which the relationship between the diameter of the permeameter (D_x) and the representative size of the particle D is less than 10 (influence of the wall effect). In such cases, the corresponding tests were not considered.

Although the author did not analyse the theoretical significance of the parameters α and β, he did confirm that over a wide range of porosities he found no relationships of α and β with the porosity.

Later, Frank Engelund (1953) [34] carried out a study to analyse the influence of turbulence in subterranean waters on uniform limestone sands, working on his own tests and those of other authors (Lindquist, 1933 [35] and Chardabellas, 1940 [36,37]). For his tests, he worked only on three samples, two of them were with D = 2.6 mm and one with D = 1.4 mm. The proposed quadratic equation Equation (25) was:

$$i = \frac{v}{g} \cdot \alpha_0 \cdot \frac{1}{D_e^2} \cdot \frac{(1-n)^3}{n^2} \cdot V + \frac{1}{g} \cdot \beta_0 \frac{(1-n)}{n^3} \cdot \frac{1}{D_e} \cdot V^2 \tag{25}$$

He used as the representative diameter of particle D the parameter D_e, which is the equivalent diameter that corresponds to the diameter of a sphere with the same volume of the particle; α_0 is the linear dimensionless coefficient of Engelund and β_0 is the quadratic dimensionless coefficient of Engelund. According to the author, "α_0 and β_0 are dimensionless numerical constants depending, for uniform soil, on the structure and the grain shape."

The structure of Equation (25) is remarkably similar to Equation (22) given by Ergun (1952) [12]. The difference in the representative diameter D of the adopted particle, D_p by Ergun (1952) [12] and D_e by Engelund (1953) [34], and the function of linear porosity f'_L obtained as:

$$f'_L = \frac{(1-n)^3}{n^2} \tag{26}$$

According to Frank Engelund (1953) [34], this porosity function f'_L of the linear expression fits better than the function f_L from Ergun Equation (23) to the measurements obtained by Rose (1953) [37] and Franzini (1951) [38], especially the latter, who worked with a wide interval of porosities $0.270 < n < 0.476$.

Ergun (1952) [12] proposed universal values for the dimensionless coefficients α and β. However, Engelund (1953) [34] did consider the shape and angularity of the particles through the dimensionless coefficients α_0 and β_0 for which he proposed the following values:

α_0 = 780 for uniform spherical particles (using data from Lindquist (1933)); α_0 = 1000 for uniform rounded sands (using data from Chardabellas (1940)), and α_0 = 1500 or above for angular sands, (using own experimental data).

β_0 = 1.8 for uniform spherical particles (using data from Lindquist (1933)); β_0 = 2.8 for uniform rounded sands (using data from Chardabellas (1940)), and β_0 = 3.6 or above for angular and uniform sands, (using own experimental data).

Finally, we must point out that it is simpler to measure as representative diameter of the particle D the value of D_e than the value of D_p, which requires the measurement of the particle specific surface area S_e, in accordance with Equation (20).

In 1979, Stephenson carried out a study whose main purpose was to research the fully developed turbulent flow regime. On the basis of an analogy with the flow in pipes (Darcy–Weisbach equation), he considered that the gradient i was proportional to the expression Equation (27):

$$i \propto \frac{1}{R_h} \cdot \frac{V_p^2}{2 \cdot g} \tag{27}$$

As the hydraulic mean radius R_h is proportional to the size of the particle D (Leps (1973)) [39] and, furthermore, this variable is more easily measured, he suggested using the exponential equation, i.e., Equation (1) with coefficient b equal to 2 for Equation (28):

$$i = \frac{1}{g} \cdot f_d \cdot \frac{1}{D} \cdot \frac{1}{n^2} \cdot V^2 \tag{28}$$

where f_d is the particle friction factor and, as the author noted, "is actually the function of the Reynolds number R_d". The author used as the representative diameter of the particle D, the parameter D_{50} for lack of other data in the bibliography. However, for conceptual reasons, we will continue working with the representative diameter of the particles D.

He considered the dimensionless groups $[R_d, f_d]$ where:

$$f_d = D \cdot g \cdot \frac{i}{V_p^2} \tag{29}$$

and R_d is defined by the Equation (15).

He represented the data of three types of porous materials: smooth spheres, river gravel, and crushed aggregates on the diagram from Stephenson (1979) [17] $[R_d, f_d]$. This data came from tests carried out by the same author and by various researchers: Dudgeon (1966) [10], Volker (1969) [40], Leps (1973) [39], and Cedergren (1977) [41]. With this experimental data and their corresponding ranges of particle sizes and hydraulic gradients, he matched three smooth curves whose lines coincided in the laminar regime. From this line, they gradually separate in the transition regime towards turbulence with three asymptotic values: f_d = 1 for smooth spheres, f_d = 2 for river gravel, and f_d = 4 for crushed aggregate (see Figure 3).

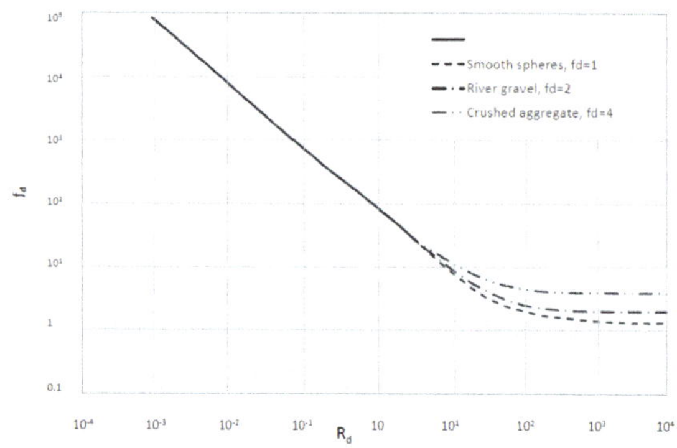

Figure 3. Schematic diagram $[R_d, f_d]$ proposed by Stephenson (1979) [17].

For laminar flow, considering the analogy with the flow in pipes and based on data from Dudgeon (1966) [10] and Cedergren (1977) [41], he proposed the relationship:

$$f_d = \frac{K_l}{R_d} \tag{30}$$

where K_l is the linear dimensionless coefficient from Stephenson (1979) [17]. The value of K_l obtained from the experimental data was 800. The limit of the Reynolds number R_d for the laminar regime proposed by Stephenson was $R_d = 10^{-4}$.

In the transition zone, he proposed Equation (31) of the type given by Equation (10):

$$f_d = \frac{K_l}{R_d} + K_t \tag{31}$$

where K_t is the quadratic dimensionless coefficient from Stephenson (1979) [17]. According to the author, Reynolds numbers of $R_d > 10{,}000$ produce the fully developed turbulent flow regime and, in this case, the author indicated that it was only a function of the shape and angularity of the particles.

Ultimately, Equation (31) (similar to the generalised equation Equation (10)), the Ergun equation, i.e., Equation (21), and by implication Equation (25) by Frank Engelund, represent continuous curves for each porous medium in a generalised diagram $[R_e, f]$ that tend to an asymptotic value for fully developed turbulent regime (see Figures 1 and 3).

Finally, substituting Equations (15), (29) and (31), we get the quadratic equation Equation (32):

$$i = \frac{v}{g} \cdot K_l \cdot \frac{1}{D^2} \cdot \frac{1}{n} \cdot V + \frac{1}{g} \cdot K_t \cdot \frac{1}{D} \cdot \frac{1}{n^2} \cdot V^2 \tag{32}$$

Equation (32), according to Li B. et al. (1998), was not initially proposed by Stephenson (1979) [17] and these authors termed it as the modified Stephenson equation.

2.2.2. Resistance Formulas Based on Intrinsic Permeability

Ward (1964) [9] obtained a quadratic formula for general application taking $\sqrt{K_0}$ as the characteristic length. For this, he worked with the dimensionless groups expressed in Equations (33) and (34):

$$f_k = \sqrt{K_0} \cdot g \cdot \frac{i}{V^2} \tag{33}$$

$$R_k = \frac{\sqrt{K_0} \cdot V}{v} \tag{34}$$

where f_k is Ward's friction factor and R_k is Ward's Reynolds number.

For the transition zone, in a $[R_k, f_k]$ diagram of the type produced by Equation (10), Ward (1964) [9] proposed Equation (35):

$$f_k = \frac{1}{R_k} + C \tag{35}$$

where C is the quadratic dimensionless coefficient of Ward which, as indicated by Ahmed and Sunada (1969) [14], "Previous investigators have interpreted the term C as a constant reflecting geometric properties of the medium."

Considering the three previous equations, the author obtained the quadratic equation, i.e., Equation (36):

$$i = \frac{v}{g} \cdot \frac{1}{K_0} \cdot V + \frac{1}{g} \cdot C \cdot \frac{1}{\sqrt{K_0}} \cdot V^2 \tag{36}$$

Ward (1964) [9] had the same aim as Ergun (1952) [12], to adjust a single value of C so that Equation (35) might be generally applicable. Engelund (1953) [34], and Stephenson (1979) [17], in fact, proposed three equations also for general application, but depending on the structure of the porous material and the shape of the particles.

Ward (1964) [9] used 20 different porous media: glass bead, ion exchange resin, sands, gravel, granular activated carbon, and anthracite. The size of each porous medium was defined through the geometric mean of the particle sizes M_g in each porous medium. The size interval was 0.27 mm < M_g < 16.10 mm. In total, 53 tests were carried out and the fluid used was water. The Reynolds number interval was $0.122 < R_k < 18.10$, with the size interval of M_g studied, he obtained an adjusted value for C of 0.55 with a standard deviation σ_s of 0.024.

The author noted that "D.K. Todd (1959) [42] showed a plot similar to Figure 1, except that an average grain diameter is used in place of the square root of the permeability in Equations (10) and (16). Because the average grain diameter is not sufficient to characterize a porous medium, there is considerable scatter in the plotted points."

Regardless of the existence of the smooth transition curve (see Figure 1), Ward (1964) [9] proposed, for engineering applications, the division into four types of flow regimes: laminar, nonlinear laminar, turbulent transition, and fully developed turbulent; with the following limits of the Reynolds number R_k among them: laminar $R_k < 0.0182$, nonlinear laminar $R_k < 1.82$, and turbulent transition $R_k < 182$.

To conclude, we must point out that Ward attributes, in some cases, the deviations seen in respect of Equation (35) to an inadequate determination of the intrinsic permeability K_0, which must be obtained with very low gradients to be in a laminar regime in accordance with Equation (6).

Continuing with Ward's study (1964) [9], Ahmed and Sunada (1969) [14] developed the same quadratic equation, i.e., Equation (36) from the Navier–Stokes' equations, based on the hypothesis that the porous medium is homogeneous and isotropic on a macroscopic scale, and that the chemical and thermodynamic effects are small. His study focused specifically on the nonlinear laminar regime.

Ahmed and Sunada (1969) [14] stated that "for flow through porous media, convective accelerations are always present, whereas turbulence is a random phenomenon dependent upon of flow velocity and space geometry." According to the authors, "This fact was demonstrated experimentally by Schneebeli (1955) [43] who used dye to identify the flow. He injected dye into the flow at various velocities (steady-state conditions) and found that, even though measurements of gradients and velocities indicated nonlinear flow, the dye assumed laminar characteristics, i.e., stream-lined flow. Increasing the flow velocities approximately four times caused the dye from one channel to mix with the dye of another, indicating that departure from Darcy's Law should be the result of convective acceleration of the fluid within the pores space."

More recently, H. Huang and J. Ayoub (2007) [30], concluded, "Derivation of the Forchheimer equation from the Navier–Stokes' equation reveals that the nature of the Forchheimer flow regime is laminar with inertial effect. The inertia resistance factor β can be used to characterize this flow regime and is therefore an intrinsic property of the porous media."

Along the same lines, Balhoff, M.T. et al. (2009) [44] indicated, "The constant, β, is referred to us as the non-Darcy coefficient and, like permeability, is an empirical value specific to the porous medium. It represents the additional inertial resistance caused by the converging/diverging and tortuous medium geometry."

The phenomenon of the turbulence in porous media is very complex. Recently, Sidiripoulou, M.G. et al. (2007) [19] referred to studies by Skjetne and Auriault (1998) [45], Panfilov et al. (2003) [46], and Fourar et al. (2004) [47] on the mechanisms of turbulence, which were related to the separation of the layer limit and the recircularisation of the vortices formed.

2.2.3. Resistance Formulas Based on the Hydraulic Mean Radius

In 1998, Li B et al. (1998) [18], based on an analogy with the flow in pipes and using as characteristic length L_c and the hydraulic mean radius R_h defined by Equation (7), proposed the dimensionless groups Equations (37) and (38):

$$f_p = 4 \cdot R_h \cdot 2g \cdot \frac{i}{V_p^2} \tag{37}$$

$$R_p' = \frac{R_h \cdot V_p}{v} \tag{38}$$

where f_p is the pore friction factor and R_p' is the Reynolds number based on R_h.

The transition curve proposed in the pore diagram $[R_p', f_p]$ was determined by Equation (39):

$$f_p = \frac{\alpha'}{R_p} + \beta' \tag{39}$$

where α' and β' were dimensionless coefficients of the pores from the linear and quadratic expressions, respectively.

Substituting Equations (37)–(39), the authors obtained the quadratic equation:

$$i = \frac{v}{8 \cdot g} \cdot \alpha' \cdot \frac{1}{R_h^2} \cdot V_p + \frac{1}{8 \cdot g} \cdot \beta' \cdot \frac{1}{R_h} \cdot V_p^2 \tag{40}$$

With the experimental data provided by the University of Ottawa (Hansen (1992)) [48], which included materials for rockfill dams in intervals of 16.0 mm < D < 40.0 mm, they obtained values of 98 for α' and 3 for β'. The permeameter used had a diameter D_x of 300 mm. The pore Reynolds number R_p', from which the total turbulent regime was developed, was 200.

They subsequently extended this data, with that supplied by Stephenson (1979) [17] and Li and Hu (1988) [18] obtaining a value of 1279 for the nonlinear dimensionless coefficient K_l and 3.84 for the quadratic dimensionless coefficient K_t which is shown in the modified Stephenson equation, i.e., Equation (32). These values are in the same order of magnitude as those obtained by Stepheson for crushed aggregate, i.e., $K_l = 800$ and $K_t = 4.00$.

More recently, Mohammad-Bagher Salahi et al. (2015) [22], working on rounded granular materials (2.10 mm < D < 17.78 mm) and aggregate (1.77 mm < D < 16.62 mm), obtained values $K_l = 488$ and $K_t = 4725$ for rounded aggregate $K_l = 588$ and $K_t = 5550$ for crushed aggregate. Both values of k_t are higher than those proposed by Stephenson (k_t = 2.00 for rounded aggregate and K_t = 4.00 for crushed aggregate). The Reynolds number interval, R_d, with which they developed the tests, was 10 < R_d < 1882.

Additionally, Li B et al. (1998) [18] represented the experimental data in the diagram from Stephenson $[R_d, f_d]$. According to an analysis of this diagram, they proposed that the fully developed turbulent regime (asymptotic curve) should be obtained for values of R_d > 2000. This value is consistent with the data previously obtained for the pore Reynolds number R_p' > 200 if we consider that the hydraulic mean radius R_h is approximately 10% of the representative size of the particle D (Parkin, 1991) [25]. However, this limit value of R_d is lower to that proposed by Stephenson of R_d = 10,000.

2.2.4. On the Physical Parameters r and s of the Forchheimer Equation

Sidiropoulou et al. (2007) [19] obtained empirical relationships for a general application to any porous medium for the coefficients r and s in the function of physical parameters such as representative particle size D and porosity n on the basis of an analysis of multiple regression using the data from various authors: Ward (1964) [9], Ahmed and Sunada (1969) [14], Arbhabhirama and Dinoy (1973) [16], Ranganadha Rao and Suresh (1979) [49], Tyagi and Todd (1970) [50] who used the data from Dudgeon (1966) [10], Venkataraman and Rao (1988) [51], and Bordier and Zimer (2000) [52]. The number from the available experimental data (N) was 115. However, in many cases, the complete data needed for the adjustment was not available, that is, r, s, D and n.

They obtained three different empirical relationships for the expressions r and s to study how the porosity influenced parameters r and s:

$$r = 0.00333 \cdot D^{-1.500403} \cdot n^{0.060350} \quad (R^2 = 0.9108;\ N = 55) \tag{41a}$$

$$s = 0.194325 \cdot D^{-1.265775} \cdot n^{-1.141417} \quad (R^2 = 0.8715;\ N = 49) \tag{41b}$$

$$r = 0.0002789 \cdot D^{-1.502361} \cdot (1-n)^{-0.216014} \quad (R^2 = 0.9142;\ N = 55) \tag{42a}$$

$$s = 1.228873 \cdot D^{-1.263314} \cdot (1-n)^{-1.532475} \quad (R^2 = 0.8762;\ N = 49) \tag{42b}$$

$$r = 6.527953 \cdot 10^{-15} \cdot D^{-1.54745} \cdot n^{-16.068711} \cdot (1-n)^{-23.157232} \quad (R^2 = 0.9188;\ N = 55) \tag{43a}$$

$$s = 1.107768 \cdot 10^{-10} \cdot D^{-1.30182} \cdot n^{-13.836369} \cdot (1-n)^{-18.365290} \quad (R^2 = 0.8806;\ N = 49) \tag{43b}$$

where D is given in metres, r in seconds per metre, and s in seconds squared per metre squared.

The best adjustment obtained was for the Equation (43a,b).

The issue with the previous empirical relationships is that they were based on porous materials made up of particles with different geometries: glass beads, granular activated carbon, ion exchange resin, sand, gravel, anthracite coal, angular gravel, round river gravel, blue metal, river gravel, marbles, and glass spheres.

In fact, these equations, were an attempt to provide a general application equation as per the studies made by Ergun (1952) [12] and Ward (1964) [9]. Frank Engelund (1953) [34] went further by proposing three general equations in the function of the shape and angularity of particles through the dimensionless coefficients α_0 and β_0. Stephenson (1979) [17] also proposed three different equations: smooth spheres, river gravel, and crushed aggregate (see Figure 3).

3. Analysis of the Relationships among Parameters of the Different Formulas of Resistance

To adequately define the filtration phenomenon through a porous medium, we use three types of equations: the generalised equation $[R_e, f]$ determined by Equation (10), the quadratic generalised equation, i.e., Equation (12), and the linear generalised equation $[R_e, \lambda]$ determined by Equation (16), all of them using the general characteristic length L_c.

The purpose of this section is to standardise the formulas described in the previous section. For this, we have chosen the quadratic generalised equation as a base in accordance with Equation (12), it includes the generalised characteristic length L_c and the generalised dimensionless coefficient A_1 (linear term) and A_2 (quadratic term).

To arrive at this standardisation, the following mathematical relationships were studied:

(a) Among the characteristic lengths, R_h, $\sqrt{K_0}$, and D;
(b) Among the Reynolds numbers R_p, R_k and R_d;
(c) Among the different laminar dimensionless coefficients α, α_0, K_l, and α', and quadratic dimensionless coefficients β, β_0, K_t, β', and C.

These relationships should allow us to compare the results proposed by different authors who use different characteristic lengths L_c and, as a result, different values for the laminar dimensionless coefficients, (α, α', α_0, and K_l), quadratic dimensionless coefficients (β, β_0, β', C, and K_t), and also different values for the limit of the Reynolds number R_e that define the zones of the flow regime: laminar, nonlinear laminar, turbulent transition, and fully developed turbulent.

The relationships that we have obtained in this section are based on uniform granular materials where the uniformity coefficient is not included C_u.

3.1. Equations with Characteristic Length Based on the Hydraulic Diameter

In accordance with the analogy of the flow in pipes, for noncircular sections, we can take the characteristic length to be L_c and the hydraulic mean diameter as D_h:

$$D_h = 4 \cdot R_h \tag{44}$$

If we substitute Equation (44) in Equation (12) we get:

$$i = \frac{v}{2 \cdot g} \cdot A'_1 \cdot \frac{1}{D_h^2} \cdot V_p + \frac{1}{2 \cdot g} \cdot A'_2 \cdot \frac{1}{D_h} \cdot V_p^2 \tag{45}$$

where A'_1 is the linear dimensionless coefficient corresponding to $L_c = D_h$ and A'_2 is the quadratic dimensionless coefficient corresponding to $L_c = D_h$.

Substituting Equation (44) in Equation (45) we obtain Equation (46):

$$i = \frac{v}{2 \cdot g} \cdot \frac{A'_1}{16} \cdot \frac{1}{R_h^2} \cdot V_p + \frac{1}{2 \cdot g} \cdot \frac{A'_2}{4} \cdot \frac{1}{R_h} \cdot V_p^2 \tag{46}$$

If we compare Equation (46) with Equation (40) proposed by Li, B. et al. (1998) [18], we get the dimensionless coefficients A'_1(linear) and A'_2(quadratic):

$$A'_1 = 4 \cdot \alpha' \tag{47}$$

$$A'_2 = \beta' \tag{48}$$

Considering Equation (7), which defines the hydraulic mean radius R_h we obtain the expression for the specific surface area S_e:

$$S_e = \frac{n}{(1-n)} \cdot \frac{1}{R_h} \tag{49}$$

Considering Equations (5) and (49), which defines the pore velocity V_p and substituting Equation (46) we get:

$$i = \frac{v}{32 \cdot g} \cdot A'_1 \cdot \frac{v}{g} \cdot \frac{(1-n)^2}{n^3} \cdot S_e^2 \cdot V + \frac{1}{8 \cdot g} \cdot A'_2 \cdot \frac{1}{g} \cdot \frac{(1-n)}{n^3} \cdot S_e \cdot V^2 \tag{50}$$

Equation (50) is the same as that developed by Sabri Ergun and A. A. Orning (1949) [8] where the dimensionless coefficients A'_1 and A'_2 have the values:

$$A'_1 = 64\alpha \tag{51}$$

$$A'_2 = \beta \tag{52}$$

Considering the Equations (47), (48), (51), and (52):

$$\alpha = \frac{\alpha'}{16} \tag{53}$$

$$\beta = \beta' \tag{54}$$

In addition, the specific surface area of the packing S is related to the average specific surface area of the particles S_e through Equation (9) with which we get:

$$S_e^2 = \frac{S^2}{(1-n)^2} \tag{55}$$

Substituting this last value in Equation (50), we obtain the quadratic equation with characteristic length L_c the surface of the packing S:

$$i = 2\alpha \cdot \frac{v}{g} \cdot \frac{1}{n^3} \cdot S^2 \cdot V + \frac{1}{8} \cdot \beta \cdot \frac{1}{g} \cdot \frac{1}{n^3} \cdot S \cdot V^2 \tag{56}$$

Alternatively, S.P. Burke and W.B. Plumer (1928) [6] developed the exponential equation with exponent (2−b):

$$i = K_b \cdot \frac{1}{g} \cdot \frac{V^2}{n^3} \cdot \left(\frac{\mu \cdot S}{\rho \cdot V} \cdot \frac{n}{(1-n)} \right)^{2-b} \tag{57}$$

where K_b is a dimensionless coefficient, which according to the authors includes the shape of the porous material and the symmetry of the packing.

The authors did not consider taking a quadratic equation from this. However, this can be achieved by doing no more than taking b = 1 for the linear component and b = 2 for the quadratic component. Through this approach we get the equation known as the quadratic equation from Burke and Plumer (1928) [6]:

$$i = [K_b]_L \cdot \frac{v}{g} \cdot \frac{1}{n^3} \cdot \frac{n}{(1-n)} \cdot S^2 \cdot V + [K_b]_T \cdot \frac{1}{g} \cdot \frac{1}{n^3} \cdot S \cdot V^2 \tag{58}$$

where $[K_b]_L$ is the linear dimensionless coefficient by Burke and Plumer $[K_b]_T$ is the quadratic dimensionless coefficient by Burke and Plumer.

Now, seeing Equation (58) and comparing it with Equation (56) we see that the dimensionless coefficients $[K_b]_L$ and $[K_b]_T$ are determined by the expression:

$$[K_b]_L = 2\alpha \cdot \frac{n}{(1-n)} \tag{59}$$

$$[K_b]_T = \frac{\beta}{8} \tag{60}$$

Accordingly, the previous equations show that a mathematical relationship exists among the formulas of the authors cited.

3.2. Equations with Characteristic Length Based on the Intrinsic Permeability

Now, if we consider, as characteristic length in Equation (12), the quadratic root of the intrinsic permeability $\sqrt{K_0}$ adopted by Ward (1964) [9], Ahmed and Sunada (1969) [14], and Arbhabhirama and Dinoy (1973) [16], among others, we obtain:

$$i = \frac{v}{2 \cdot g} \cdot A_1'' \cdot \frac{1}{K_0} \cdot V_p + \frac{1}{2 \cdot g} \cdot A_2'' \cdot \frac{1}{\sqrt{K_0}} \cdot V_p^2 \tag{61}$$

where A_1'' is the laminar dimensionless coefficient corresponding to $L_c = \sqrt{K_0}$ and A_2'' is the quadratic dimensionless coefficient corresponding to $L_c = \sqrt{K_0}$.

If we compare Equation (61) with Equation (30) by Ward (1964), we get the dimensionless coefficients A_1'' and A_2'':

$$A_1'' = 2 \cdot n \tag{62}$$

$$A_2'' = 2 \cdot C \cdot n^2 \tag{63}$$

3.3. Equations with Characteristic Length Based on the Representative Size of Particles

We are going to consider in general the representative size of the particle D, as obviously, the formulas developed in this case can be applied for any representative size such as $D_e, D_a, D_p, D_{50}, M_g$, etc.

For this general case of parameter D and applying Equation (12) we obtain the equation:

$$i = \frac{v}{2\cdot g}\cdot A_1''' \cdot \frac{1}{D^2}\cdot V_p + \frac{1}{2\cdot g}\cdot A_2''' \cdot \frac{1}{D}\cdot V_p^2 \tag{64}$$

where A_1''' is the linear dimensionless coefficient corresponding to $L_c = D$ and A_2''' is the quadratic dimensionless coefficient corresponding to $L_c = D$.

If we observe Equation (64) and compare it with modified Stephenson equation, i.e., Equation (36), we get the dimensionless coefficients A_1''' and A_2''':

$$A_1''' = 2K_l \tag{65}$$

$$A_2''' = 2K_t \tag{66}$$

Once the quadratic equations have been obtained for each characteristic length L_c based on Equation (12), we can determine the mathematical relationships between the main parameters.

3.4. Relationships among Characteristic Lengths

Next, we determine the relationships between R_h and $\sqrt{K_0}$ and between R_h and D. With regard to the first relationship, being the linear components of Equations (45) and (61):

$$\frac{1}{2}\cdot A_1' \cdot \frac{v}{g}\cdot \frac{1}{D_h^2} = \frac{1}{2}\cdot A_1'' \cdot \frac{v}{g}\cdot \frac{1}{K_0} \tag{67}$$

Substituting the values of A_1' defined in Equation (51) and A_1'' defined in Equation (62) into Equation (67) we obtain:

$$32\cdot\alpha\cdot\frac{1}{D_h^2} = n\cdot\frac{1}{K_0} \tag{68}$$

$$D_h^2 = \frac{32\cdot\alpha\cdot K_0}{n} \tag{69}$$

Substituting the value of the hydraulic mean diameter D_h defined in Equation (44) into Equation (69), we obtain the relationship between the hydraulic mean radius R_h and the intrinsic permeability K_0:

$$R_h = \frac{\sqrt{2\cdot\alpha}}{n^{0.5}}\cdot\sqrt{K_0} \tag{70}$$

We determine the relationship between the hydraulic mean radius R_h and the representative size of the particle D, by equaling the linear components of the Equation (45) and the Equation (64):

$$\frac{1}{2}\cdot A_1' \cdot \frac{v}{g}\cdot \frac{1}{D_h^2} = \frac{1}{2} A_1''' \cdot \frac{v}{g}\cdot \frac{1}{D^2} \tag{71}$$

Substituting the values of A_1' defined in Equation (51) and A_1''' defined in Equation (65) in the previous expression we obtain:

$$32\cdot\alpha\cdot\frac{1}{D_h^2} = K_l\cdot\frac{1}{D^2} \tag{72}$$

$$D_h^2 = \frac{32\cdot\alpha}{K_l}\cdot D^2 \tag{73}$$

With which we obtain the relationship:

$$R_h = \frac{\sqrt{2 \cdot \alpha}}{\sqrt{K_l}} \cdot D \tag{74}$$

We determine the relationship between the hydraulic mean radius R_h and the representative size of the particle D by applying Equation (7). We know that, for a sphere, the specific surface area S_e has a value of $6/D$, where D is the diameter of the sphere. To consider the shape, angularity, and surface roughness of the particles that may affect its specific surface area S_e (Crawford, C.W. et al. (1986) [3]; Sabin G. C. W. and Hansen D. (1994) [4], we can define a coefficient F that is determined by the expression:

$$S_e = \frac{6 \cdot F}{D} \tag{75}$$

In accordance with Equation (7) that defines hydraulic mean radius R_h and considering Equation (72) we get the relationship:

$$R_h = \frac{n}{6 \cdot F \cdot (1-n)} \cdot D \tag{76}$$

Loudon (1953) [53] used a coefficient similar to F to determine the permeability of sands. The author used D as the representative size of the particle, (D_g) defined as the average geometry between two consecutive sieves:

$$S_e = F \cdot \frac{6}{D_g} \tag{77}$$

where:

$$D_g = \sqrt{D_n \cdot D_{n-1}} \tag{78}$$

where D_n and D_{n-1} are the apertures in two consecutive sieves.

Loudon proposed the following values for the coefficient F:

(a) Round sand, F = 1.10;
(b) Semi angular sand, F = 1.25;
(c) Angular sand, F = 1.40.

In accordance with Martins (1990) [54] and considering the coefficient of the shape c' we can obtain the relationship:

$$c' = 6 \cdot F \tag{79}$$

In accordance with the values of c' proposed by Linford, A. and Saunders, D. (1967) [55] and Martins, R. and Escarameria, M. (1989) [56] (according to Martins), the equivalent values of the coefficient F are:

(a) Angular Particles c' = 8.5 (F = 1.47).
(b) Round particles c' = 6.3 (F = 1.05).

These values are similar to those proposed by Loudon (1953) [53].

With these values and applying Equation (76), we can estimate the value of the hydraulic mean radius R_h.

Finally, we determine the relationships between $\sqrt{K_0}$ and D considering the linear components of Equations (61) and (64):

$$\frac{1}{2} \cdot A_1'' \cdot \frac{v}{g} \cdot \frac{1}{K_0} = \frac{1}{2} \cdot A_1''' \cdot \frac{v}{g} \cdot \frac{1}{D^2} \tag{80}$$

Substituting the values of A_1'' defined in Equation (62) and A_1''' defined in Equation (65) in the previous expression we get:

$$n \cdot \frac{v}{g} \cdot \frac{1}{K_0} = K_l \cdot \frac{v}{g} \cdot \frac{1}{D^2} \tag{81}$$

Developing:

$$n \cdot D^2 = K_l \cdot K_0 \tag{82}$$

With which we finally get the relationship:

$$\sqrt{K_0} = \frac{n^{0.5}}{\sqrt{K_l}} \cdot D \tag{83}$$

Summarizing above, Table 1 shows the relationships among the characteristic lengths R_h, $\sqrt{K_0}$ and D.

Table 1. Relationships among characteristic lengths L_c.

L_c	A_1	A_2	R_e	f	R_h	$\sqrt{K_0}$	D
D_h	64α	β	$\frac{4 \cdot R_h \cdot V_p}{v}$	$8R_h \cdot g \cdot \frac{i}{V_p^2}$	$4 \cdot R_h$	$4 \cdot \frac{\sqrt{2\alpha}}{n} \cdot \sqrt{K_0}$	$4 \cdot \frac{\sqrt{2\alpha}}{\sqrt{K_l}} \cdot D$
$\sqrt{K_0}$	$2n$	$2 \cdot C \cdot n^2$	$\frac{\sqrt{K_0} \cdot V_p}{v}$	$\sqrt{K_0} \cdot 2g \cdot \frac{i}{V_p^2}$	$\frac{n^{0.5}}{\sqrt{2\alpha}} \cdot R_h$	1	$\frac{n^{0.5}}{\sqrt{K_l}} \cdot D$
D	$2K_l$	$2 \cdot K_t$	$\frac{D \cdot V_p}{v}$	$D \cdot 2g \cdot \frac{i}{V_p^2}$	$\frac{\sqrt{K_l}}{\sqrt{2\alpha}} \cdot R_h$	$\frac{\sqrt{K_l}}{n^{0.5}} \cdot \sqrt{K_0}$	1

3.5. Relationships among Reynolds Numbers

First, we determine the relationships between the pore Reynolds number R_p and the Ward Reynolds number R_k. Considering the pore Reynolds number D_h:

$$R_p = \frac{4 \cdot R_h \cdot V_p}{v} \tag{84}$$

And substituting Equation (70) in Equation (84) we obtain:

$$R_p = 4 \cdot \left(\frac{\sqrt{2 \cdot \alpha}}{n^{0.5}} \cdot \sqrt{K_0} \right) \cdot \frac{V}{n \cdot v} \tag{85}$$

If we compare the last equation with Equation (28) that defines the Reynolds number from Ward (R_k) we obtain the relationship:

$$R_p = \frac{4 \cdot \sqrt{2 \cdot \alpha}}{n^{1,5}} \cdot R_k \tag{86}$$

Now, we determine the relationships between the pore Reynolds number R_p based on D_h and the Reynolds number of the particles R_d.

Considering Equation (74) that relates to the hydraulic mean radius R_h with the representative size of the particles D, and substituting in Equation (84) that defines pore Reynolds number R_p based on D_h we obtain:

$$R_p = 4 \cdot \left(\frac{\sqrt{2 \cdot \alpha}}{\sqrt{K_l}} \cdot D \right) \cdot \frac{V_p}{v} = 4 \cdot \frac{\sqrt{2 \cdot \alpha}}{\sqrt{K_l}} \cdot \frac{D}{v} \cdot \frac{V}{n} \tag{87}$$

Alternatively, in accordance with Equation (15) that defines the Reynolds number of the particles, and substituting this last expression into the previous expression, we obtain the relationship between R_p and R_d:

$$R_p = \frac{4 \cdot \sqrt{2 \cdot \alpha}}{\sqrt{K_l}} \cdot R_d \tag{88}$$

In addition, in accordance with Equation (76) that relates to the hydraulic mean radius R_h and the representative size of the particles D, and substituting in Equation (84) that defines R_p as:

$$R_p = 4 \cdot [n/(6 \cdot F \cdot (1-n)) \cdot D] \cdot \frac{V_p}{v} \tag{89}$$

In accordance with Equation (15) we finally obtain another relationship between R_p and R_d:

$$R_p = \frac{4 \cdot n}{6 \cdot F \cdot (1-n)} \cdot R_d \tag{90}$$

Finally, we determine the relationships between the Reynolds number of Ward R_k and the Reynolds number of the particles R_d.

In accordance with Equation (28) that defines the Reynolds number of Ward and considering Equation (83) that relates the intrinsic permeability K_0 with the representative size of the particles D, and substituting this last equation into the definition of the Reynolds number of Ward R_k, we obtain:

$$R_k = \frac{D}{\sqrt{K_l}} \cdot n^{0.5} \cdot \frac{V}{v} = \frac{D}{\sqrt{K_l}} \cdot n^{0.5} \cdot \frac{n}{n} \cdot \frac{V}{v} \tag{91}$$

Finally, in accordance with the definition of the Reynolds number of the particles R_d defined by the Equation (15) we obtain the relationship:

$$R_k = \frac{n^{1.5}}{\sqrt{K_l}} \cdot R_d \tag{92}$$

In accordance with the above, Table 2 shows the relationships among the Reynolds numbers R_p, R_k and R_d.

Table 2. Relationships among the Reynolds numbers R_e.

L_c	R_e	R_p	R_k	R_d
$4R_h$	R_p	1	$\frac{4\sqrt{2\alpha}}{n^{1.5}} \cdot R_k$	$\frac{4\sqrt{2\alpha}}{\sqrt{K_l}} \cdot R_d$
$\sqrt{K_0}$	R_k	$\frac{n^{1,5}}{4\sqrt{2\alpha}} \cdot R_p$	1	$\frac{n^{1,5}}{\sqrt{K_l}} \cdot R_d$
D	R_d	$\frac{\sqrt{K_l}}{4\sqrt{2\alpha}} \cdot R_p$	$\frac{\sqrt{K_l}}{n^{1.5}} \cdot R_k$	1

3.6. Relationships among the Laminar Dimensionless Coefficients α, α_0, K_l and α'

Considering Equations (74) and (76) and equaling both equations, we obtain:

$$\frac{\sqrt{2\alpha}}{\sqrt{K_l}} = \frac{n}{6 \cdot F \cdot (1-n)} \tag{93}$$

Developing Equation (93), we obtain:

$$2\alpha \cdot 36 \cdot F^2 (1-n)^2 = K_l \cdot n^2 \tag{94}$$

Solving the linear dimensionless coefficient α, we finally determine the relationship between the linear dimensionless coefficients α and K_l Equation (95).

$$\alpha = \frac{n^2}{72 \cdot F^2 \cdot (1-n)^2} \cdot K_l \tag{95}$$

Next, we relate the coefficients α and α_0. If we equal the linear components of Engelund Equation (25) considering the representative diameter of the particle D instead of D_e and Equation (64), we obtain:

$$\frac{v}{g} \cdot \alpha_0 \cdot \frac{1}{D^2} \cdot \frac{(1-n)^2}{n^2} = \frac{v}{2 \cdot g} \cdot (2 \cdot K_l) \cdot \frac{1}{D^2} \cdot \frac{1}{n} \tag{96}$$

We get:

$$K_l = \alpha_0 \cdot \frac{(1-n)^3}{n} \tag{97}$$

Substituting the values of K_l in Equation (95) we obtain Equation (98):

$$\alpha = \frac{\alpha_0 \cdot (1-n) \cdot n}{72 \cdot F^2} \tag{98}$$

Coefficient α' relates to coefficient α through Equation (53)

Summarizing, Table 3 shows the relationships among the laminar dimensionless coefficients α, K_l and α_0.

Table 3. Relationships among the laminar dimensionless coefficients.

L_c	Laminar	α	K_l	α_0
$4R_h$	α	1	$\frac{n^2}{72 \cdot F^2 \cdot (1-n)^2} \cdot K_l$	$\frac{(1-n) \cdot n}{72 \cdot F^2} \cdot \alpha_0$
$\sqrt{K_0}$	1			
D	K_l	$\frac{72 \cdot F^2 \cdot (1-n)^2}{n^2} \cdot \alpha$	1	$\frac{(1-n)^2}{n} \cdot \alpha_0$
D	α_0	$\frac{72 \cdot F^2}{(1-n) \cdot n} \cdot \alpha$	$\frac{n}{(1-n)^3} \cdot K_l$	1

3.7. Relationships among Turbulent Dimensionless Coefficients β, β_0, K_t, β' and C

If we equal the quadratic components of Equations (45) and (61):

$$\frac{1}{2} \cdot A'_2 \cdot \frac{1}{g} \cdot \frac{1}{4R_h} \cdot \frac{1}{n^2} = \frac{1}{2} \cdot A''_2 \cdot \frac{1}{g} \cdot \frac{1}{\sqrt{K_0}} \cdot \frac{1}{n^2} \tag{99}$$

Considering the values of A'_2 and A''_2 defined by Equations (52) and (63) and substituting them in the first expression we get:

$$\beta \cdot \frac{1}{4R_h} = 2 \cdot C \cdot n^2 \cdot \frac{1}{\sqrt{K_0}} \tag{100}$$

In accordance with Equation (70) that defines the relationship between the hydraulic mean radius R_h and the intrinsic permeability K_0 and substituting this equation in Equation (100):

$$\frac{\beta}{4} \cdot \frac{n^{0.5}}{\sqrt{2\alpha}} \cdot \frac{1}{\sqrt{K_0}} = 2 \cdot C \cdot n^2 \cdot \frac{1}{\sqrt{K_0}} \tag{101}$$

We finally get:

$$\beta = 8 \cdot \sqrt{2 \cdot \alpha} \cdot C \cdot n^{1.5} \tag{102}$$

This last derived equation shows the dependence between the linear dimensionless coefficient α and the quadratic dimensionless coefficient β. As indicated by Huang H. and Ayoub J. (2007), "both expressions are intrinsically related, and the division is not arbitrary."

If we equal the quadratic components of Equations (45) and (64) we obtain the relationship turbulent coefficients β and K_t:

$$\frac{1}{2}\cdot A_2'\cdot\frac{1}{g}\cdot\frac{1}{4\cdot R_h}\cdot\frac{1}{n^2}=\frac{1}{2}\cdot A_2'''\cdot\frac{1}{g}\cdot\frac{1}{D}\cdot\frac{1}{n^2} \tag{103}$$

If we consider Equation (76) that relates the characteristic lengths R_h and D the values A_2' and A_2''' given by Equations (52) and (66), substituting in the Equation (103) we obtain:

$$\beta\cdot\frac{6\cdot F\cdot(1-n)}{4\cdot n}\cdot\frac{1}{D}=2\cdot K_t\cdot\frac{1}{D} \tag{104}$$

Therefore, we finally obtain the relationship:

$$\beta = 86\cdot nF\cdot(1-n)\cdot K_t \tag{105}$$

Finally, equaling Equations (104) and (105) that define the coefficient β:

$$8\sqrt{2\cdot\alpha}\cdot C\cdot n^{1,5}=\frac{8}{6}\cdot\frac{n}{F\cdot(1-n)}\cdot K_t \tag{106}$$

By clearing C in Equation (106) we obtain the relationship between the turbulent dimensionless coefficients C and K_t:

$$C=\frac{1}{6}\cdot\frac{1}{\sqrt{2\alpha}}\cdot\frac{1}{F\cdot n^{0.5}(1-n)}\cdot K_t \tag{107}$$

Finally, if we equal the quadratic components of Equation (25) of Engelund (1953) and Equation (64):

$$\frac{1}{g}\cdot\beta_0\cdot\frac{(1-n)}{n^3}\cdot\frac{1}{D}=\frac{1}{2\cdot g}\cdot(2\cdot K_t)\cdot\frac{1}{D}\cdot\frac{1}{n^2} \tag{108}$$

We obtain the relationship:

$$K_t=\beta_0\cdot\frac{(1-n)}{n} \tag{109}$$

And substituting in Equation (108) we obtain:

$$\beta=\frac{8}{6}\cdot\frac{1}{F}\cdot\beta_0 \tag{110}$$

Coefficient β' is equal to coefficient β in accordance with Equation (54).

Accordingly, Table 4 shows the relationships among the dimensionless coefficients β, C, K_t and β_0.

Table 4. Relationships among the turbulent coefficients.

L_c	Turbulent	β	C	K_t	β_0
$4R_h$	β	1	$8\cdot\sqrt{2\alpha}\cdot n^{1.5}\cdot C$	$\frac{8}{6}\cdot\frac{n}{F(1-n)}\cdot K_t$	$\frac{8}{6}\cdot\frac{1}{F}\cdot\beta_0$
$\sqrt{K_0}$	C	$\frac{1}{8\cdot\sqrt{2\alpha}\cdot n^{1.5}}\cdot\beta$	1	$\frac{1}{6\cdot\sqrt{2\alpha}\cdot Fn^{0.5}(1-n)}\cdot\beta$	$\frac{1}{6\cdot\sqrt{2\alpha}\cdot F\cdot n^{1.5}}\cdot\beta_0$
D	K_t	$\frac{6\cdot F(1-n)}{8n}\cdot\beta$	$6\cdot\sqrt{2\alpha}\cdot F\cdot n^{0.5}\cdot(1-n)\cdot C$	1	$\frac{(1-n)}{n}\cdot\beta_0$
D	β_0	$\frac{6}{8}\cdot F\cdot\beta$	$6\cdot\sqrt{2\alpha}\cdot F\cdot n^{1.5}\cdot C$	$\frac{n}{(1-n)}\cdot K_t$	1

4. Conclusions

The seepage process in a coarse porous medium can be represented by three types of equation: a generalized equation $[R_e, f]$, i.e., Equation (10), a generalized quadratic equation, i.e., Equation (12), and a generalized linear equation $[R_e, \lambda]$, i.e., Equation (16); all these equations consider a characteristic length L_c.

The linearized equation, i.e., Equation (16), allows one to check whether the coefficients A_1 and A_2 remain constant or, on the contrary, vary throughout the three non-Darcy flow regimes: nonlinear laminar, turbulent transition, and fully developed turbulent.

On the basis of the analysis of the different relationships between the gradient and the seepage velocity developed by different authors, all of them based on Equation (1) proposed by Forchheimer (1901) [1], it can be concluded that all the physical parameters considered in the different formulations are related to each other. Such parameters are the characteristic length (L_c, Table 1), the Reynolds number (R_e, Table 2), the dimensionless coefficient of the linear term r (A_1, Table 3), and the dimensionless coefficient of the quadratic term s (A_2, Table 4).

In this paper, we establish the equations that relate these parameters, thus, facilitating comparisons among the main studies carried out to date by different authors.

Author Contributions: Conceptualization, J.C.L.; formal analysis, J.C.L.; investigation, J.C.L.; resources, J.C.L.: data curation, J.C.L.; writing-original draft preparation, J.C.L., R.M.; writing-review and editing, R.M., M.Á.T.; visualization, J.C.L.; supervision, R.M., M.Á.T. All authors have read and agreed to the published version of the manuscript.

Funding: This research received no external funding.

Institutional Review Board Statement: Not Applicable.

Informed Consent Statement: Not Applicable.

Data Availability Statement: Some of the data used in the report comes from the mentioned references.

Conflicts of Interest: The authors declare no conflict of interest.

Nomenclature

a	Coefficient of the exponential equation that depends on the characteristics of the porous medium
A_1	Generalised dimensionless coefficient of the linear expression r
A_2	Generalised dimensionless coefficient of the quadratic expression
A_1'	Linear dimensionless coefficient corresponding to $L_c = D_h$
A_2'	Quadratic dimensionless coefficient corresponding to $L_c = D_h$
A_1''	Laminar dimensionless coefficient corresponding to $L_c = \sqrt{K_0}$
A_2''	Quadratic dimensionless coefficient corresponding to $L_c = \sqrt{K_0}$
A_1'''	Laminar dimensionless coefficient corresponding to $L_c = D$
A_2'''	Quadratic dimensionless coefficient corresponding to $L_c = D$
c'	Coefficient from Martins
b	Exponent of the exponential equation function of the flow conditions
C	Quadratic dimensionless coefficient of Ward
C_u	Coefficient of uniformity
D	Representative size of the particles in uniform materials
D_{50}	Sieve opening through which 50% of the material passes
D_a	Average size of sieve openings
D_e	Diameter equivalent or diameter of a sphere with the same volume as the particle
D_g	Geometric mean between the two consecutive sieves
D_h	Hydraulic mean diameter
D_m	Particle mean diameter
D_p	Effective diameter or diameter of a sphere with the same specific surface area as the particle
D_x	Diameter of the permeameter
F	Coefficient of Loudon which considers the shape and angularity of the particles
f	Generalised friction factor, by Darcy–Weisbach
f_L'	Function of porosity, by Engelund

f_d	Particle friction factor
f_E	Friction factor of Ergun
f_k	Friction factor of Ward
f_p	Function of linear porosity
f_p	Pore friction factor
f_T	Function of quadratic porosity
g	Gravitational acceleration
i	Hydraulic gradient
K_0	Intrinsic permeability of the porous medium
K_b	Coefficient of Blake that considers the shape of the porous material and the symmetry of the packing
$[K_b]_L$	Linear dimensionless coefficient, S. P. Burke and W. B. Plummer
$[K_b]_T$	Quadratic dimensionless coefficient, S. P. Burke y W. B. Plummer
K_l	Linear dimensionless coefficient, by Stephenson
K_t	Quadratic dimensionless coefficient, by Stephenson
L_c	Characteristic length
M_g	Geometric mean of the size of the particles that constitute the porous medium
n	Porosity
r	Linear coefficient of the Forchheimer equation of function of the characteristics of the porous medium and fluid.
R_e	Generalised Reynolds number
R_d	Particle Reynolds number
R_E	Reynolds number, by Ergun
R_h	Hydraulic mean radius
R_k	Reynolds number, by Ward
R_p	Pore Reynolds number of Dh
s	Quadratic coefficient of the Forchheimer equation of function of the characteristics of the porous medium.
S	Surface area per volume unit of the packed porous medium
S_e	Average specific surface area of solid particles
S_P	Average surface area of the particles
V	Average fluid velocity based on the transversal section
v	Kinematic viscosity
V_P	Pore velocity
v_p	Average particle volume
α	Linear dimensionless coefficient of the expression r, by Sabri, Ergun, and A. A. Orning
α'	Linear dimensionless coefficient of pores
α_0	Linear dimensionless coefficient r, by Engelund
β	Quadratic dimensionless coefficient of the expression r, by Sabri, Ergun and A. A. Orning
β'	Quadratic dimensionless coefficient of pores
β_0	Quadratic dimensionless coefficient of the expression s, by Engelund
λ	Linearised generalised friction factor
ρ	Fluid density
σs	Geometric standard desviation of the size distribution of the porous medium

References

1. Forchheimer, P.H. Wasserbewegung durch Boden. *Z. Des Ver. Dtsch. Ing.* **1901**, *50*, 1781–1788.
2. Taylor, D.W. *Fundamentals of Soil Mechanics*; John Wiley & Sons. Inc.: New York, NY, USA, 1948.
3. Crawford, C.W.; Plumb, O.A. The influence of surface roughness on resistance to flow through packed beds. *J. Fluids Eng.* **1986**, *108*, 343–347. [CrossRef]
4. Sabin, G.C.; Hansen, D. The effects of particle shape and surface roughness on the hydraulic mean radius of a porous medium consisting of quarried rock. *Geotech. Test. J.* **1994**, *17*, 43–49.
5. Blake, F.C. The Resistance of Packing to Fluid Flow. American Institute of Chemical Engineers, Ed.; Transcription at the Richmond Meeting: Richmond, VA, USA, 1922; pp. 415–421.
6. Burke, S.P.; Plummer, W.B. Gas flow through packed columns. *Ind. Eng. Chem.* **1928**, *20*, 1196–1200. [CrossRef]

7. Bear, J. Dynamics of Fluid in Porous Media. Dover Publications, Inc.: Mineola, NY, USA, 1988; pp. 176–185.
8. Ergun, S.; Orning, A.A. Fluid flow through randomly packed columns and fluidized beds. *Ind. Eng. Chemistry* **1949**, *41*, 1179–1184. [CrossRef]
9. Ward, J.C. Turbulent flow in porous media. *J. Hydraul. Div.* **1964**, *5*, 1–12.
10. Dudgeon, R.D. An experimental study of the flow of water through coarse granular media. *Houllie Blanche* **1966**, *7*, 785–802. [CrossRef]
11. Morcom, S.R. Transactions of the Institution of Chemical Engineers. The Institution (Great Britain): London, UK, 1946; Volume 24, p. 30.
12. Ergun, S. Fluid through packed columns. *Chem. Eng. Prog.* **1952**, *18*, 89–94.
13. Kadlec, H.R.; Knight, L.R. *Treatment Wetlands*; Lewis Publishers: Boca Raton, FL, USA, 1996.
14. Ahmed, N.; Sunada, D.K. Nonlinear flow in porous media. *J. Hydraul. Div.* **1969**, *95*, 1847–1858. [CrossRef]
15. Kovacks, G. Relationship between velocity of seepage and hydraulic gradient in the zone of high velocity. In Proceedings of the XIII Congress International Association for Hydraulic Research, Research Institute for Water Resources Development, Budapest, Hungary, 9 December 1969; Volume 14.
16. Arbhabhirama, A.; Dinoy, A.A. Friction factor and Reynolds number in porous media flow. *J. Hydraul. Div.* **1973**, *6*, 901–911. [CrossRef]
17. Stephenson, D.J. Flow through Rockfill. *Dev. Geotech. Eng.* **1979**, *27*, 19–37. [CrossRef]
18. Li, B.; Garga, V.K.; Davies, M.H. Relationships for Non-Darcy flow in rockfill. *J. Hydraul. Eng.* **1998**, *124*, 206–212. [CrossRef]
19. Sidiropoulou, M.G.; Moutsopoulos, K.N.; Tsihrintzis, V.A. Determination of Forchheimer equation coefficients a and b. *Hydrol. Process.* **2007**, *21*, 534–554. [CrossRef]
20. Moutsopoulos, K.N.; Papaspyros, I.N.E.; Tsihrintzis, V.A. Experimental investigation of inertial flow processes in porous media. *J. Hydrol.* **2009**, *374*, 242–254. [CrossRef]
21. Sedghi-Asl, M.; Rihimi, H.; Salehi, R. *Non-Darcy Flow of Water through a Packed Colum*; Springer Science: Berlin/Heidelberg, Germany, 2013; pp. 215–227.
22. Salahi, M.-B.; Sedghi-Asl, M.; Parvizi, M. Nonlinear flow through a packed-column experiment. *J. Hydraul. Eng.* **2015**, *20*, 04015003. [CrossRef]
23. White, F.W. *Fluid Mechanics*, 5th ed.; McGraw-Hill, Inc: New York, NY, USA, 2003.
24. Wilkins, J.K. Flow of water through rockfill and its application to the design of dams. *Soil Mech. Conf. Proc.* **1956**, *10*, 141–149.
25. Parkin, A.K. Through and overflow rockfill dams. In *Advances in Rockfill Structures*; Springer: Berlin/Heidelberg, Germany, 1991; pp. 571–592.
26. Ferdos, F.; Wörman, A.; Ekström, I. Hydraulic conductivity of coarse Rockfill used in hydraulic structures. *Trans. Porous. Med.* **2015**, *108*, 367–391. [CrossRef]
27. Wright, D.E. Nonlinear flow through granular media. *J. Hydraul. Div.* **1968**, *4*, 851–872. [CrossRef]
28. Dybbs, A.; Edwards, R.V. *Workshop on Heat and Mass Transfer in Porous Media*; Case Western Reserve University; N° 75 117. Department of Fluid, Thermal and Aerospace Sciences; Springfield, VA, USA, 1975; p. 228.
29. Fand, R.M.; Kim, B.Y.K.; Lam, A.C.C.; Phan, R.T. Resistance to the flow of fluids through simple and complex porous media whose matrices are composed of randomly packed spheres. *J. Fluid Eng.* **1987**, *109*, 268–273. [CrossRef]
30. Huang, H.; Ayoub, J. Applicability of the Forchheimer equation for Non-Darcy flow in porous media. *SPE J.* **2007**, *13*, 112–122. [CrossRef]
31. McCorquodale, J.A.; Hannoura, A.A.; Nasser, S. Hydraulic conductivity of rockfill. *J. Hydraul. Res.* **1978**, *16*, 123–137. [CrossRef]
32. Kozeny, J. Über Kapillare Leitung des Wassers im Boden; 1927. Available online: https://www.zobodat.at/pdf/SBAWW_136_2a_0271-0306.pdf (accessed on 14 July 2021).
33. Leva, M.; Grimmer, M. American Institute of Chemical Engineers, AiChE (www.aiche.org). *Chem. Eng. Progress* **1947**, *43*, 549-54, 633-8, 713-18.
34. Engelund, F. On the laminar and turbulent flows of ground water through homogeneous sand. *Hydraul. Lab.* **1953**, *4*, 356–361.
35. Lindquist, E. *On the Flow of Water through Porous Soil; Reports to the First Congress of Large Dams*; Communications Diverses: Stockholm, Sweden, 1933; pp. 81–101.
36. Chardabellas, P. *Durchflusswiderstände im Sand und ihre Abhängigkeit von Flüssigkeits und Bodenkennziffern. Mitteilungen der Preussischen Versuchsanstalt für Wasser Erd und Schiffbau*; Preussischen Versuchsanstalt für Wasser Erd und Schiffbau: Berlin, Germany, 1940; Volume 40.
37. Rose, H.E. An Investigation into the Laws of Flow of Fluids through Beds of Granular Materials. *Proc. Inst. Mech. Eng.* **1945**, *153*, 141. [CrossRef]
38. Franzini, J.B. Porosity factor for case of laminar flow through granular media. *Trans Am. Geophys. Union* **1951**, *32*, 443. [CrossRef]
39. Leps, T.M. *Flow through Rockfill*; John Wiley and Sons: Hoboken, NJ, USA, 1973; pp. 87–107.
40. Volker, R.E. Nonlinear flow in porous media be finite elements. *J. Hydraul. Div.* **1963**, *6*, 2093–2114.
41. Cedergren, H.R. *Seepage, Drainage and Flows Nets*; John Willey and Sons Inc.: Hoboken, NJ, USA, 1977.
42. Todd, D.K. *Ground Water and Hidrology*; John Wiley and Sons, Inc.: Hoboken, NJ, USA, 1959.
43. Schneebeli, G. Experiences sur la limite de validite de la turbulence dams un ecoulement de filtration. *Houille Blanche* **1955**, *2*, 141–149. [CrossRef]

44. Balhoff, M.T.; Wheeler, M.F. A predictive pore-scale model for non-Darcy flow in porous media. *SPE J.* **2009**, *110838*, 579–587.
45. Skjetne, E.; Auriault, J.L. High-velocity laminar and turbulent flow in porous media. *Trans. Porous Media* **1998**, *36*, 131–147. [CrossRef]
46. Panfilov, M.; Oltean, G.; Panfilova, I.; Bues, M. Singular nature of nonlineal manoscale effects in high-rate flow through porous media. *Comtes Rendus Mecc.* **2003**, *331*, 41–48. [CrossRef]
47. Fourar, M.; Radilla, G.; Lenormand, R.; Moyne, C. On the non-linear behavior of a laminar single-phase flow through two and three dimensional porous media. *Adv. Water Resour.* **2004**, *27*, 669–677. [CrossRef]
48. Hansen, D. The Behaviour of Flow through Rockfill Dams. Ph.D. Thesis, Department of Civil Engineering, University of Ottawa. Ottawa, ON, Canada, 1992.
49. Ranganadha Rao, R.P.; Suresh, C. Discussion of "Nonlinear Flow in Porous Media". *J. Hydraul. Div.* **1970**, *96*, 1732–1734. [CrossRef]
50. Tyagi, A.K.; Todd, D.K. Discussion of "Nonlinear Flow in Porous Media". *J. Hydraul. Div.* **1970**, *124*, 1734–1738. [CrossRef]
51. Venkataraman, P.; Rao, P.R.M. Darcian, trasitional and turbulent flow through pororus media. *J. Hydraul. Eng.* **1998**, *124*, 840–846. [CrossRef]
52. Bordier, C.; Zimmer, D. Drainage equationsand non-Darcian modelingin coarse porous media or geosynthetic materials. *J. Hydraul.* **2000**, *228*, 174–187. [CrossRef]
53. Loudon, A.G. The comptation of permeability from simple soil test. *Georechnique* **1952**, *3*, 165–183.
54. Martins, R. Turbulent seepage flow through rockfill structures. *Water Power Dam Constr.* **1990**, *90*, 41–45.
55. Linford, A.; Saunders, D. *A Hydraulic Investigation of through and over Flow Rockfild Dams*; British Hydromechanics Research Association: Cranfield, UK, 1967.
56. Martins, R.; Escarameria, M. *Caracterizaçao dos Materiais para o Estudo em Laboratorio do Escoamento de Percolaçao Turbulento*; Springer: Lisbon, Portugal, 1989.

water

Article

Numerical Modeling of the Effects of Toe Configuration on Throughflow in Rockfill Dams

Nils Solheim Smith [1,*], Ganesh H. R. Ravindra [1,2] and Fjóla Guðrún Sigtryggsdóttir [1]

[1] Department of Civil and Environmental Engineering, Norwegian University of Science and Technology (NTNU), S.P. Andersens veg 5, 7491 Trondheim, Norway; ganesh.hiriyanna.rao.ravindra@trondheim.kommune.no (G.H.R.R.); fjola.g.sigtryggsdottir@ntnu.no (F.G.S.)
[2] Department of Water and Wastewater, Kommunalteknikk, Trondheim Kommune, Erling Skakkes Gate 14, 7013 Trondheim, Norway
* Correspondence: nils.solheim.smith@ntnu.no

Abstract: The rockfill toe structure situated within the downstream slope of rockfill dams is an integral part of a defense mechanism safeguarding the dam structure in throughflow situations. Recent studies have concluded that the rockfill toe structure can have significant impacts on throughflow development and stability of rockfill dams under scenarios of accidental throughflow caused by overtopping of the dam core. The ability to numerically model the effect of various toe configurations on flow through rockfill dams can support the design of effective toe drainage structures for rockfill dams. Development and calibration of a reliable numerical modeling tool in this regard has been challenging owing to lack of availability of extensive datasets from physical modeling investigations. This study further employs datasets gathered by a recent physical modeling study investigating the effects of various toe configurations on throughflow development in rockfill dam models. A commercial numerical seepage modeling tool with an option for non-Darcy flow was calibrated against the datasets with good calibration metrics. The study is novel in providing a rare report on the usage of this option. The calibrated tool can further be employed to carry out a wide array of simulations to arrive at an ideal design for a toe structure for rockfill dams and for assessment of hydraulic performance of toe structures.

Keywords: rockfill dams; dam safety; throughflow; numerical modeling; non-Darcy flow

Citation: Smith, N.S.; Ravindra, G.H.R.; Sigtryggsdóttir, F.G. Numerical Modeling of the Effects of Toe Configuration on Throughflow in Rockfill Dams. *Water* **2021**, *13*, 1726. https://doi.org/10.3390/w13131726

Academic Editors: Miguel Á. Toledo and Rafael Moran

Received: 5 May 2021
Accepted: 14 June 2021
Published: 22 June 2021

1. Introduction

Embankment dams, constructed with locally excavated earth or rockfill represent 78% of the total number of existing dams worldwide [1]. Embankment dams comprising of coarse rockfill materials in more than 50% of the dam volume are defined as rockfill dams and represent 13% of the worldwide dam population [1]. A rockfill dam structure generally consists of an impervious element, filter zones, support fill and some means of controlling the development of phreatic surface and seepage through the dam structure.

The issue of dam safety has gained much attention in the recent past. Stringent measures are being put in place by the respective national dam safety authorities to ensure safety of dams. Although there exists significant amount of accumulated scientific literature within the research discipline of embankment dams in general, technical literature describing throughflow behavior of rockfill dams is scarce. This article aims at adding to the research discipline of rockfill dam safety. Dam safety assessment is a complex task, as it is influenced by multitudes of internal and external factors [2]. It is essential to determine the most common causes of dam failures to identify probable factors which commonly contribute to dam instability. Statistics from the International Commission on Large Dams (ICOLD) state overtopping as the main cause of embankment dam failure appearing as the primary factor in 31% of the total number of failures, and is further involved in another 18% of failures as a secondary agent [3]. Hence, equipping embankment dams with defense mechanisms against unanticipated overtopping or leakage events is of importance from a dam safety perspective. This includes safeguarding against accidental throughflow conditions arising when the core of a rockfill dam is overtopped, resulting in turbulent flow within the downstream dam shoulder, as shown in Figure 1b.

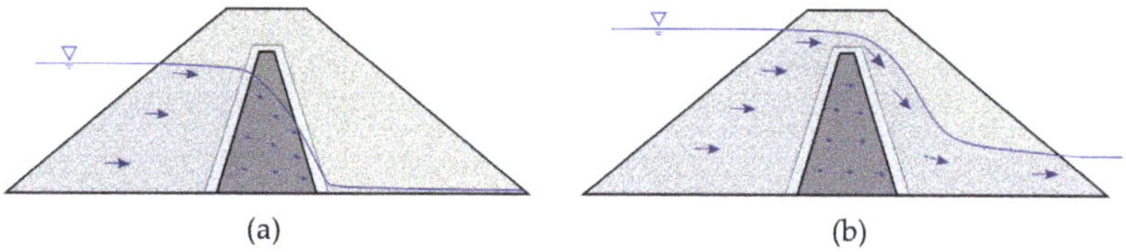

Figure 1. Sketch of throughflow situations for an embankment dam with a central core. (**a**) Normal conditions with seepage through the core. (**b**) Accidental load situation with overtopping of the core leading to large throughflow.

The rockfill toe structure situated within the downstream slope of rockfill dams can be considered as an integral part of a defense mechanism installed to protect the dam structure under normal seepage situation as described by Figure 1a, as well as under accidental overtopping situations leading to extreme throughflow conditions as shown in Figure 1b. In fact, some previous studies into embankment dam failures describe the downstream toe as a critical location for failure initiation under throughflow scenarios [4–8]. Furthermore, findings of Toledo and Morera [9] and Moran and Toledo [10] suggest that rockfill toes may be used as effective protection against throughflow in rockfill dams. Furthermore, Moran et al. [11] present a procedure for the design of external toe protection for rockfill embankments.

A recent investigation conducted by Ravindra [12] studied the effects of various configurations of rockfill toes on throughflow development within hydraulic scale models of rockfill dams (Figure 2). Findings from the experimental studies, presented by Kiplesund et al. [13], highlight the fact that toe configuration can have significant impact on the development and progression of phreatic surfaces within rockfill dam models subjected to incremental overtopping scenarios. The toe configurations were also found to influence the stability of the downstream slope. The study gave valuable insight into the significance of rockfill toes with regard to rockfill dam safety. The present study builds numerical models on the data accumulated through the experimental investigations [12,13].

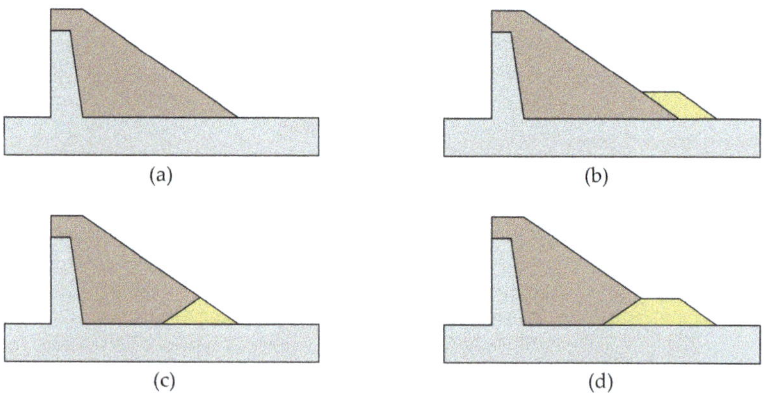

Figure 2. Displaying the investigated toe configurations, where (**a**) shows no toe configuration, (**b**) external toe, (**c**) internal toe and (**d**) combined toe configuration.

The present study is a part of an ongoing research program into dam overtopping and throughflow. Figure 3 visualizes the modeling strategies of this overarching research program from full-scale to model-scale investigations, as well as numerical modeling efforts. The full-scale part embraces consideration of the design of existing dams [14,15], as well as analysis of data accumulated from full-scale tests [16]. The hydraulic scale models that relate to the numerical modeling of the present study are those presented by Kiplesund et al. [13]. These considered the scaling of previous hydraulic scale models for investigating riprap erosion protection on the downstream slope of embankment dams [17–19]. The combined application of the different modeling strategies is for enhanced applicability and relevance of the research for full-scale dam cases.

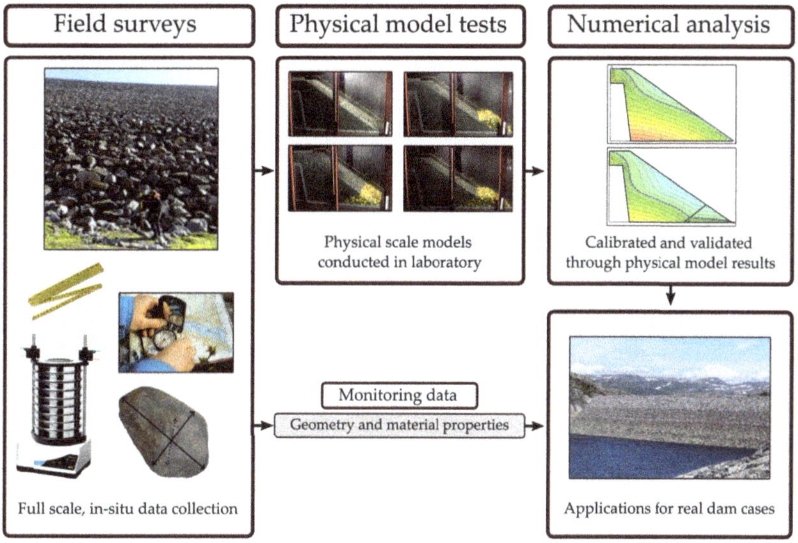

Figure 3. Schematic overview of the overarching research program.

The overarching goal of the present part of the research program is to evaluate the hydraulic response of rockfill dams exposed to accidental throughflow scenarios (Figure 1b) and to study the effects of rockfill toes on throughflow hydraulic properties of rockfill dams. For this purpose, a numerical model is developed for replicating results from physical model tests considering turbulence of the flow. Hence, an important aspect of the present study is the implementation of a geotechnical software [20] commonly employed in dam engineering for practical applications as well as in research [21,22]. However, the modeling usually assumes laminar flow or Darcy flow conditions, suitable for cases as in Figure 1a. Thus, the present study aims at investigating the ability of a tool provided within such software [23] to model turbulent or non-Darcy flow commonly encountered in the physical rockfill dam models. This has a relevance when proceeding to numerical models of real dam cases considering non-Darcy flow for the accidental overtopping situation. Moreover, numerical modeling of the effect of various toe configurations on flow through rockfill dams has not been looked into in the past. The datasets gathered through the previously mentioned physical modeling investigations [12,13] are used to calibrate numerical models employing the numerical seepage software SEEP/W [20] with a non-Darcy tool. The aim is to predict the development of throughflow within rockfill dam structures and to numerically model the effect of a drainage component within the downstream dam slope on non-linear throughflow development.

2. Background

Flow through porous media is generally characterized as either Darcy or non-Darcy type based on flow properties. The linear Darcy flow theory is widely implemented in soil mechanics and is described by the following equation:

$$v = k\, i, \tag{1}$$

where the velocity of flow, v, is described by a linear relationship between hydraulic conductivity, k, and the hydraulic gradient, i.

Darcy's law is only valid at low velocities, i.e., laminar flow. At higher velocities, the inertial forces distort the streamlines and turbulent flow occurs, removing the linear relationship. In rockfill material, the voids are of a magnitude that turbulent flow is expected [16]. An illustration is made in Figure 4. Where Darcy's law is applicable, flow is evenly distributed and laminar. In the rockfill case, the voids are bigger and velocities vary along with the grains redirecting the flow.

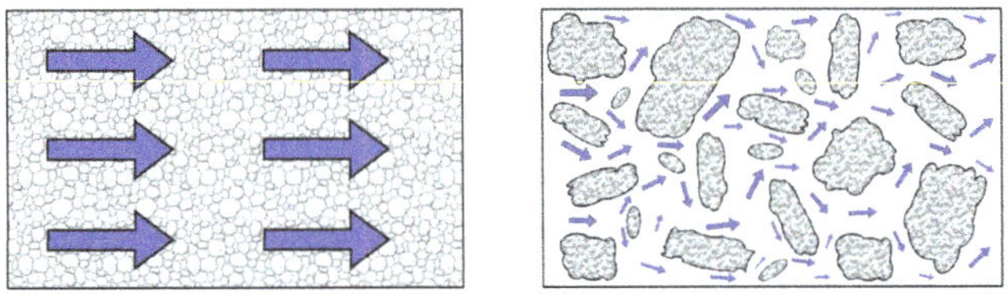

Figure 4. Representation of different flow regimes through porous media. Where (**a**) shows laminar flow through uniform small-grained material and (**b**) demonstrates turbulent flow condition through material with coarser grains and larger voids.

Non-Darcian or turbulent flow through porous media is generally represented as a power-law function:

$$v = a\, i^b, \tag{2}$$

where, a and b represent empirical coefficients to be determined experimentally. Coefficient a depends on the properties of fluid and porous media such as porosity, particle shape, particle size, roughness, tortuosity of void structure and viscosity of fluid. Parameter b is dependent upon the state of flow or the level of flow turbulence [8].

Until recently, the performance of the general non-linear flow law of the form presented in Equation (2) was verified only through experimental studies conducted in permeameters. Past studies have investigated non-linear flow through rockfill medium through elaborate permeameter experimental testing conducted on rockfill with sizes ranging from $d_{50} = 10$ mm to 240 mm [8,24–29]. Several empirical relationships describing the non-linear i-v flow properties have also been proposed as a result of these investigations. Although a considerable number of investigations have investigated non-Darcian flow through rockfill material in permeameters of varying sizes, experimental validation of these past findings in rockfill embankments exposed to throughflow conditions can be stated as quintessential for validation of past research findings in terms of relevance of application in rockfill dam engineering. A recent study put forth by Ravindra et al. [16] has made attempts at validating some of the widely employed non-linear flow equations from the past and have also further proposed a new equation applicable for non-linear flow through homogeneous rockfill dams.

Dealing with soil or rockfill, which is generally heterogeneous and discontinuous in nature, approximate solutions are normally pursued [30]. The finite element method is a powerful tool for approximating complex field problems. The domain in which the analysis is being conducted is divided into finite elements creating a mesh. For each node in the mesh, the field variable is explicitly calculated through a mass balance approach.

The functions that define how the field variable varies in the domain are controlled through the material properties. The mass balance approach for the utilized software relevant for this study can be summarized by the following general equation [20]:

$$\frac{dM_{st}}{dt} = \dot{m}_{in} - \dot{m}_{out} + \dot{M}_S \tag{3}$$

where M_{st} is the stored mass in the control volume, the inflow and outflow terms, $\dot{m}_{in}$ and $\dot{m}_{out}$, represent flow in and out of the control volume and M_S is the source term, with dot-notation representing rates.

For seepage problems, the governing differential equation utilized by the software in a 2D case is defined by:

$$\frac{\partial}{\partial x}\left(k_x \frac{\partial H}{\partial x}\right) + \frac{\partial}{\partial y}\left(k_y \frac{\partial H}{\partial y}\right) + Q = \frac{\partial \theta}{\partial t} \tag{4}$$

where k is the hydraulic conductivity in x- and y-direction, H is the total head, Q is the mass source or sink term. The right side of the equation is the change in volumetric water content, θ, with respect to time. The simulations in this study are conducted in a steady-state condition, which yields that there are no time-dependant variables and the right side of Equation (4) becomes zero.

The finite element method then entails that the governing differential equation must be satisfied at every node; assembled in matrix form, this can be summarized as the finite element equation:

$$[\mathbf{K}]\{\mathbf{h}\} = \{\mathbf{q}\} \tag{5}$$

where $\mathbf{K}$ represents the global element matrix, defining each element's geometry and material properties; $\mathbf{h}$ is the primary unknown vector consisting of the total head at each node. Lastly, $\mathbf{q}$ is the resultant vector also called the nodal flow vector, defined by the boundary conditions. This system of equations is iteratively solved so that each element in addition to the whole domain satisfies the governing equation.

To account for non-linearity of the flow, or non-Darcy flow, an added feature is usually required. Professional packages are available that employ a flux approach where the nonlinear nature of Equation (2) is relegated to an apparent hydraulic conductivity term, $k_{w,a}$, by rearranging the equations as follows with the hydraulic gradient expressed in vector notation as $\nabla h = \left(\frac{\partial h}{\partial x}, \frac{\partial h}{\partial y}, \frac{\partial h}{\partial z}\right)$ and velocity expressed as a flux vector, q_w [23]:

$$q_w = -k_{w,a}\, \nabla h \tag{6}$$

where the apparent hydraulic conductivity, $k_{w,a}$ can be expressed in terms of total head as [23]:

$$k_{w,a} = \frac{-1 + \sqrt{1 + 4\, C_F\, k^{3/2} \sqrt{\frac{\rho_w}{g\mu_w}}\, |\nabla h|}}{2\, C_F \sqrt{k} \sqrt{\frac{\rho_w}{g\mu_w}} |\nabla h|} \tag{7}$$

where C_F is the form drag constant, k is the hydraulic conductivity of the porous media, ρ_w is fluid density, g is gravitational acceleration and μ_w is the dynamic viscosity of fluid.

It can be seen from Equation (7) that the apparent conductivity will decrease with increasing velocity, providing non-linear behavior of the rockfill material. Definition of the parameters that govern the non-linear behavior are based on the closed form equation for hydraulic conductivity derived by Van Genuchten [31]. The input parameters include, saturated and residual water content, α, n, l, C_F and fluid temperature. Originally developed for agrophysical purposes, the equation builds on the soil water retention curve, which can be established through laboratory testing, to find the relative conductivity between saturated and unsaturated material.

Listing some parameters found for clay to sandstone soils, n-values ranges from 1.2 to 10 [31]. In a later study, typical values are presented as 1.2 for fine soils and 2.7 for coarse soil [32]. For the α-parameter variation lies between 0.01 and 1 for fine material including clay [33,34]. There exist multiple studies with varying values for the form drag constant, C_F, and there are no input limitations in the add-in of the software used [20]. As a selected limitation for the present study, the drag constant can vary between 0.5 and 1.5 for coarse granular material [35]. The l-parameter represents the inter connectivity and tortuosity of the voids in the material, with values ranging from −1 to 2 in different solutions [32].

Several past studies available in the international literature perform the function of defining turbulent flow through uniform and homogeneous rockfill materials [8,16,24–29]. However, the validity of these equations as applied to zoned rockfill structures comprising several different materials with varying properties has not been investigated. This can be attributed to the fact that hydraulic throughflow properties in zoned rockfill dam models can be very complex and deriving general results/relationships to describe such behavior can be challenging. Hence, numerical modeling can be considered as a well suited method for investigating such complex hydraulic aspects in rockfill dams. This study aims at employing a numerical model to obtain a representative description of flow through rockfill dam models with two individual zones. This can form a strong launchpad for further developments to the model which can help improve our capabilities to model complex hydraulic behaviors within large scale rockfill dams.

3. Materials and Methods

Methodology, instrumentation and material properties adopted for the physical modeling studies are succinctly explained in the following chapter. The process relating to the numerical analysis is then explained, covering both the design of the model and analysis.

3.1. Physical Model

The physical models and results are described in detail by Kiplesund et al. [13] and only the main features are presented herein. The rockfill dam models (Figure 5) were built in a 25 m long, 1 m wide and 2 m high flume at the hydraulics laboratory of NTNU, Trondheim. The effects of various configurations of rockfill toes (no toe, internal toe, external toe and combined toe configurations) on throughflow development within rockfill dam models were studied. The model only consisted of the downstream half of a rockfill dam structure with an aluminum core built on a horizontal platform elevated 350 mm from the flume floor. The reasoning being that under throughflow conditions, behavior of the downstream shoulder is of specific interest from a dam safety standpoint in comparison to the upstream embankment as the downstream slope of rockfill embankments are exposed to higher degree of destabilizing forces under turbulent throughflow conditions.

The physical model of a rockfill dam with a toe structure is presented in Figure 5. Region (a) within Figure 5 represents the internal toe configuration, Region (b) the external configuration and Regions (a) + (b) represent the combined configuration. The individual setups are visualized in Figure 2. The instrumentation for the physical model is comprised of 8 pressure sensors installed along the dam foundation (P3–P10 seen in Figure 5) for measurements of the pore pressure distributions within the dam models under throughflow conditions. The pressure measurements in this study are shown as the piezometric head, using the origin in Figure 5 as the datum. Additional pressure sensors were installed (i) on top of the metallic dam core (P2) and (ii) at the upstream section of the model (P1) for measurements of water levels over the core and the upstream reach of the model, respectively. Discharge to the flume was fed by two pumps with a combined capacity of about 0.4 m^3/s regulated through a digital discharge meter. The physical tests were conducted so that the discharge was fixed and the resulting water level upstream of the dam variable.

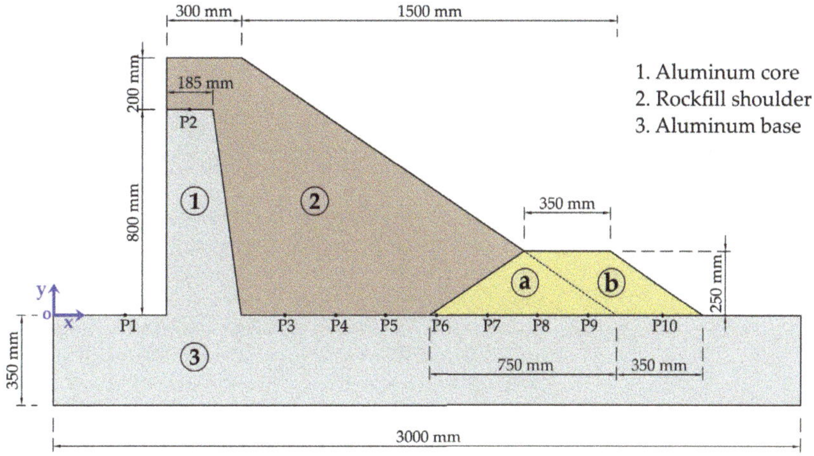

Figure 5. Sketch of the model displaying the rockfill shell, along with the base and core, with two regions for (**a**) internal and (**b**) external toe configuration. The coordinate system, drawn in blue, is placed at the origin. The locations of the installed pressure sensors are listed as P1–P10 (see Table 1).

Table 1. Sensor position along the dam body, represented as the x-axis in Figure 5.

Sensor	P1	P2	P3	P4	P5	P6	P7	P8	P9	P10
Position (m)	0.29	0.54	0.93	1.13	1.33	1.53	1.73	1.93	2.14	2.44

The selected rockfill material grain size was based on data analysis from existing rockfill dams in Norway. The gradation curves were down-scaled by a ratio of 1:10, barring some of the finest materials due to limitations of the flume pumping system. The selected gradation curve was thus slightly narrower, and lies on the coarser boundary of Norwegian standards. In total, 1800 kg of shell material was mixed in order to complete the model. Some key material parameters for the rockfill shoulder and toe materials are presented in Table 2. Presented are density, key grain sizes and the coefficient of conformity, $c_u = d_{60}/d_{10}$. The resulting gradation curves for the well-graded shell material and uniform toe material can be seen in Figure 6.

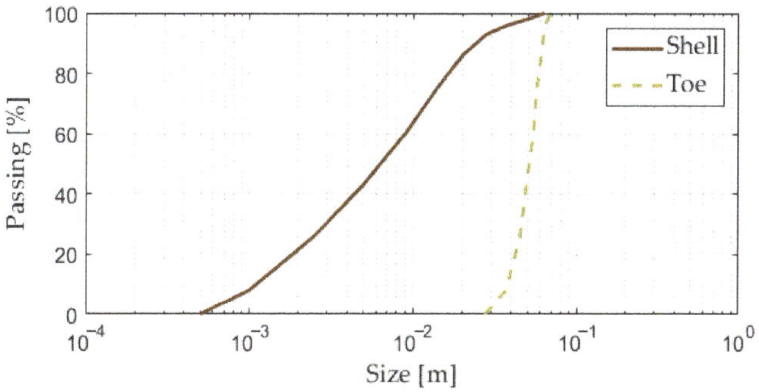

Figure 6. Gradation curve shown for shell and toe material.

Table 2. Material properties for shell and toe.

	ρ	d_{10}	d_{50}	d_{60}	c_u
Shell	2720	1.2	6.5	9.0	7.50
Toe	2860	37	52	55	1.42
	(kg/ m^3)	(mm)	(mm)	(mm)	(-)

A total of twelve physical tests were conducted, comprising three tests on each individual toe configuration. The testing methodology consisted of exposure of the rockfill dam models to incremental throughflow magnitudes. The discharge intervals were adopted as $\Delta q = 0.5 \times 10^{-3}$ m^3/s over N discharge steps with initial exposure set to $q_i = 1 \times 10^{-3}$ m^3/s. The discharge levels were maintained constant over regular time periods of $\Delta t = 1800$ s to allow for flow stabilization at each overtopping interval. Reference is made to Kiplesund et al. [13] for detailed summary of the tests and the respective testing protocols. The tests were visually documented through video recordings.

3.2. Numerical Model

Independent of which software one is interested in utilising, the modeling procedure begins with defining the domain to investigate, i.e., drawing the geometry. This can be imported from CAD software or can be defined within the selected software for numerical analysis. In the present study, the 2D geometry is drawn within the software used [20]. The different sections of the model with varying material properties are drawn as separate regions. The pressure sensor positions along the dam body are defined as nodes within the numerical model to allow for juxtaposition of results from the numerical and physical modeling efforts.

3.2.1. Material Properties

Definition of the hydraulic properties is an important and challenging configuration of the model. The multiphase nature of throughflow with water displacing air in the voids requires consideration of saturation. When defining the model case, it can be considered a fully submerged case or a unsaturated/saturated case. The unsaturated/saturated case is selected for the present model. The water content and hydraulic conductivity can consequently be defined. Volumetric water content functions, are preinstalled in the professional software used in this study [20], based on grain size distribution and saturated water content. These parameters can be established by laboratory sieving analysis and porosity measurements. For the shell, the saturated water content is defined as 0.15, based on in-situ cone porosity measurements. From sieving analysis, the d_{60} and d_{10} were found to be 9 mm and 1.2 mm, respectively. The toe material grain size analysis yielded d_{60} and d_{10} to be 36 mm and 11 mm. Being coarser and uniformly graded, the toe material volumetric water content was estimated to be 0.4.

By using the non-Darcy add-in [23] the hydraulic conductivity function is replaced to replicate a non-Darcy condition. To enable this function, the add-in is selected as function type. For the shell, the saturated water content is set to 0.15, residual water content is set to 10% of saturated levels, 0.015, α is set to 8, n-parameter is set to 2, hydraulic conductivity is set to 0.003 m/s, the l-parameter is set to -0.5, the form-drag coefficient is set to 1.5 and finally temperature is defined as 20 ° C. For the toe material the saturated water content is set to 0.4, residual water content is set to 0.04, again based on 10% of the saturated values. α is set to 15, n-parameter is set to 4, hydraulic conductivity is set to 0.1 m/s, the l-parameter is set to -1, the form-drag coefficient is set to 0.75 and temperature is again defined as 20 ° C. A summary of the calibrated parameter set can be seen in Table 3.

Table 3. Summary of numerical model parameters.

	Vol. Water Content		Fitting Param.		Hydr. Cond.	Tortuoisity & Connectivity	Form Drag Constant	Fluid Temp.
	Sat.	Res.	α	n	k (m/s)	l	FDC	T (°C)
Shell	0.150	0.015	8	2	0.003	−0.5	1.50	20
Toe	0.40	0.040	15	4	0.100	−1.0	0.75	20
Limit	[0–1]	[0–1]	-	>1	-	-	-	-

3.2.2. Mesh

The automated mesh is drawn based on a global element size of 0.02 m, shown in Figure 7. This mesh size was selected as it provides good resolution and manageable calculation times. Specific nodes were placed to match the position of the pressure sensors, where node 2 and node 16 correspond to sensors P1 and P2, respectively. Nodes 5–11 correspond to sensors P3–P9, node 12 corresponds to sensor P10. However, for the numerical analysis, P1 is located outside the dam domain, which makes it unavailable for comparison. Similarly, for no toe and internal configuration, P10 is located outside the dam domain, meaning that values cannot be compared for these two toe configurations. Upon completion of the meshing, the analysis is ready to be run. The results produce the phreatic surface and pore pressure development through the dam. To compare results between the numerical analysis and the physical tests, data can be exported from the numerical model. The points of interest being the pore pressure development along the points matching the positions of the pressure sensors, the positions and corresponding sensors are listed in Table 1.

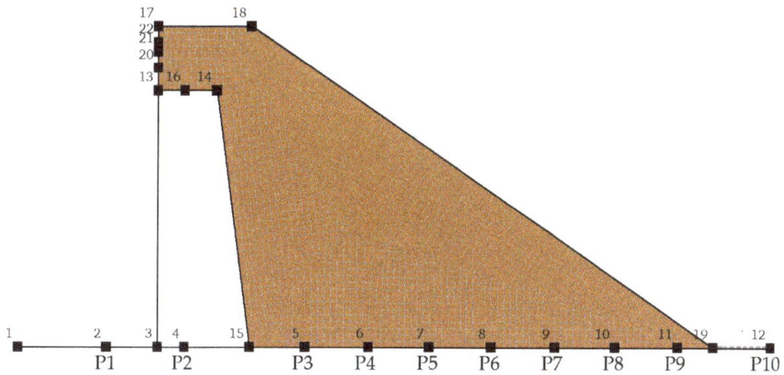

Figure 7. Design of the numerical model, showing the mesh and numbered nodes.

3.2.3. Boundary Conditions

To run a simulation, the boundary conditions must be defined. First, a total head or pressure head boundary must be added. In this case, a zero-pressure point is added to the downstream toe. Furthermore, the drainage boundary was set along the upper edge of the dam, set as a water-rate of 0 m^3/s with potential seepage face review—this allows water to escape the domain upon reaching the boundary. Lastly, the input is defined through a water flux for each discharge level. The flux is entered as $m^3/s/m^2$, which requires the correct area of flux to be defined. This is done by selecting the appropriate water level from the physical test results. For example, for an applied discharge of $q = 1 \times 10^{-3}$ m^3/s, the average entry water level was measured at 0.87 m (P1) over the horizontal platform for all the tested models. The water flux is therefore defined as 1×10^{-3} m^3/s for the 0.07 m above the crest of the core. Similarly, for $q = 2 \times 10^{-3}$ m^3/s the inflow occurs along 0.15 m of the face. Lastly, for $q = 4 \times 10^{-3}$ m^3/s the inflow boundary is 0.2 m, the height of the crest. The setup is shown in Figure 8.

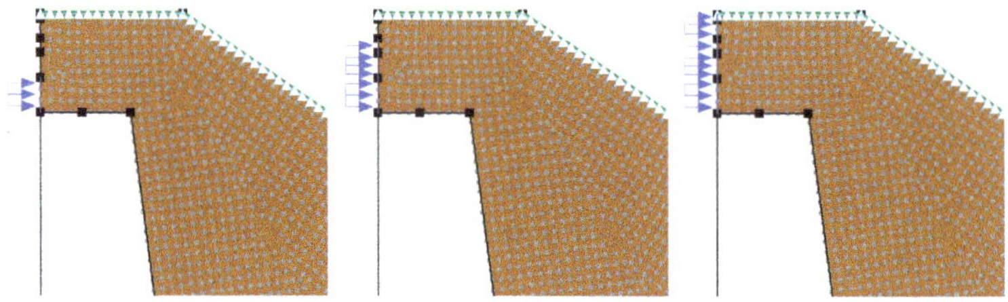

(a) (b) (c)

Figure 8. Input boundary flux shown for (**a**) $q = 1 \times 10^{-3}$ m^3/s, (**b**) $q = 2 \times 10^{-3}$ m^3/s and (**c**) $q = 4 \times 10^{-3}$ m^3/s.

3.2.4. Calibration and Evaluation

The parameter values that were selected for the numerical analysis were attained through trial and error. It was decided to utilize one parameter set which would provide the best fit for all configurations. Firstly, the results from numerical modeling of the no toe configuration were calibrated with the physical results. Upon arriving at a parameter-set, the parametric assumptions were further tested on the external, internal and combined configurations to evaluate the model performance. Necessary modifications/fine adjustments were made to the assumptions to achieve better fit with physical observations. The iteration process was repeated until an overall satisfactory fit was obtained. To evaluate the accuracy of the parameter-set, the main metric used was the root-mean-square error (RMSE), which calculates the standard deviation of the error between modelled and observed results. The root-mean-square value is favorable due to the resulting error being directly readable in the same unit as the modelled variable.

4. Results

The outcomes from the numerical modeling efforts to simulate observations and measurements from the physical modeling investigations are discussed within this section of the article. To enable visual comparison of the results, graphical plots of the dam models are depicted with pressure contours (pore pressure distributions) for incrementally applied discharge levels (Figure 9). The numerical and physical results are then collated through the comparison of the phreatic line for each individual discharge magnitude (Figure 10). Analysis of the modeling accuracies is further quantitatively described employing a statistical methodology. To aid in reading the results between numerical and physical models, the pressure sensors located along the dam body are given on the x-axis, with the positions provided in Table 1.

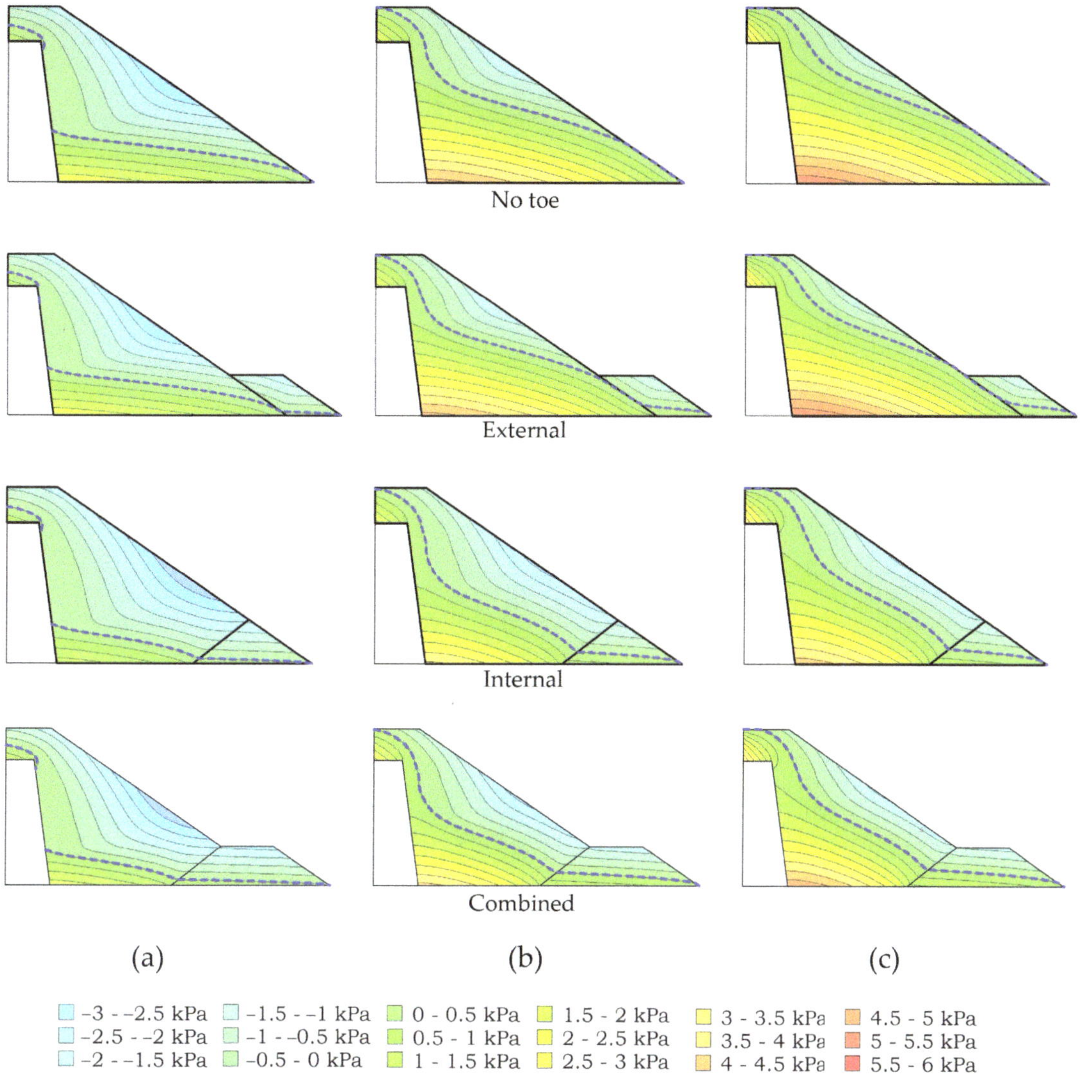

Figure 9. Numerical modeling results for the four toe configurations; column (**a**) shows results for $q = 1.0 \times 10^{-3}$ m^3/s, column (**b**) $q = 2.0 \times 10^{-3}$ m^3/s and column (**c**) $q = 4.0 \times 10^{-3}$ m^3/s. The colored contours display the water pressure shown in the legend, while the phreatic surface is displayed as the dotted blue line.

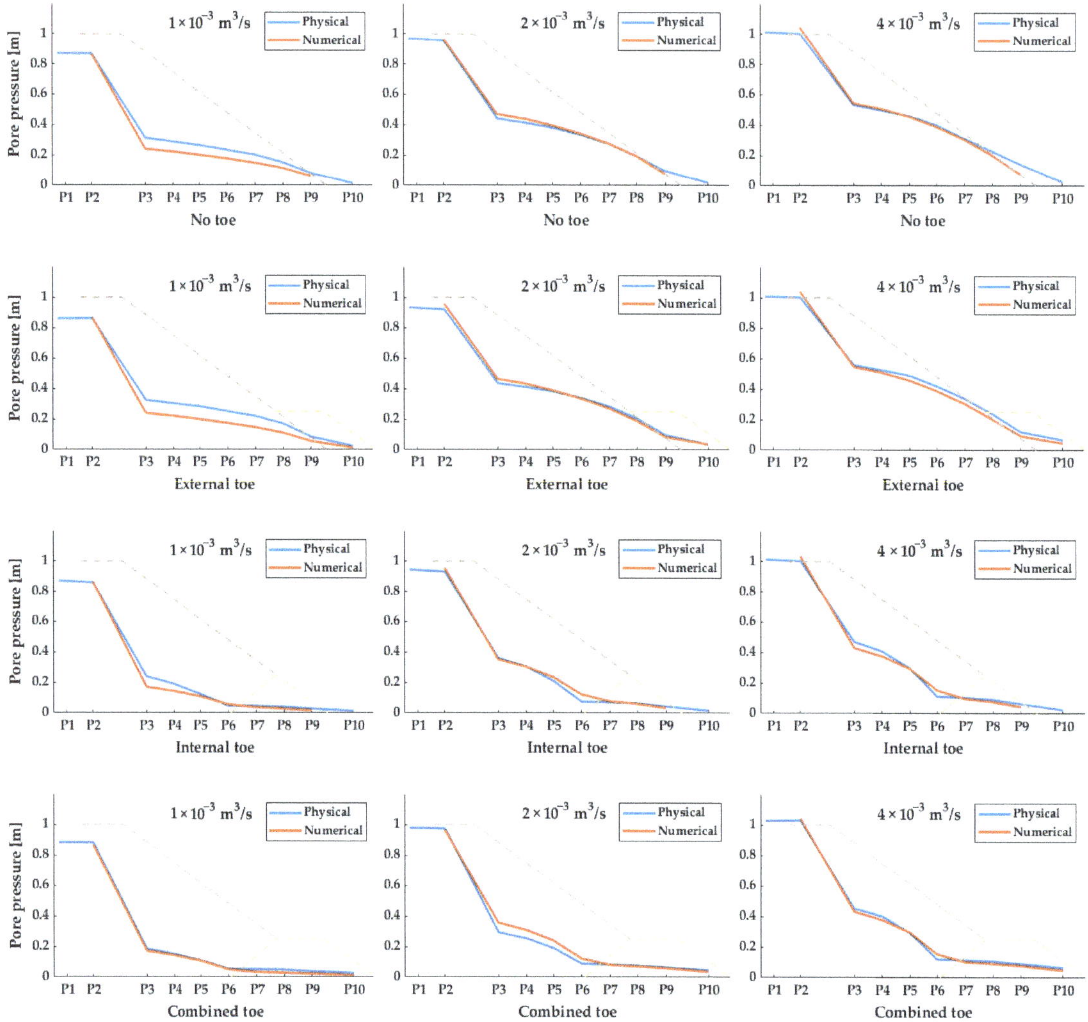

Figure 10. Numerical modeling results juxtaposed to physical modeling results. The stippled line displays the outline of the dam and toe configuration.

4.1. General Results from Numerical Modeling

The outcomes of the numerical analysis are summarized in Figure 9 as images of various model configurations as a function of the applied discharge. The pore pressure distributions are displayed as color contours with zero pressure contours marked as dotted blue lines representing the phreatic surface. In general, the calibrated numerical model demonstrates the good ability of the model to simulate throughflow conditions within the various dam models.

Results from the numerical analysis are overlaid with measurements from the physical modeling studies as depicted in Figure 10 for discharge magnitudes, $q = 1.0 \times 10^{-3}$ m^3/s, 2.0×10^{-3} m^3/s and 4.0×10^{-3} m^3/s. Juxtaposition of the results in Figure 10 illustrates that the numerical model accurately predicts the throughflow patterns within the various rockfill dam models subjected to various throughflow levels. The influence of the high permeability zones, i.e., the various toe configurations, on the development of phreatic lines within the dam models is also accurately modelled. The internal and the combined toe configurations lead to reduced pore pressure levels within the dam structure for a given applied throughflow magnitude in comparison with rockfill dam models with no toe. Furthermore, the external toe configuration appears to have minimal impact on throughflow development within the dam structure. Both these findings are in alignment with the documented findings from the physical studies [13]. The numerical model also appears to be capable of simulating the flow transitions from (a) the dam core towards the shoulder and (b) from the downstream shoulder towards the toe structure. This was verified by visual inspection of the video footage from the physical tests. Further detailed descriptions of qualitative results from comparative evaluations conducted between various physical and numerical model configurations are detailed in subsequent sections.

4.1.1. No Toe

In the depictions of Figure 10 for the no toe model, the phreatic surfaces undergo transitions as they enter the downstream shoulder structure. The numerical simulations appear to closely resemble the physical observations in this regard. Further, the precited pore pressure development trends in general confirm well the observations for sensor locations P2 to P8. From P8 to P9, the numerical predictions follow the dam surface, which is defined as a drainage boundary. In the numerical model, the zero-pressure boundary point is located at the toe, whereas in the physical model the water column at the downstream end of the dam, resulting from the throughflow exiting the dam, is measured. This leads to divergence in the results at the downstream end of the dam structure.

4.1.2. External Toe

The comparative evaluation results for rockfill dams with external toe configuration are shown in Figure 10. The pressure developments through the dam structure, as previously stated, appear very similar to the results from simulations for dams without toe structures. Comparison of measurements from sensor P10 with the simulations demonstrates good fit for flow within the external toe structure.

4.1.3. Internal Toe

Results for the internal toe configuration can be seen in Figure 9. The phreatic line can be seen to undergo gradual decrements along the dam length and experiences marked lowering of pore pressure prior to entering the toe structure. Furthermore, the phreatic surface within the toe structure drops significantly and flattens towards the exit boundary. Figure 10 shows a comparison of the physical and numerical results. The results in general correlate well with each other as documented by the depictions for the internal toe. Within the toe structure, the two results have a similar trajectory, however the numerical model analysis translates towards zero at the end of the toe (defined boundary condition), whereas the physical measurement is slightly higher due to the throughflow water exiting the dam.

4.1.4. Combined Toe

The numerical results in Figure 9, for dam model coupled with the combined toe configuration, closely resemble the results from the internal configuration case. However, with a longer flat decrease in pore pressures within the toe. For the combined toe configuration, a near perfect fit is obtained to the physical model results for $q = 1.0 \times 10^{-3}$ m^3/s.

4.2. *Performance Evaluation*

This section aims at quantitative evaluation of the model performance as pertained to ability of the numerical model to simulate phreatic surface developments within the model dam structures. To accomplish this task, a statistical evaluation is conducted adopting a relative changes approach. Relative changes are computed adopting the methodology put forth by Kiplesund et al. [13], wherein percentage pore pressure differences were computed for the different models with the no toe case as the baseline. Similar computations were conducted here for results from the numerical modeling investigations producing Table 4. This gives a good picture of quantitative conformity between results from the numerical and physical models.

Table 4 does not include P9 due to the boundary conditions in place, as the phreatic line of the numerical model will automatically follow the dam surface and thus is not comparable with measurements at location P9 in the physical models. The table in general demonstrates conformity between the percentage pressure changes with the physical modeling results. Hence, the numerical model is able to accurately simulate the pressure profiles at various locations within the dam structure. The model also accounts for the influence of the high permeability zones to a good degree.

The root-mean-squared error (RMSE) was further calculated for each model configuration exposed to various discharge magnitudes. The results show that the RMSE in general increases with increasing discharge. The cumulative RMSE for all models was found to be 0.023 m, and the largest error is computed for $q = 1.0 \times 10^{-3}$ m^3/s.

Table 4. Comparison of pressure reduction for different toe configurations relative to no toe configuration for physical and numerical models.

		Physical Model Rel. Pressure Reduction (%)							Numerical Model Rel. Pressure Reduction (%)						
Toe Config.	**q (L/s)**	**P2**	**P3**	**P4**	**P5**	**P6**	**P7**	**P8**	**P2**	**P3**	**P4**	**P5**	**P6**	**P7**	**P8**
	1.0	−1	3	4	7	8	9	12	0	−1	−1	−1	−1	−2	−3
	1.5	−2	2	2	3	4	5	7	0	−1	−1	−1	−1	−2	−3
	2.0	−4	−1	−1	1	1	3	6	−1	−1	−2	−2	−2	−2	−3
External	2.5	−3	0	1	2	3	5	8	0	0	0	−1	−1	−1	−1
	3.0	−2	1	2	3	4	6	7	0	0	0	0	0	0	0
	3.5	−1	5	6	6	5	8	4	0	0	0	0	0	0	1
	4.0	0	5	5	6	5	8	4	−1	0	0	0	0	0	0
	1.0	−1	−24	−34	−53	−80	−77	−73	0	−29	−36	−46	−68	−75	−75
	1.5	−1	−17	−26	−46	−78	−75	−69	−1	−27	−33	−43	−66	−73	−72
	2.0	−3	−18	−26	−44	−77	−74	−67	−1	−25	−31	−40	−64	−72	−69
Internal	2.5	−1	−16	−24	−42	−76	−72	−64	0	−20	−26	−35	−59	−66	−59
	3.0	0	−14	−21	−39	−74	−70	−61	−1	−22	−27	−37	−62	−70	−65
	3.5	0	−11	−18	−36	−73	−68	−61	−1	−22	−27	−36	−61	−69	−64
	4.0	0	−11	−18	−36	−72	−67	−59	−1	−21	−26	−35	−60	−68	−62
	1.0	2	−41	−48	−59	−77	−74	−67	0	−29	−35	−46	−71	−77	−74
	1.5	2	−35	−41	−53	−75	−72	−64	0	−26	−32	−43	−67	−74	−68
	2.0	2	−33	−39	−50	−73	−70	−60	0	−24	−30	−40	−64	−71	−63
Combined	2.5	2	−32	−36	−47	−71	−68	−56	0	−20	−25	−35	−60	−65	−54
	3.0	2	−26	−31	−43	−69	−65	−53	0	−22	−27	−37	−62	−68	−58
	3.5	3	−15	−20	−39	−71	−64	−54	0	−21	−26	−36	−61	−67	−56
	4.0	3	−15	−19	−36	−70	−63	−53	0	−21	−26	−35	−60	−66	−55

Legend	0–20%	21–40%	41–60%	61–80%

4.3. *Laminar versus Turbulent Flow in Numerical Models*

The numerical results presented are all models that consider a non-linear flow regime through the use of the non-Darcy tool [23]. However, the traditional seepage modeling, considering laminar flow, was also investigated. It was found that the seepage through the shell material, representing the rockfill shoulder, could be reliably modelled in a laminar regime for the no toe model. Difficulties occurred, in models with a toe configurations, when the flow transitioned into the toe region. Firstly, there were convergence issue with the analysis.

Secondly, to achieve comparable results to the non-Darcy modeling, it was necessary to alter the material parameter set between the models of different toe configurations.

5. Discussion

The results from the numerical analysis demonstrate that turbulent non-Darcy flow through rockfill dam structures can be modelled with good calibration metrics. However, some challenges with the numerical modeling work were encountered. This section discusses these challenges and aims at putting forth recommendations and insights that can potentially supplement further research in this regard.

5.1. Boundaries

The upstream boundary condition is simplified by assuming that the discharge is evenly distributed along the corresponding water level (see Figure 8). Thus, the velocity profile will be homogeneously distributed along the face. This can be stated as a simplification. However, due to the low entry velocities at the entry surface, the variability in the velocities with depth (0.2 m high crest) can be considered as insignificant. Additionally, the effect of this simplification on the results of the study are deemed to be minimal.

Definition of a drainage boundary with a potential seepage face results in that water is able to escape the domain along the boundary. In the numerical model, the dam surface is a drainage boundary, including the surface at the crest. Thus, for the high upstream phreatic line, as occurs for the highest discharges, a minor amount of water exits the dam at the crest where the inflow enters the dam, as shown in Figure 11. The effect was observed in the numerical model results for discharges, $q = 2.5 \times 10^{-3}$ m^3/s and higher. To ensure that drainage out of the system does not distort the results in the downstream dam shoulder, data were extracted to calculate the water loss at the highest applied discharge, $q = 4.0 \times 10^{-3}$ m^3/s. Considering the external toe case as an example, the total loss was computed to be 5.7×10^{-8} m^3/s. Furthermore, for the no toe configuration, the total loss was calculated as 6.8×10^{-8} m^3/s. The magnitude of losses was therefore deemed negligible for all models. This further entails that the same applies to lower applied discharges, $q < 4.0 \times 10^{-3}$ m^3/s.

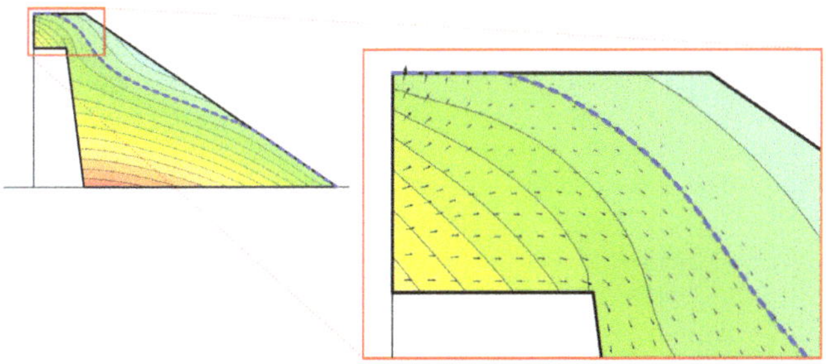

Figure 11. Detail showing flux vectors at the crest of the no toe configuration for $q = 4.0 \times 10^{-3}$ m^3/s. The maximum flux shown in the top left corner is 0.0068 m^3/s/m^2; the vectors are magnified by a factor of two to increase visibility.

5.2. Pressure Development

Investigating the pressure development within the dam models, for $q = 1.0 \times 10^{-3}$ m^3/s, the numerical results for the no toe and external toe models show somewhat lower pressure values than physical modeling observations. The reasoning for this could be that the selected parameter set and successive numerical results were better fitted to the higher discharges. However, for both the internal toe and combined toe cases, the fit is similar for all the discharges, and even best for $q = 1.0 \times 10^{-3}$ m^3/s in the case of the combined toe. Furthermore, the observed discrepancy does not affect the general outcome of the study relating to investigating the different toe configurations. Moreover, it is of value to be able to use the same material properties for all the models and discharges. Comparison of the numerical and physical model results in Figure 10 is a validation of the parameter set used and supports further investigations using the numerical model.

The numerical results for the internal and combined toe configurations are in good agreement with the physical results. However, for location P6, close to the interface of the rockfill material and the toe, the pore pressure is slightly lower for the physical models for discharges $q = 2.0 \times 10^{-3}$ m^3/s and larger. This can be due to increased permeability at this location in the physical model, explained by the methodology adopted for construction of the dam. The toe is first placed in its position in layers and the dam is built adjacent to the toe. When compacting the dam layers, the shell material in contact with the toe needs to be carefully tamped, so as to not cause filling of the voids or disturbing the shape of the toe. This can cause the interface and surrounding area to have greater permeability than the rest of the dam. For the numerical model, the material properties are defined for the respective region, and the interface has no effect outside of the contact. A similar limitation lies at the crest of the dam, which can be seen as the numerical results yielding higher pressure values at P2 for increasing discharge. Construction of the physical model requires a metal mesh at the face in order to withhold the materials from sliding into the upstream reservoir. This causes some of the finer material to slide through, as well as diminishing the compactability. Ultimately this can have an increasing effect on the permeability compared to the main body of the shell, which is not represented in the numerical model, as it remains completely homogeneous. These effects are important to note when proceeding to real dam cases, considering that the method of construction introduces regions of different permeability.

From Table 4, the physical model results show the relative reduction in pore pressure decreases with increasing discharge. The same overall trend can be seen for the numerical results. However, the results for $q = 2.5 \times 10^{-3}$ m^3/s, do not agree with that trend. These peculiarities can be linked to issues with the input boundary. For $q = 2.5 \times 10^{-3}$ m^3/s, the water level was measured at 0.99 m at P1, meaning that the mesh size used, of 0.02 m, is split for this edge.

A consequence of lowering the phreatic surface, due to introduction of a high permeability zone such as the toe installations, is increased flow velocities within the dam structure. Increased velocity can have a detrimental effect on stability through internal erosion processes, if it is not accounted for in the dam design. To investigate the velocity increase, the numerical model was used, extracting data at a horizontal line from the top of the core into the dam. Comparing the internal toe configuration to no toe, the data showed a 10% average increase in velocity fluxes for $q = 4.0 \times 10^{-3}$ m^3/s.

5.3. Calibration

The parameter set that was obtained for the dam and the toe structure was arrived at through trial and error. When using the non-Darcy add-in, there are multiple physical properties and fitting parameters that affect the flow patterns and subsequent phreatic line developments. The critical component being the hydraulic conductivity, defining the permeability of the dam. Minute adjustments of the hydraulic conductivity will have great influence on the phreatic line and pore pressure development. Upon setting an agreeable hydraulic conductivity, the Van Genuchten fitting parameters and tortuosity parameters were dialed in.

The α-parameter will alter the fit at the upstream end of the dam, showing influence on the early sensor locations. It was found that increasing the α-value in general raised the phreatic surfaces.

The l-parameter had significant influence on the flow pattern through the downstream dam shell and the toe structure. It was found that, when decreasing the l-parameter flow, velocity vectors were seen to a larger degree above the phreatic lines. This effect has the largest influence on the toe, where there is high inter-connectivity of the voids. The l-parameter was also found to affect the pressure in the main body of the dam.

Some limitations on the parameters are hard to estimate, as the user interface of the non-Darcy add-in of the software used [20] does not limit any input, but incorrect values will cause failure to find a solution during analysis. For example, the n-parameter is said to have a limit yet, with increasing hydraulic conductivity, the limit changes. In this sense, the add-in calibration can be slightly cryptic, prolonging the process. The calibration process aimed at finding a parameter set best fitting P2-P10, as the phreatic lines on top of the core remained largely unaffected by toe design in the physical model. However, for internal and no toe configurations, the last measurement points, P9 and P10 were affected by the end of the domain.

In examining the available literature regarding the Van Genuchten input parameters, it should be clarified that the described parameters are detailed for soils of different compositions. Another limitation to bring up is that the apparent conductivity used in the non-Darcy add-in is designed for groundwater aquifers, and is valid when velocities remain low to intermediate [23]. This could pose issues with upscaling the model to larger dams where velocities can be considerably larger. There is no available research utilizing the non-Darcy add-in, which further adds some uncertainty.

Equifinality of parameters is an additional point of discussion for the non-Darcy modeling. As calibration is done on a trial and error basis, with multiple fitting parameters, the results could possibly be reproduced with another parameter-set.

5.4. Application and Future Recommendations

The study demonstrates how numerical models can be useful for deeper apprehension of the results from physical tests. The numerical model enables detailed investigation of flow through the dam structure at every specific location, not just discreet positions determined by, e.g., installed pressure sensor locations. Moreover, the numerical model has the advantageous possibility of investigating different parameters at specific locations that the physical test cannot, such as velocity. In addition, through a calibrated numerical model, one can experiment with modifications to the physical model which are more resource intensive than modifications to numerical models. Hence, prior to customization of the physical model, it is highly recommended to utilize the numerical model for planning of future experiments. The numerical model requires very little resources for alteration of the design, which can then be used to design the changes and hypothesize the results of alterations to the physical model.

It is important to proceed from physical scale models to full scale dam cases, preferably with relevant data for calibration and validation. In general, to optimize the calibration process, it is recommended to investigate the usage of optimization algorithms to find optimally fitted parameter sets. With adoption of an optimization algorithm or machine learning method, higher precision calibrations could be achieved, but one must be cautious of unrealistic parameters. Before expanding the numerical model to other cases, it is recommended to verify the numerical model on a prototype rockfill dam with well-documented throughflow data. Calibrating the model to a full-scale embankment dam can provide verification of the applicability and validity of the non-Darcy add-in within prototype dams.

6. Conclusions

The goal of this study was to investigate the applicability of a professional software to numerically model turbulent flow through rockfill dams. Moreover, it was to further understand the effects of different rockfill toe configurations within the downstream dam shoulder. Numerical models were successfully calibrated against results from physical model tests, employing one set of material parameters for different model setups. Through this, the present article makes a strong case highlighting the potential for numerical modeling of turbulent non-Darcy flow through rockfill dam structures. The numerical analysis results further support findings of the physical study [13] relating to effectiveness of the different toe configurations. In comparison to the dam without a toe, the external toe protects the exit zone on the downstream side from eroding supporting the findings and recommendations of Moran et al. [11]. Additionally, the internal and combined toe configurations are effective in lowering the phreatic line within the dam, for enhanced slope stability compared to the cases without a toe or an external toe.

The numerical modeling study presented demonstrates the efficacy of the model with regard to predicting throughflow in rockfill dams to a high degree of accuracy. Numerical modeling tools over the years have become increasingly reliable and robust and the trend appears to extend into the future. Development and calibration of numerical modeling tools to assist in the design and evaluation of rockfill dams can be stated as an effective method for enhancing dam safety. Since a significant number of iterations can be run in such a numerical model, a wide variety of material properties and loading conditions could be evaluated leading to better practical decision-making. Such numerical models can also be invaluable for research and development. A small number of physical modeling studies can lead to reliably calibrated numerical models and these models can further be employed to carry out a spectrum of investigations on model variations. Further research into development, calibration and, in turn, validation of numerical modeling tools within the research discipline of rockfill dam engineering is highly recommended.

Author Contributions: Conceptualization: F.G.S., G.H.R.R.; methodology: N.S.S., G.H.R.R., F.G.S.; formal analysis: N.S.S.; investigation: N.S.S., G.H.R.R., F.G.S.; original draft preparation: N.S.S., F.G.S., G.H.R.R.; writing: N.S.S., F.G.S., G.H.R.R.; visualization: N.S.S.; supervision: F.G.S.; project administration: F.G.S. All authors have read and agreed to the published version of the manuscript.

Funding: Financial support for the research venture was provided by Hydrocen, Norway.

Institutional Review Board Statement: Not applicable

Informed Consent Statement: Not applicable.

Data Availability Statement: Data presented in this study can be made available upon request from the corresponding author.

Acknowledgments: The authors acknowledge the support and co-operation offered by Geir H. Kiplesund, NTNU and Marius Rokstad, NTNU with this research project. Appreciation also goes to Livia Pitorac for allowing the use of her photo (real dam cases) within Figure 3.

Conflicts of Interest: The authors declare no conflict of interest.

References

1. International Commission on Large Dams (ICOLD). World Register on Dams. General Synthesis. 2020. Available online: http://www.icold-cigb.org/GB/World_register/general_synthesis.asp (accessed on 25 April 2021).
2. Sigtryggsdóttir, F.G.; Snæbjörnsson, J.T.; Grande, L.; Sigbjörnsson, R. Interrelations in multi-source geohazard monitoring for safety management of infrastructure systems. *Struct. Infrastruct. Eng.* **2016**, *12*, 327–355. [CrossRef]
3. International Commission on Large Dams (ICOLD). *Dam Failures Statistical Analysis*; Bulletin 99-3; ICOLD: Paris, France, 1995.
4. Solvik, O. Throughflow and stability problems in rockfill dams exposed to exceptional loads. In Proceedings of the 17th International Congress on Large Dams, Vienna, Austria, 17–21 June 1991.
5. Marulanda, A.; Pinto, N.L.S. Recent Experience on Design, Construction and Performance of CFRD Dams, J. Barry Cooke volume. In Proceedings of the International Symposium on Concrete Face Rockfill Dams and 20th International Commission on Large Dams Congress, Beijing, China, 18 September 2000; pp. 279–299.
6. Cruz, P.T.; Materón, B.; Freitas, M. *Concrete Face Rockfill Dams*; CRC Press: Leiden, The Netherlands, 2009.
7. Javadi, N.; Mahdi, T.F. Experimental investigation into rockfill dam failure initiation by overtopping. *Nat. Hazards* **2014**, *74*, 623–637. [CrossRef]
8. Siddiqua, S.; Blatz, J.; Privat, N. Evaluating turbulent flow in large rockfill. *J. Hydraul. Eng.* **2011**, *137*. [CrossRef]
9. Toledo, M.; Morera, L. *Design of Overtopping-Resistant Rockfill Dams*; CRC Press: Leiden, The Netherlands, 2015; pp. 133–142.
10. Morán, R.; Toledo, M. Research into protection of rockfill dams from overtopping using rockfill downstream toes. *Can. J. Civ. Eng.* **2011**, *38*, 1314–1326. [CrossRef]
11. Morán, R.; Toledo, M.A.; Larese, A.; Monteiro-Alves, R. A procedure to design toe protections for rockfill dams against extreme through-flows. *Eng. Struct.* **2019**, *195*, 400–412. [CrossRef]
12. Ravindra, G.H.R. Hydraulic and Structural Evaluation of Rockfill Dam Behavior When Exposed to Throughflow and Overtopping Scenarios. Ph.D. Thesis, Norwegian University of Science and Technology, Trondheim, Norway, 2021.

13. Kiplesund, G.H.; Ravindra, G.H.R.; Rokstad, M.M.; Sigtryggsdóttir, F.G. Effects of toe configuration on throughflow properties of rockfill dams. *J. Appl. Water Eng. Res.* **2021**. [CrossRef]
14. Hiller, P.H. *Survey of Placed Riprap on the Downstream Slopes of Rockfill Dams*; Report B1-2016-1; Norwegian University of Science and Technology: Trondheim, Norway, 2016.
15. Ravindra, G.H.R.; Sigtryggsdóttir, F.G.; Asbølmo, M.; Lia, L. Toe support conditions for placed ripraps on rockfill dams—A field survey. *VANN* **2019**, *3*, 185–199.
16. Ravindra, G.H.R.; Sigtryggsdóttir, F.G.; Høydal, O. Non-linear flow through rockfill embankments. *J. Appl. Water Eng. Res.* **2019**, *7*, 247–262. [CrossRef]
17. Hiller, P.H.; Aberle, J.; Lia, L. Displacements as failure origin of placed riprap on steep slopes. *J. Hydraul. Res.* **2018**, *56*, 141–155. [CrossRef]
18. Ravindra, G.H.R.; Sigtryggsdóttir, F.G.; Lia, L. Buckling analogy for 2D deformation of placed ripraps exposed to overtopping. *J. Hydraul. Res.* **2021**, *59*, 109–119. [CrossRef]
19. Ravindra, G.H.; Gronz, O.; Dost, J.B.; Sigtryggsdóttir, F.G. Description of failure mechanism in placed riprap on steep slope with unsupported toe using smartstone probes. *Eng. Struct.* **2020**, *221*, 111038. [CrossRef]
20. GeoSlope. *Heat and Mass Transfer Modeling with GeoStudio*, 2nd ed.; Geoslope International Ltd.: Calgary, AB, Canada, 2017.
21. Kheiri, G.; Javdanian, H.; Shams, G. A numerical modeling study on the seepage under embankment dams. *Model. Earth Syst. Environ.* **2020**, *6*, 1075–1087. [CrossRef]
22. Papagiannakis, A.; Fredlund, D. A steady state model for flow in saturated–unsaturated soils. *Can. Geotech. J.* **1984**, *21*, 419–430. [CrossRef]
23. GeoSlope. *Add-In: Non-Darcy Flow, Theoretical Background*; Geoslope International Ltd.: Calgary, AB, Canada, 2018.
24. Engelund, F. On the laminar and turbulent flows of groundwater through homogeneous sand. *Akad. Tek. Vidensk.* **1953**, *3*, 1–105.
25. Wilkins, J.K. Flow of water through rockfill and its application to the design of dams. *N. Z. Eng.* **1955**, *10*, 382–387.
26. Dudgeon, C.R. An experimental study of the flow of water through coarse granular media. *Houille Blanche* **1966**, *7*, 785–801. [CrossRef]
27. Ferdos, F.; Yang, J. Characterization of Hydraulic Behaviours of Coarse Rock Materials in a Large Permeameter. *J. Geosci. Environ. Prot.* **2013**, *1*, 1–6. [CrossRef]
28. Ferdos, F.; Ekström, I. Hydraulic Conductivity of Coarse Rockfill used in Hydraulic Structures. *Transp. Porous Media* **2015**, *108*, 367–391. [CrossRef]
29. Parkin, A.K.; Trollope, D.H.; Lawson, J.D. Rockfill Structures Subject to Water Flow. *J. Soil Mech. Found. Div.* **1966**, *92*, 135–151. [CrossRef]
30. López-Acosta, N.P. Numerical and Analytical Methods for the Analysis of Flow of Water Through Soils and Earth Structures. In *Groundwater*; Javaid, M.S., Ed.; IntechOpen: Rijeka, Croatia, 2016; Chapter 5. [CrossRef]
31. Van Genuchten, M. A Closed-form Equation for Predicting the Hydraulic Conductivity of Unsaturated Soils1. *Soil Sci. Soc. Am. J.* **1980**, *44*, 892–898. [CrossRef]
32. Van Genuchten, M.; Pachepsky, Y. Hydraulic Properties of Unsaturated Soils. In *Encyclopedia of Earth Sciences Series*; Springer: New York, NY, USA, 2011; pp. 368–376._69. [CrossRef]
33. Mace, A.; Rudolph, D.L.; Kachanoski, R.G. Suitability of Parametric Models to Describe the Hydraulic Properties of an Unsaturated Coarse Sand and Gravel. *Ground Water* **1998**, *36*, 465–475. [CrossRef]
34. Yang, X.; You, X. Estimating Parameters of Van Genuchten Model for Soil Water Retention Curve by Intelligent Algorithms. *Appl. Math. Inf. Sci.* **2013**, *7*, 1977–1983. [CrossRef]
35. Hoffmans, G.J.C.M. *The Influence of Turbulence on Soil Erosion*; Eburon Academic Publishers: Delft, The Netherlands, 2012; pp. 55–56.

water

MDPI

Article

Hydrodynamic Performance and Design Evolution of Wedge-Shaped Blocks for Dam Protection against Overtopping

Francisco Javier Caballero [1,2,*], Miguel Ángel Toledo [1], Rafael Moran [1,3] and Javier San Mauro [3]

1 SERPA Dam Safety Research Group, Department of Civil Engineering: Hydraulics, Energy and Environment, E.T.S. de Ingenieros de Caminos, Canales y Puertos, Universidad Politécnica de Madrid (UPM), Profesor Aranguren s/n, 28040 Madrid, Spain; miguelangel.toledo@upm.es (M.Á.T.); r.moran@upm.es (R.M.)
2 ACIS Innovation+Engineering S.L. (ACIS2in), Planeta Urano 13, P18 2°A, 28983 Parla, Madrid, Spain
3 Centre Internacional de Metodes Numerics en Enginyeria (CIMNE), Campus Norte UPC, Gran Capitán s/n, 08034 Barcelona, Spain; jsanmauro@cimne.upc.edu
* Correspondence: franciscojavier.caballero@upm.es; Tel.: +34-616-978-633

Abstract: Dam safety requirements have become stronger in recent years, highlighting, among other issues, the need to increase the discharge capacity of existing spillways and the protection of embankment dams against potential overtopping, which are particularly threatened by the hydrological consequences of climate change. The current economic situation requires solutions that ensure the safety of these infrastructures at an affordable cost. Wedge-shaped blocks (WSBs) are one of these solutions. A more detailed understanding of the performance of WSBs was the objective of this work and, based on this, the evolution of WSB design. An extensive empirical test program was performed, registering hydrodynamic pressures on the block faces and leakage through the joints between blocks and their air vents. A new WSB (named ACUÑA) with a different design of air vents was tested in comparison to Armorwedge™, which was used as a reference case. Moreover, the hydraulic behavior of the WSB was analyzed according to the saturation state of the granular drainage layer. The ACUÑA unit was designed with air vents in the upper part of the riser where the registered negative pressures were higher. Negative pressures were also measured at the base of the block when the granular drainage layer was not fully saturated. Finally, the beneficial effect of sealing some of the joints between blocks was quantified.

Keywords: wedge-shaped block; WSB; overtopping; dam protection; dam spillway; dam safety; ACUÑA

Citation: Caballero, F.J.; Toledo, M.Á.; Moran, R.; San Mauro, J. Hydrodynamic Performance and Design Evolution of Wedge-Shaped Blocks for Dam Protection against Overtopping. *Water* **2021**, *13*, 1665. https://doi.org/10.3390/w13121665

Academic Editor: Giuseppe Pezzinga

Received: 30 April 2021
Accepted: 11 June 2021
Published: 15 June 2021

Publisher's Note: MDPI stays neutral with regard to jurisdictional claims in published maps and institutional affiliations.

1. Introduction and Background

Wedge-shaped blocks (WSBs) are modular elements made of precast concrete. They are intended to prevent the erosion and scour in soils caused by water flowing at high velocity. Such blocks are considered one of the feasible alternatives to protect embankment dams against erosion caused by overtopping [1]. They are installed in overlapping rows on the downstream shell of embankment dams or levees. Typically, the WSBs are manufactured with high-strength concrete, although initially, steel-reinforced blocks were also used [2].

The former idea of protecting dams against overtopping by overlapping concrete blocks comes from Gordienko, from the Moscow Institute of Civil Engineering in the late 1960s [3]. Subsequently, new advances were carried out by Pravdivets [2,4], Bramley, May and Baker [3,5–9], Clopper [10], Slovensky [11], Gaston [12], Frizell [13,14], Thornton et al. [15] and Relvas and Pinheiro [16–20], among others [21]. From this knowledge, the first technical guide to build spillways using wedge-shaped blocks was published [3], and the Armorwedge™ patent was developed by the U.S. Bureau of Reclamation (US5544973A). Additionally, several dams in operation have been built with this technology since 2007. Barriga dam in Spain [22] (Figure 1 and Figure S1 in Supplementary Materials), Bruton dam [23], Ogden dam [24,25] and Norton-Fitzwarren dam [26] were built in the United Kingdom, and Friendship Village auxiliary spillway was built in Missouri, USA [1].

(a) (b)

Figure 1. (**a**) The Barriga dam (Burgos, Spain). (**b**) The upper area detail of the WSB chute spillway of the Barriga dam.

The vents' surface and their position in the blocks are essential for the correct operation of WSB. A recent investigation in Spain, which is presented here, has focused on this subject. Earlier, in the 1990s, researchers from the University of Salford obtained the critical drainage surface in the block to achieve a satisfactory reduction of uplift pressure [3] and also performed a comparison between the drainage of the WSBs through slots located in the lower area of the overlapping zone between blocks and orifices located in the block tread [6,8]. In most of the experimental research, the setup of the tests was developed by placing the blocks over a drainage layer, which may become saturated at a certain inlet flow rate (q) value [5–9,11,17,19]. The uplift developed at the base of the blocks was a key aspect of their stability. As every block overlaps the ones located at the downstream row, the revetment works hydraulically as a stepped spillway. A fraction of the inlet flow leaks through the joints between adjacent blocks and inlets at the drainage layer under the blocks, usually formed of gravel material. The resulting seepage flow at such a drainage layer is critical for the stability of the block. [15,16]. The hydraulic stability of WSBs is mainly achieved due to three effects: the positive pressure of the main discharge flow, which impinges on the upper face of the next blocks downstream; the overlapping of the different rows; and the development of negative pressures on the block tread. Such stability is enhanced by the effect of the vents (also termed as "air vents" or "holes") transmitting the suction generated on the block tread towards the block base when the drainage layer is not saturated (Figure 2a). If the underlay is saturated, such suction may cause the return of a fraction of the drainage flow to the spillway chute, reducing the uplift pressure under the blocks (Figure 2b). In supplementary materials, some videos are included where this behavior can be appreciated (folders "01_No_saturation_*d2*" and "02_Suction"). WSBs have proven to be highly stable even in very unfavorable conditions [3].

There are references for this protection system in dams under 18 m high, with a maximum unit discharge of 3.9 $m^2 s^{-1}$ [1]. Nonetheless, good and likely better behavior should be expected for a greater velocity due to the positive effect of suction.

Since 2011, new research efforts have aimed to complement the theoretical and practical knowledge of WSB technology [27–31]. Such work aims to deepen our knowledge on an alternative technology in order to improve dam safety, which is particularly threatened by the hydrological consequences of climate change [32–39]. Specifically, the main goal of the research was to increase the cost efficiency of the discharge capacity of existing spillways and the protection of earth dams against potential overtopping. One of the results obtained is the development of a new design of WSB, ACUÑA (patented on 8 May 2017 ES2595852), which aims to improve the behavior of pre-existing blocks that have air vents in the lower part of the riser (i.e., Armorwedge™). The new design aims to improve the transmission of negative pressures and, therefore, the stability of the block. The new design also aims to achieve other additional construction improvement objectives for the implementation of the blocks.

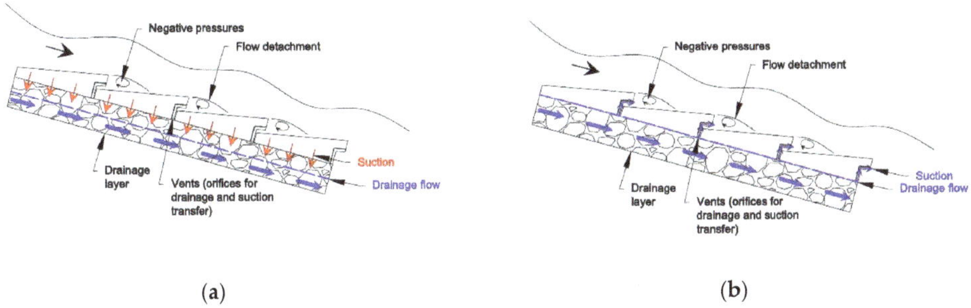

Figure 2. The hydraulic performance of a WSB spillway: (**a**) the drainage layer not saturated and (**b**) the drainage layer saturated.

2. Materials and Methods

2.1. Experimental Facility

The laboratory flume (located at CEDEX Hydraulics Lab, Madrid, Spain) includes a 0.50 m wide steel and methacrylate chute. The vertical slope is 2H:1V, and the maximum vertical drop is 4.7 m (Figure 3a). The side walls of the chute are 0.85 m high in the direction normal to the bottom. The maximum inlet flow rate (q) available is 0.24 $m^2\ s^{-1}$. The water supply is carried out through water pumps from the lower tank to the inlet tank with a maximum elevation of 6 m from the lab floor and a horizontal area of $2.5 \times 2.5\ m^2$. The tank has a 0.5 m wide and a 0.75 m high lateral opening, which connects to the chute by a 1.5-m-long horizontal inlet (Figure 3b). The WSBs tested in each trial are laid over the bottom of the chute and placed in 47 horizontal rows (Figure 3c). At the downstream end of the chute, there is a rectangular stilling basin that dissipates the energy of the flow. The measurement of the discharge is performed at a rectangular thin-plate weir at the end of the stilling basin before the flow is conveyed back to the lower tank. Next to the chute, on one of the side walls (in row 32), there is an outlet pipe of the seepage discharge in order to measure the drainage flow under the rows of WSBs by means of a triangular thin-plate weir. Additional photographs, schemes and videos of the experimental facility have been included in supplementary materials (Figure S2 in Supplementary Materials and 'folder 00_Experimental_set_up').

The instrumentation setup is able to measure the water level, the inlet flow (q) and drainage flow (q_d) discharge and the pressures on different positions of the blocks. Two methacrylate WSBs were designed as measuring blocks built with methacrylate sheets (Figure 4a). On such blocks, pressure sensors were installed on the block faces at different positions (Figure 4b). The methacrylate measuring blocks were installed on rows 5, 10, 15, 25, 30 and 35 (colored black in Figure 3c) to achieve measurements at different positions along the longitudinal profile of the chute. The measuring devices can be grouped as follows:

1. Measuring devices for water levels and discharge of skimming and seepage flows:
    - Electromagnetic flowmeter to measure the pumped flow rate.
    - Triangular thin-plate weir to measure the flow that leaks through the open joints between adjacent blocks and seeps through the granular layer.
    - Electromagnetic limnimeters (4) for measuring the water level at the following points: the inlet tank, the upstream end of the chute, the abovementioned triangular thin-plate weir and the rectangular thin-plate weir at the end of the stilling basin.

2. Pressure measurement system to register the water pressures at several points on the block tread, base and the riser step of the WSBs [27], formed by:
    - A set of 12 Messtech submersible XA-700 pressure transducers connected to measuring tubes installed on one of the measuring blocks.
    - A pressure gauge (Scanivalve DSA3207 Corp. model) with 12 sockets measuring tubes installed on a second measuring block.

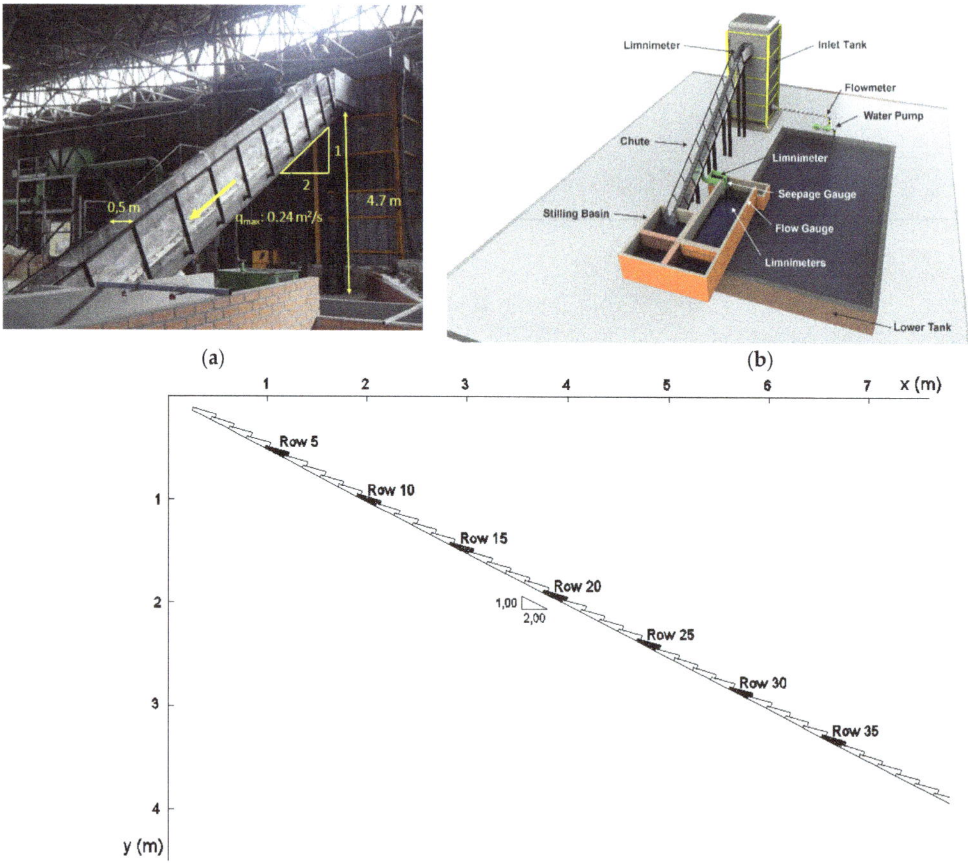

Figure 3. The experimental setup: (**a**) a photo of the testing facility, (**b**) scheme and (**c**) a pthe instrumented rows.

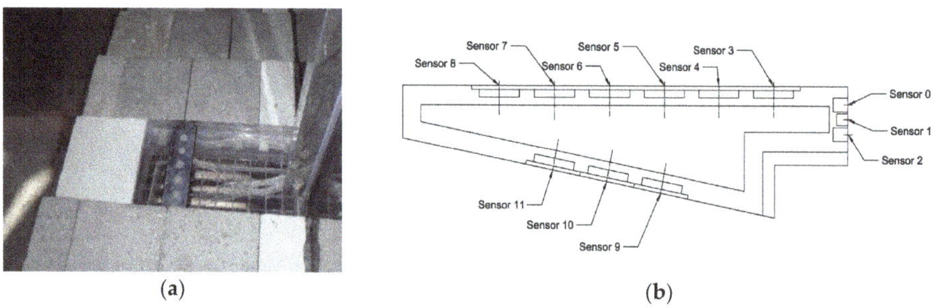

Figure 4. The measuring methacrylate WSB (**a**) and a block section with the locations of pressure sensors indicated (**b**) [28].

The data acquisition was performed at a time interval of 5 min per test for each inlet flow rate (q). Such acquisition was conducted using the Messtech submersible XA-700 pressure transducers (National Instruments) data collection equipment (cDAQ), which allows one to obtain 30 items of data per second and channel, and an Ethernet connection system by Scanivalve DSA3207 Corp., which allows one to obtain 70 items of data per second and channel. Furthermore, a conventional video camera was used for recording the tests.

2.2. *Flow Test Characterization*

2.2.1. Flow Regimes

The type of flow regime (nappe, transition or skimming) determines the pressure pattern of the stepped chutes. For example, with a transition flow, the suction on the base of the block is lower than with skimming flow [18]. Several authors [2,13,18] relate the slope of the pseudo-bottom (i.e., the straight line connecting the step edges) to the ratio between the flume critical depth (h_c) and the height of the block riser (h_s), h' (Figure 5, Table 1). Some of them [18] were used here to predict the flow regimes for each tested unit discharge (q).

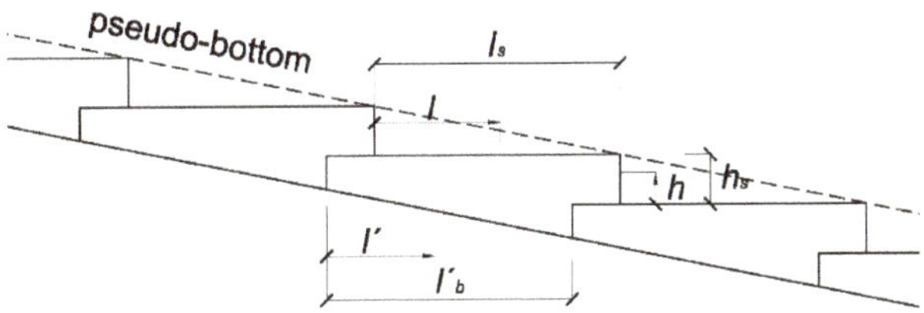

Figure 5. The geometric parameters of the block.

Table 1. Values of h' and q used in the tests.

h' ($h_c\ h_s{}^{-1}$)	4.51	3.99	3.44	2.84	2.52	2.17	1.37
q ($m^2\ s^{-1}$)	0.24	0.20	0.16	0.12	0.10	0.08	0.04

In order to compare results with other authors, dimensionless distances were used; l is the distance measured from the upstream end of the exposed tread (position of the pressure measurements on each sensor), and l_s is the total exposed length of the upper surface of the block. At the base of the block, l' is the distance measured from the upstream end of the base (position of the pressure measurements on each sensor), and l'_b is the total length of the base of the block. The tread length (l) refers to the unit length (l_s) (Figure 5) to obtain the dimensionless parameter ($l\ l_s{}^{-1}$).

Figure 6 shows that every unit discharge used during the tests corresponded to skimming flow according to every author, except for Chamani-Ratjaratnam [40] and André [41]. According to the criteria proposed by Chamani-Ratjaratnam, the two lower discharges of the present study correspond to the transition flow, and all others correspond to the skimming flow. According to André, only the lower discharge corresponds to the transition flow.

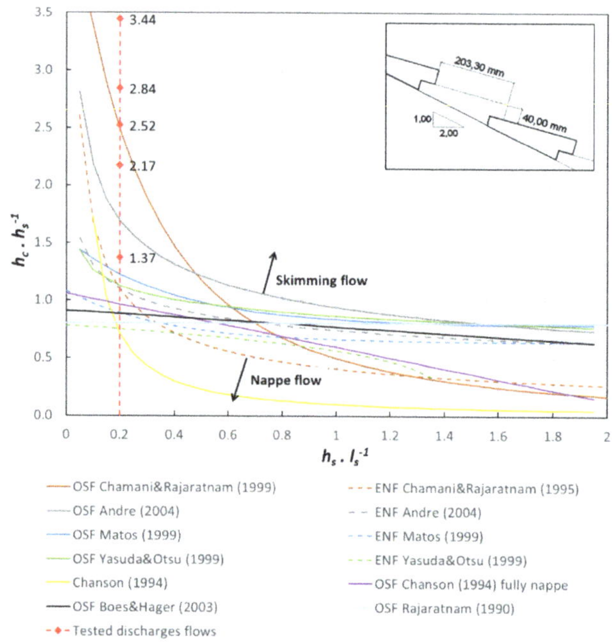

Figure 6. Flow regime (skimming, transition and nappe flow) for the tested discharges according to the criteria of [40–45].

2.2.2. Inception Point

The inception point is located at the position where the boundary layer intersects with the free surface of the flow. Such location is important to establish the upstream limit of air entrainment through the flow surface.

The location of the inception point (Table 2) was determined by visual observation according to the criteria of Mateos and Elviro [46]; i.e., the location was established where a permanent presence of air bubbles was observed for all the tested discharges. The observed location was compared with the results of the empirical formulas proposed by Relvas [18], Chanson [43,47] and Matos [44]. Results are included in Figure S3 of the Supplementary Materials.

Table 2. The inception point location of tested flows according to different authors.

Flow Rate q ($m^2\ s^{-1}$)	H' ($h_c\ h_s^{-1}$)	Row Number	Distance of the Inception Point L_i			
		Observed	Observed Interval	Relvas [18]	Chanson [43]	Matos [44]
0.04	1.37	3	673–880	779	684	617
0.08	2.17	4–5	1035–1293	1283	1121	1035
0.10	2.52	5–6	1190–1500	1507	1314	1223
0.12	2.84	6	1293–1500	1718	1496	1401
0.16	3.44	6–7	1293–1707	2114	1837	1737
0.20	3.99	7–8	1500–1914	2482	2154	2053
0.24	4.51	9–10	1914–2328	2831	2453	2352

The results obtained in the research carried out by Gaston [12] presented a good agreement with the formulation of Relvas [18] for slopes of the chute of 2.5H:1V and 2H:1V.

However, the values of the roughness Froude number (F^*, Equation (1)) used by Gaston were much higher than the ones used by Relvas and ourselves in the present study.

$$F^* = q / \sqrt{g \cdot sin\theta \cdot (h_s \cdot cos\theta)^3} \quad (1)$$

defined in terms of h_s and θ, where θ is the angle formed by the flume with the horizontal.

Some previous formulations, such as that of Matos [44], do not take into account the influence of the slope of the channel, and the results are relatively close (but higher) to those obtained here by visual observation for the highest flow rates (from 0.16 $m^2 s^{-1}$).

2.2.3. Uniform Flow Area

If the channel is long enough, the uniform regime is reached. Then, the amount of entrained and aspirated air would be equal. Flow can be considered stable in this area. The upstream limit of the uniform flow area (Table 3) was estimated by means of empirical formulas [3,18,48]. Results are also included in Figure S4 of the Supplementary Materials.

Table 3. The uniform flow depth location (distance and row) of tested flows according to different authors.

Flow Rate q ($m^2 s^{-1}$)	**h' ($h_c h_s^{-1}$)**	**Uniform Flow Depth Location (m)**					
		CIRIA Guide Hewlett et al. [3]		**Boes and Minor [48]**		**Relvas [18]**	
		Distance (m)	**Row**	**Distance (m)**	**Row**	**Distance (m)**	**Row**
0.04	1.37	1.09	5	1.83	8	1.56	7
0.08	2.17	1.73	8	2.91	13	2.57	12
0.10	2.52	2.01	9	3.38	16	3.01	14
0.12	2.84	2.27	10	3.81	18	3.44	16
0.16	3.44	2.75	13	4.62	22	4.23	20
0.20	3.99	3.20	15	5.36	25	4.96	23
0.24	4.51	3.61	17	6.05	28	5.66	27

2.3. *Testing Program*

The tests were carried out for different inlet flow rates (q) from a minimum value of 0.04 to a maximum of 0.24 $m^2 s^{-1}$ (Table 1). As has been noted, the goal was to define a new, more stable block. The experimental methodology was divided into several phases. Firstly, we carried out a comprehensive analysis of the performance of the Armorwedge™ block (Figure 7a), which was used as a reference as it has been successfully applied in actual cases thus far [1,22]. Specifically, the research focused on the effect of both the hydrodynamic pressures and the leakage flow through the joints among the blocks and the aeration vents. In addition to this, the uplift pressure generated by the seepage flow through the drainage layer was also measured. Initially, the Armorwedge™ block was tested without a granular support layer in free drainage conditions. Thus, every leakage flow was conducted separately over the bottom of the chute and below the metallic grid which supported the blocks (Figure 8a). In several trial tests, the rapid stabilization of the hydrodynamic pressures on the different sensors was verified, an aspect that was achieved in a few seconds. All the tests for the determination of hydrodynamic pressures were carried out at least twice to corroborate the results obtained.

Based on the obtained results and additional numerical research [27,30], the proposed WSB, ACUÑA, was designed for testing in the phases of the test program described below (Table 4 and Figure 8b). Air vents were located in the upper part of the riser, where the greatest negative pressure was achieved according to previous research [9,11,17], and the experimental results are discussed below.

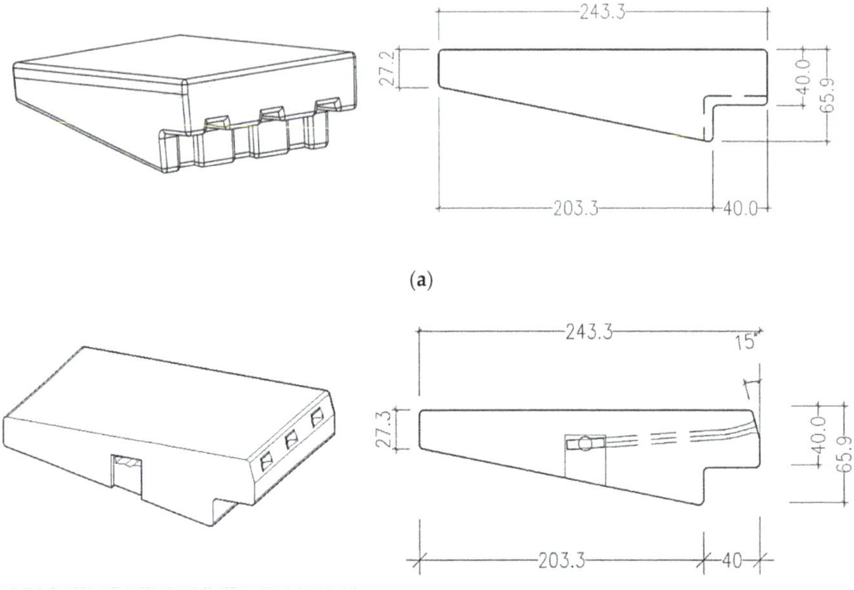

(a)

(b)

Figure 7. A 3D view and longitudinal section of the WSBs tested during the experimental research: (**a**) WSB Armorwedge™ and (**b**) WSB ACUÑA (dimensions in mm, the width of both block types: 165 mm) [30].

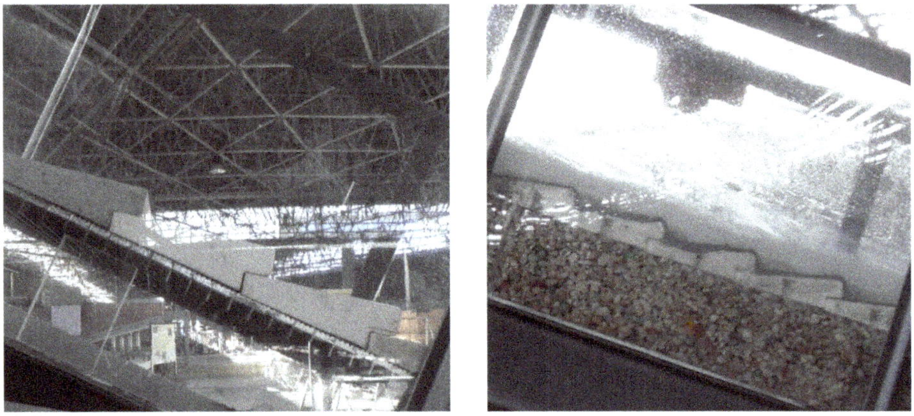

(a) (b)

Figure 8. WSBs placed over (**a**) a non-slip metallic grid and (**b**) a granular drainage layer.

Table 4. A summary of tests carried out (chronologically ordered): Armorwedge™ block ($w1$), ACUÑA block ($w2$), Free drainage conditions ($d1$) and Granular layer drainage conditions ($d2$). Measured variables: P, hydrodynamic pressures and L, leakage flow. Joints and air vents: longitudinal joints (LJ), transversal joints (TJ) and air vents (AV).

WSB	Drainage Conditions	q ($m^2\ s^{-1}$)	Measured Variables	Joints and Air Vents Conditions	Number of Tests Performed
$w1$	$d1$	0.04–0.24	P on rows 5 and 25/L	Without sealing	3
			P on rows 10 and 30/L	Without sealing	2
			P on rows 15 and 35/L	Without sealing	2
			P on rows 15 and 43/L	Without sealing	1
			L	Without sealing	1
			L	LJ sealing	1
			L	LJ and TJ sealing	1
$w2$	$d1$	0.04–0.24	L	Without sealing	1
				LJ sealing	1
				LJ and TJ sealing	1
	$d2$	0.04–0.20	P on rows 10 and 25/L	Without sealing	2
				LJ sealing	2
				LJ and TJ sealing	2
$w1$	$d2$	0.04–0.20	P on rows 10 and 25/L	Without sealing	1
				LJ sealing	1
				LJ and TJ sealing	1
$w2$	$d2$	0.04–0.20	L	Without sealing	1
				LJ and TJ sealing of rows 1 to 8.	1
				$LJ/TJ/AV$ sealing of rows 1 to 8.	1
				LJ and TJ sealing of rows 1 to 16.	1
				$LJ/TJ/AV$ sealing of rows 1 to 16.	1
				LJ and TJ sealing of rows 1 to 24.	1
				$LJ/TJ/AV$ sealing of rows 1 to 24.	1
				LJ and TJ sealing of rows 1 to 32.	1
				$LJ/TJ/AV$ sealing of rows 1 to 32.	1

Then, a set of laboratory tests was performed in order to compare the behavior of ACUÑA and Armorwedge™ blocks in two different drainage conditions. First, the free drainage condition ($d1$) was maintained. Thus, the blocks were placed over a fixed metallic grid located 0.2 m over the channel bottom so that there was a free space between the blocks and the base of the channel with the purpose of simulating the conditions of a high permeability underlay (for example, a clean, highly permeable rockfill; Figure 8a). These tests were considered representative of an underlay condition where uplift pressures are not expected. In the second stage, ACUÑA and Armorwedge™ blocks were placed over a layer of 0.20 m thick, homogeneous gravel (D_{50} of 12.6 mm, D_{10} of 8.4 mm, C_u of 1.54; Figure 9) layer ($d2$). This granular layer was extended on the impervious bottom of the channel (Figure 8b).

The second set of tests aimed to simulate the hydraulic performance of the blocks over impervious soil, such as clay or sandy clays, with an intermediate permeable bedding layer of gravels.

Conceptually, the presence or absence of saturation of the drainage layer is of special importance in the operation of WSBs. For this reason, additional tests were performed to determine the origin of the drainage flow. The first four rows of WSBs were sealed with the aim of reducing leaks and simulating the first section of a real spillway chute with WSBs and an overlapping slab in the upper area (Figure 1b). The drainage flow was firstly measured with the free drainage condition, $d1$, and then the blocks were placed over a granular drainage layer, $d2$, in order to determine the pattern of the seepage through the blocks towards the drainage layer along the chute with both drainage conditions. Next, the origin of the leakage was investigated with the aim of discriminating between the leaks through the contact joints, which were longitudinal joints and transverse joints in the area of block overlap, as well as through the vents (Figure 10). These types of tests were carried out both in the Armorwedge^{TM} and the ACUÑA block. It should be remembered that the blocks were placed in the channel without any type of waterproofing between them.

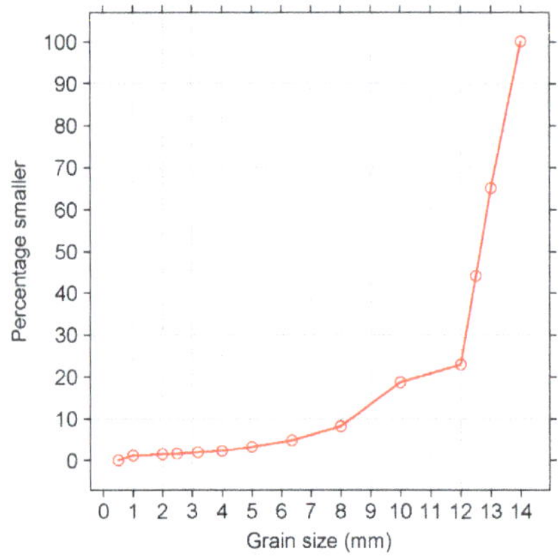

Figure 9. The grain size of the drainage layer.

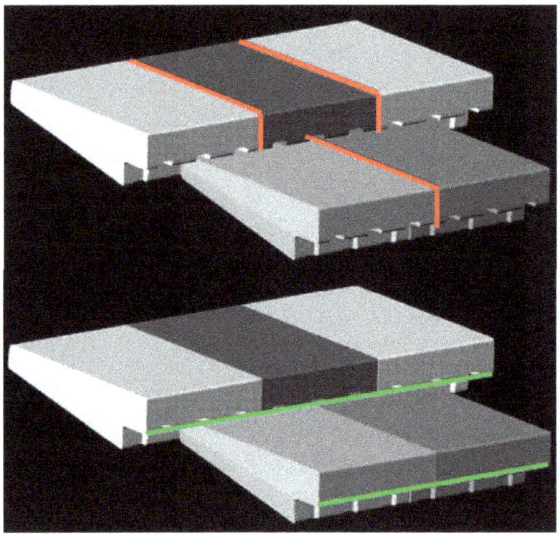

Figure 10. Longitudinal joints (red) and transverse joints (green) between blocks. Adapted from [14].

Finally, a new test program was carried out with the ACUÑA block to find out how the leakage flow was distributed among the spillway sectors along the chute. The spillway was completely waterproofed sequentially in the upstream–downstream direction in each of the four sectors of eight rows. The location of the triangular sharp-crested weir allowed the measuring of the drainage discharge up to row 32. A test was also performed with no sealing. The purpose of these tests was to verify whether, as is evident, the greatest leaks were produced by the upper zone of the flume and to obtain the data regarding the percentage. Table 4 presents a summary of the tests carried out for the research described in this article.

Additional photographs and videos of the tests have been included in supplementary materials in the following folders: '03_*w*1_*d*1_2017', '04_*w*2_*d*1_2017', '05_*w*1_*d*2_2018', '06_*w*2_*d*2_2018', '07_*w*2_*d*2_2019_no_sealing' and '08_*w*2_*d*2_2019_sealing_8rows'.

3. Results and Discussion

The results obtained in the research regarding hydrodynamic pressures on the different blocks tested ($w1$, Armorwedge™ and $w2$, ACUÑA) and with different drainage situations ($d1$: free drainage conditions and $d2$: blocks placed over a layer of 0.20 m thick homogeneous gravel), as well as the seepage from the flume to the different drainage layers ($d1$ and $d2$), are summarized in the following sections.

3.1. Hydrodynamic Pressures

3.1.1. Pressures on the Block Tread

Armorwedge™ Block ($w1$) Tests with Free Draining Conditions ($d1$)

Caballero et al. [28] found good agreement when comparing the registered records of the average pressures with those observed by Bramley et al. [8], Slovensky [11] and Relvas and Pinheiro [19] for a uniform flow regime.

Figure 11 shows additional results for the Armorwedge™ block regarding the average pressure on the sensorized block, which was located in different rows of the stepped chute. The pressure head ($p\,\gamma^{-1}$) refers to the riser height of the block (h_s) to obtain a dimensionless parameter ($p\,\gamma^{-1}\,h_s{}^{-1}$).

In agreement with previous research ([8,11,19]), two pressure zones with positive and negative pressure were found in the block tread. The boundary was located between 30% and 40% of the tread length. The maximum positive pressure heads were registered between 52% and 82% of the tread length. Slovensky registered these as between 52% and 67% of the tread length. However, the magnitude of the standard deviation must be considered when comparing these results. For the skimming uniform flow, the maximum ranges were between 67% and 82% of the tread length. We systematically observed for the skimming flow an increase in the mean maximum pressures on the block tread up to rows 25–30 and a decrease from these rows onwards (Figure 11). In the upper part of the channel, before reaching the uniform flow, the velocity increased, which might explain the increase in the impact pressures in the downstream direction. However, once the uniform flow was reached, the velocity remained constant. The decrease in maximum pressure in this area might be due to the increase in air entrainment [11]. However, this explanation given by Slovensky contradicts the definition of uniform flow. It is possible that uniform flow may not yet be fully achieved, although this fact contradicts existing empirical formulations [3,18,49]. Another possibility would be a measurement error, but as shown in the experimental program (Section 2.3. Table 4), the tests in rows 25, 30 and 35 have been repeated at least 2 times, always obtaining very similar results.

In accordance with [8,11,19], the maximum mean pressures usually increased with an increasing flow rate discharge in all sections in the channel (Figure 11). Some exceptions were observed; for example, in row 25 (Figure 11d), likely due to a measurement error, and in row 10 (Figure 11c), where the average pressures did not increase monotonically, the flow was still accelerated there. Furthermore, as shown in Figure 11, it can be concluded that as the flow moves downstream, there is also a displacement of the mean maximum pressure downstream in the block tread. Thus, for higher flow rates, from 0.16 $m^2\,s^{-1}$ upwards, there is a displacement of the average maximum pressure measured from sensor 5 in row 5 to sensor 3 (Figure 3) in rows 30 and 35. This can be explained by a more distant jet flow launch from the edge of the upper step as a consequence of the acceleration and increase in the speed of the flow. Nevertheless, it should be noted that the exact position of the maximum pressure could not be specified due to the limited number of sensors available.

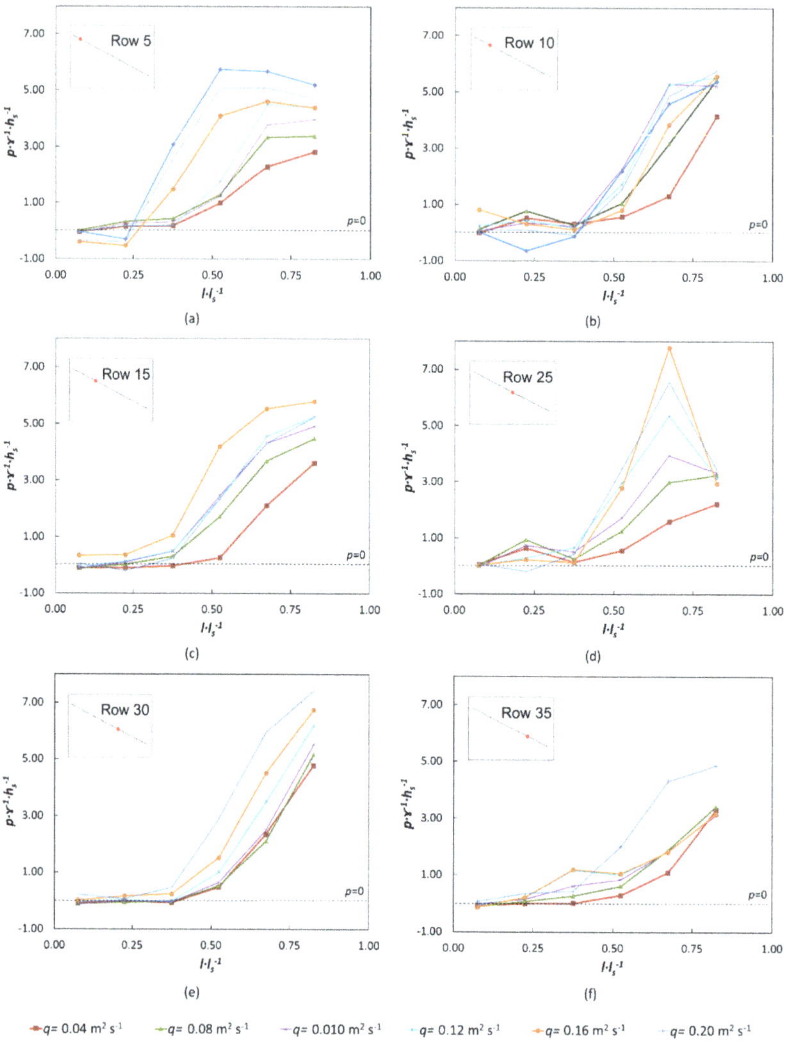

Figure 11. Mean pressures on the tread of the Armorwedge™ block ($w1$) in free draining conditions ($d1$). (**a**) Row 5; (**b**) row 10; (**c**) row 15; (**d**) row 25; (**e**) row 30; (**f**) row 35.

A slight displacement of the point between the zone of clearly positive pressures and the zone of negative or close to zero pressures was also observed once the flow accelerated. This aspect also seems quite logical as the impact of the flow on the block tread shifted downstream as the flow accelerated.

Mean pressures and standard deviation on the tread of Amorwedge^{TM} block at different rows of the chute for skimming flow rates have been included in supplementary materials (Figure S5 of Supplementary Materials).

Effect of Drainage Layer ($d2$) on Armorwedge^{TM} Block ($w1$)

The pressures on the block tread are usually similar for free drainage ($d1$) and drainage layer ($d2$) conditions; this was observed on the first four sensors available on the block tread (sensors 8–5 in Figure 3) in row 25 (Figure 12) for all the tested discharge flow rates.

However, the last sensor (sensors 3 in Figure 3) showed greater pressures with up to a 300% increase for the *d2* conditions (Figure 12b–d). In row 10 (Figure 13), there was good agreement between the values obtained for the *d1* and *d2* scenarios in the case of the highest discharge flows on all sensors, except sensor number 3, the last one (Figure 3).

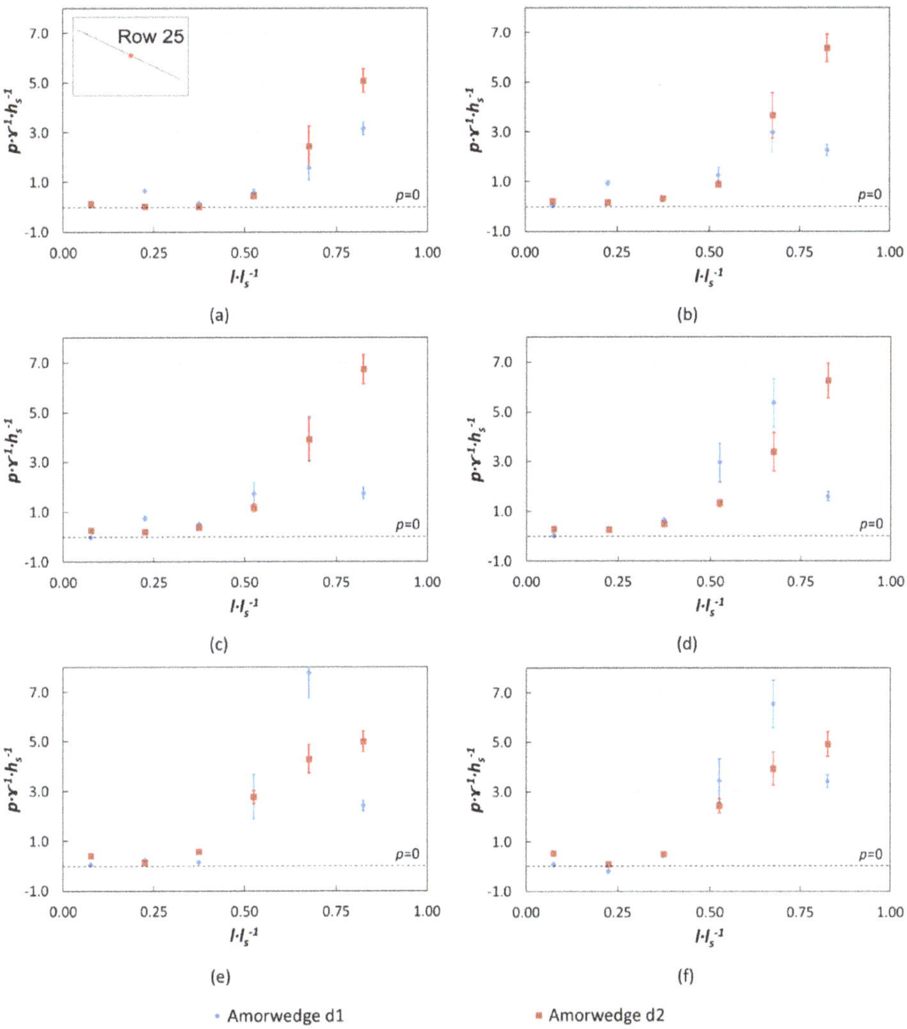

Figure 12. Average and standard deviation of the hydrodynamic pressures registered on the tread of Armorwedge^{TM} block ($w1$) in row 25 of the chute in different drainage conditions: free drainage, *d1*, and gravel drainage layer, *d2*, with unit flow: (**a**) 0.04 m^2 s^{-1}, (**b**) 0.08 m^2 s^{-1}, (**c**) 0.10 m^2 s^{-1}, (**d**) 0.12 m^2 s^{-1}, (**e**) 0.16 m^2 s^{-1} and (**f**) 0.20 m^2 s^{-1}.

The next step was to compare the hydrodynamic pressures on the block tread for the Armorwedge^{TM} and ACUÑA blocks with free drainage and with a granular drainage layer (Figures 14 and 15).

Comparison of the Armorwedge^{TM} ($w1$) and ACUÑA ($w2$) Blocks

The comparison was performed with both types of blocks on a drainage layer, which is the usual layout. As a general rule, similar pressures on the tread were observed for the Armorwedge^{TM} block ($w1$) and the ACUÑA block ($w2$) in rows 10 and 25 for all flow rates (Figures 14 and 15). Nonetheless, the differences were very modest and barely noticeable; in some sensors, the pressure was slightly higher in the ACUÑA block, in others, it was higher in the Armorwedge^{TM} block, and there were also cases where they could be considered almost coincident.

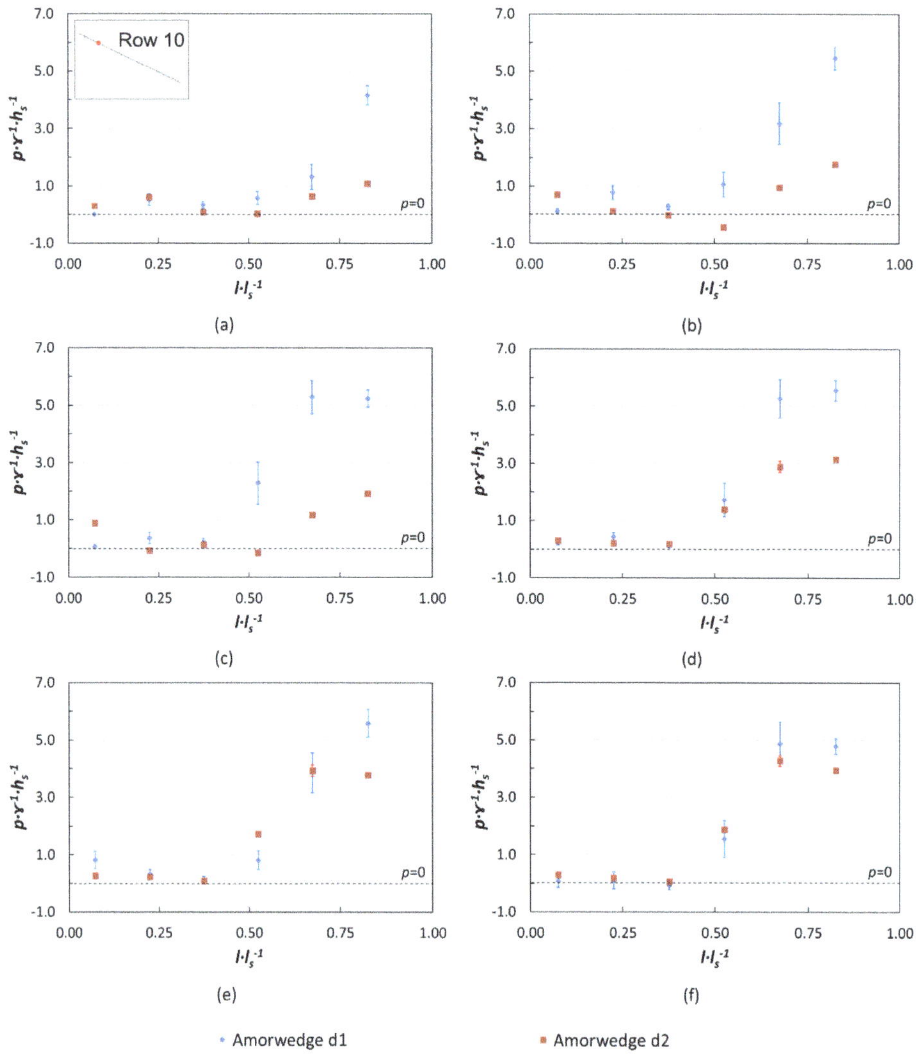

Figure 13. Average and standard deviation of the hydrodynamic pressures registered on the tread of Amorwedge^{TM} block ($w1$) in row 10 of the chute in different drainage conditions: free drainage, *d1*, and gravel drainage layer, *d2*, with unit flow: (**a**) 0.04 m^2 s^{-1}, (**b**) 0.08 m^2 s^{-1}, (**c**) 0.10 m^2 s^{-1}, (**d**) 0.12 m^2 s^{-1}, (**e**) 0.16 m^2 s^{-1} and (**f**) 0,20 m^2 s^{-1}.

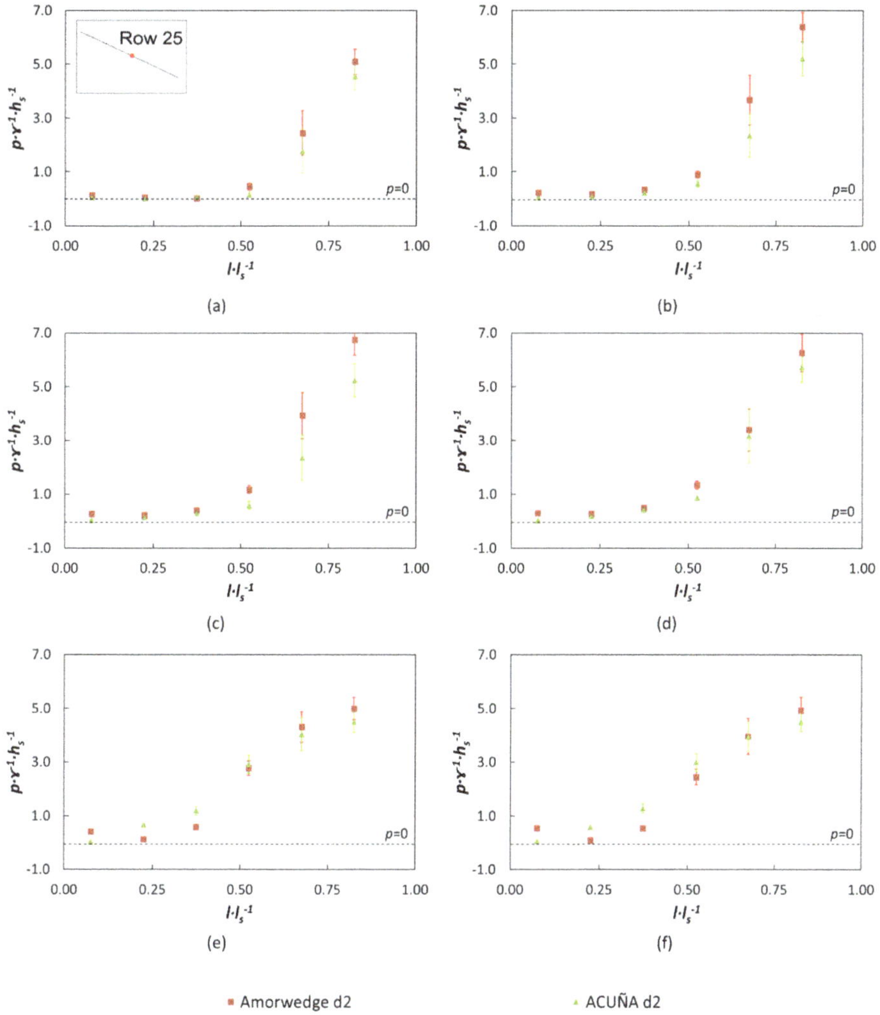

Figure 14. Average and standard deviation of the hydrodynamic pressures registered on the tread of Armorwedge^{TM} ($w1$) and ACUÑA ($w2$) block in row 25 of the chute for different unit discharges simulating blocks laid over an impervious soil with an intermediate and permeable bedding granular layer ($d2$): **(a)** 0.04 $m^2 s^{-1}$, **(b)** 0.08 $m^2 s^{-1}$, **(c)** 0.10 $m^2 s^{-1}$, **(d)** 0.12 $m^2 s^{-1}$, **(e)** 0.16 $m^2 s^{-1}$ and **(f)** 0.20 $m^2 s^{-1}$.

However, a very slight increase in the maximum pressures for the highest discharged flows (Figure 14) was observed in row 25 for the Armorwedge^{TM} block, as well as in row 10 (Figure 15); the point that separates positive and negative, or close to zero, pressures was usually located more upstream in the ACUÑA block. The chamfer of this type of block might move the impact point slightly upward and cause the described effect. This increases the positive pressures on the block tread, which is favorable for block stability.

Effect of Sealing Joints between Blocks

Additional tests were performed for both types of blocks with a drainage layer ($d2$), having previously sealed the joints between the blocks, first the longitudinal ones and then the transverse joints. Although the aim was to discriminate the preferential leakage areas and quantify the leakage (shown later in this paper), the hydrodynamic pressures were also registered. We observed that there was little variation in the pressures on the block tread compared with the unsealed scenario. Two examples are shown in Figure 16.

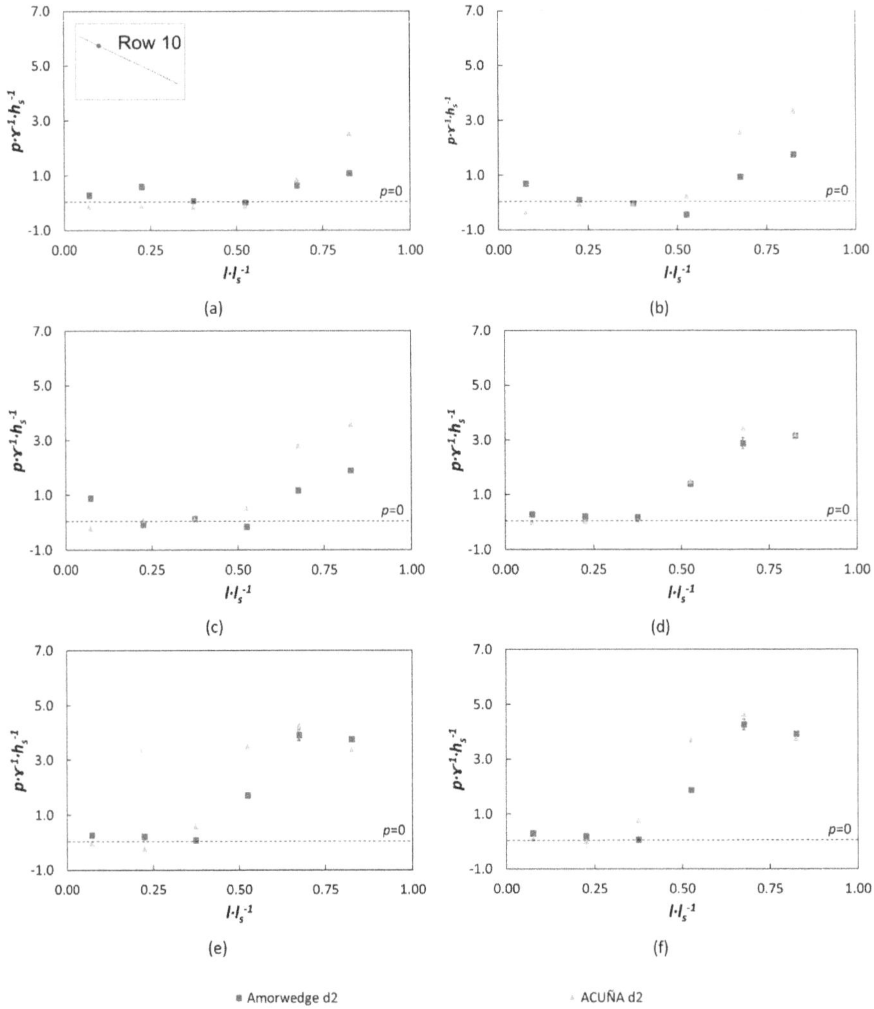

Figure 15. Average and standard deviation of the hydrodynamic pressures registered on the tread of the Armorwedge^{TM} ($w1$) and ACUÑA ($w2$) block in row 10 of the chute for different unit discharges simulating the blocks laid over an impervious soil with an intermediate and permeable bedding granular layer ($d2$): (**a**) 0.04 m^2 s^{-1}, (**b**) 0.08 m^2 s^{-1}, (**c**) 0.10 m^2 s^{-1}, (**d**) 0.12 m^2 s^{-1}, (**e**) 0.16 m^2 s^{-1} and (**f**) 0.20 m^2 s^{-1}.

3.1.2. Pressures on the Block Riser

Pressure distribution along the block riser was measured. Several authors ([8,11]) measured the pressure on the block riser at a certain location, but there was a lack of data regarding the pressure distribution along the riser. Thus, the pressure was measured at three points along the riser. As expected, pressures were negative or close to zero at the top of the riser, where there was higher suction (Figure 17).

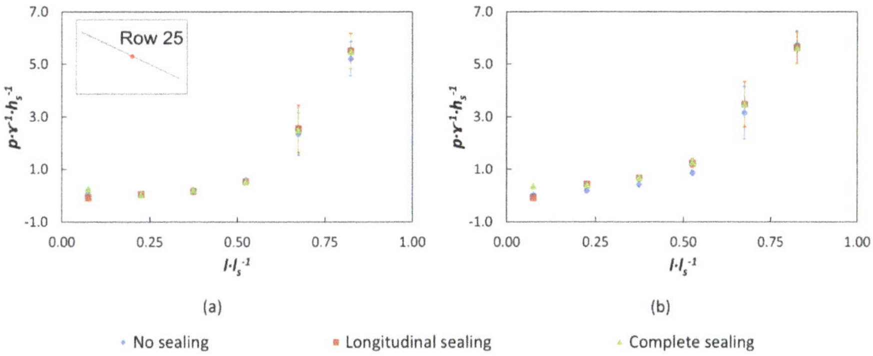

Figure 16. Pressure on the block tread of the ACUÑA block (*w*2) in row 25 in scenario *d2*: without joint sealing, with longitudinal sealing and with total sealing. (**a**) 0.08 $m^2\ s^{-1}$ and (**b**) 0.12 $m^2\ s^{-1}$.

Similar behavior was observed in the riser of row 25 for both drainage scenarios (Figure 18 and [24], Figure 19), but with different absolute pressure values. Thus, the minimum pressure was systematically located in the upper sensor for all of the tested cases with skimming flow, which was from 0.10 $m^2\ s^{-1}$ upwards (Figures 18 and 19). Finally, it was also observed that in row 10, the pressure pattern and values were highly coincident for the tests with free drainage (*d*1) and with a drainage layer (*d*2). Figures showing the pressures obtained in the riser are included in supplementary materials (Figures S6–S9 in Supplementary Materials).

The results of the experimental research agree with the results of previous numerical models. Velocity and pressure fields obtained by numerical modeling showed the development of a slightly oscillating vortex with a horizontal axis near the concave junction of the block tread and riser [27,30]. This vortex generates a zone of negative pressure on the surface of the riser, with minimum values in the upper third [27,30] near the edge of the riser. This flow pattern was also observed in the experimental test. The characteristics of the vortex determine the pressure distributions in the tread and riser of the block as well as the operating conditions of the vents. The vents of the Armorwedge™ block are located at the base of the riser. Positive pressure in that area, although low, might cause a flow circulation towards the drainage layer through the vents. Moving the position of the air vents on the ACUÑA block to the upper part of the riser was proposed after observing the described flow pattern and the values of the pressures on the base of the block, shown in the next section.

3.1.3. Pressures on the Base of the Block

Three pressure gauges were installed at the base of the block (Figure 3). Two of them registered similar values in agreement with [8], while the third seemed to have not been measured correctly. Therefore, we decided to evaluate the pressure on the base of the block, assuming a uniform distribution with an average value as registered by one of the two sensors that measured similar pressures.

Armorwedge™ Block (*w*1) Tests in Free Draining Scenario (*d*1)

The free drainage conditions allow the complete evacuation of leakage, avoiding the saturation of the drainage layer and facilitating the transmission of suction to the base of the block, which was measured here for the first time. Hydrodynamic pressures at the base of the Armorwedge™ block decreased as the unit discharge of skimming flow over the block increased (Figure 20). At the end of the test, the flow discharge decreased to zero. As shown in Figure 20, the base of the block was mainly subjected to sub-atmospheric pressures.

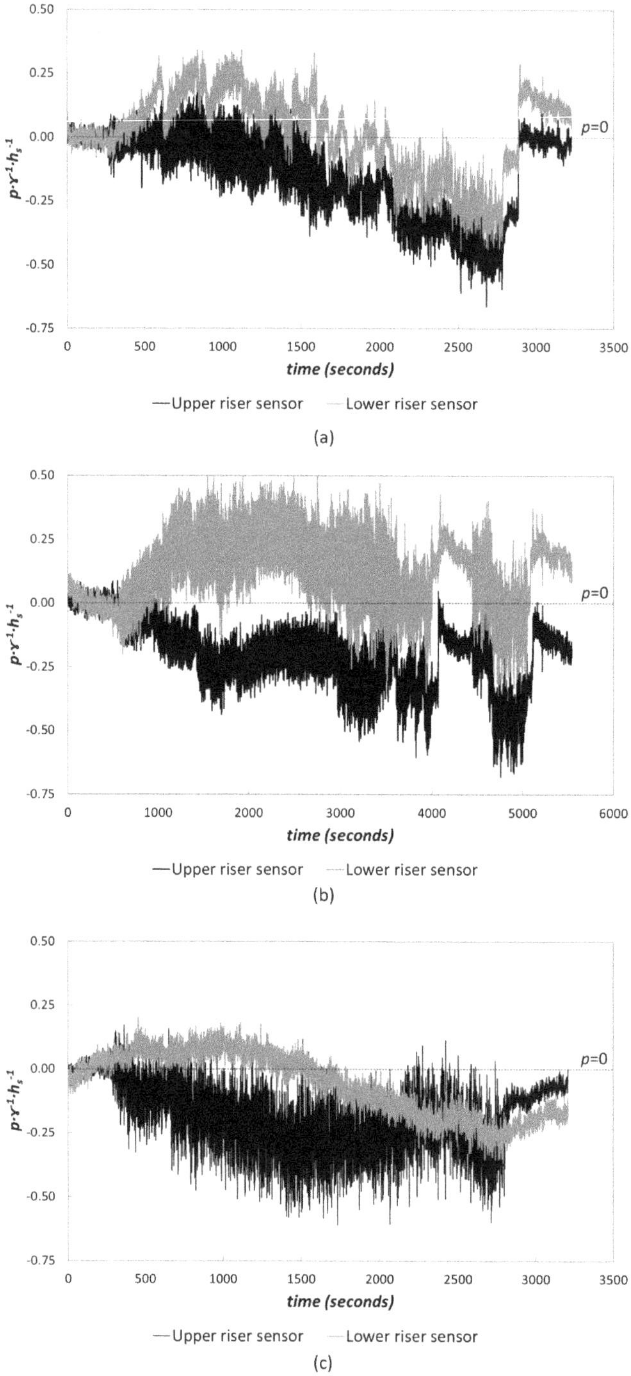

Figure 17. Hydrodynamic pressures on the Armorwedge^{TM} block (w1) riser in rows: (**a**) 15; (**b**) 25; (**c**) 35 for unit flow rates (q). They range from 0.04 $m^2 s^{-1}$ to 0.24 $m^2 s^{-1}$ and a decrease to zero discharge.

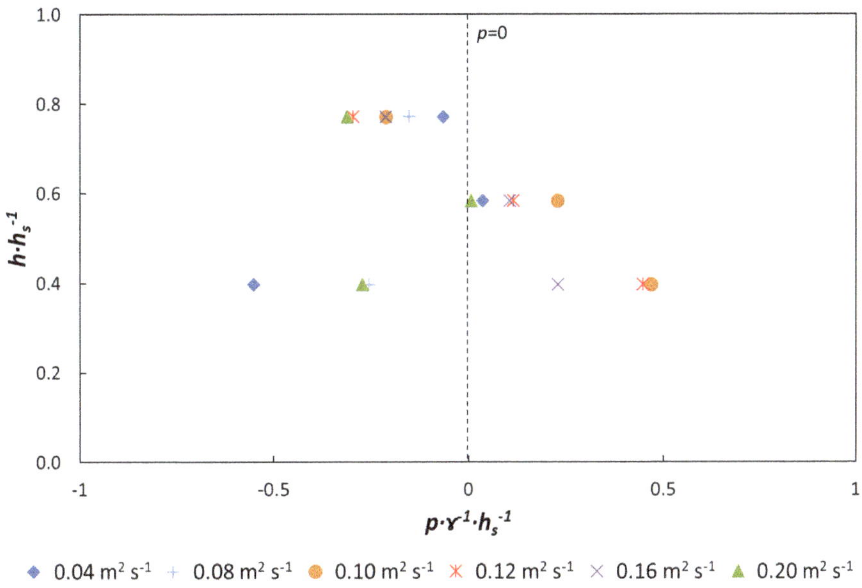

Figure 18. The average of the hydrodynamic pressures registered on the block riser of the Armorwedge™ block (*w*1) in row 25 of the chute for different unit discharges simulating free drainage conditions (*d*1).

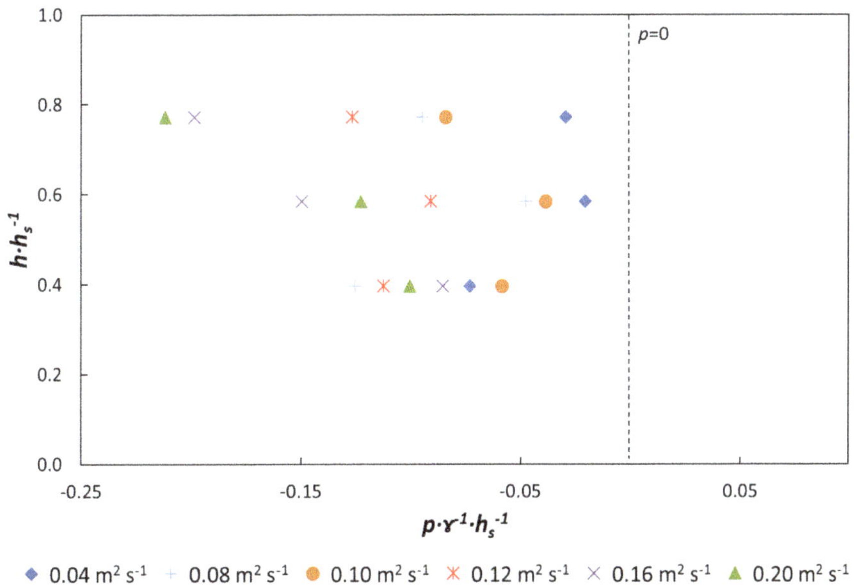

Figure 19. The average of the hydrodynamic pressures registered on the block riser of the Armorwedge™ block (*w*1) in row 25 of the chute for different unit discharges simulating the blocks laid over an impervious soil with an intermediate and permeable bedding granular layer (*d*2).

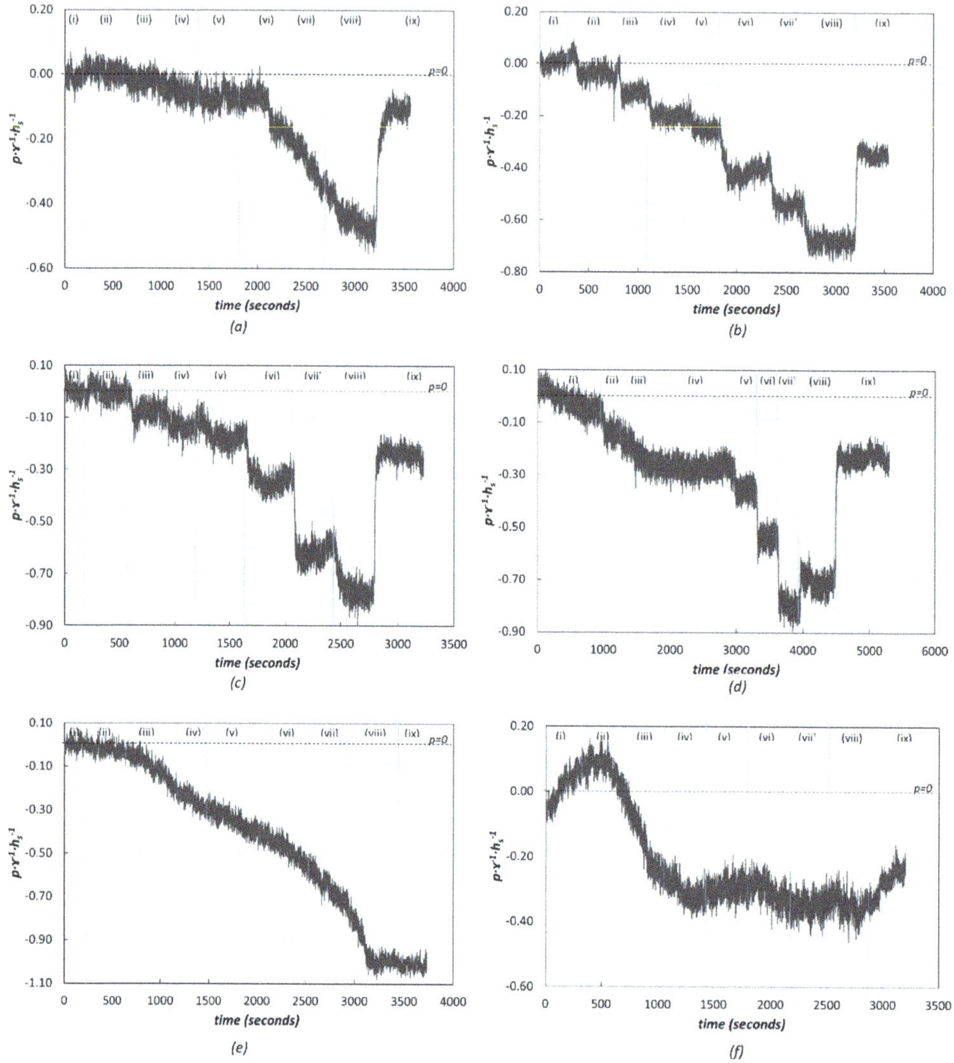

Figure 20. Hydrodynamic pressures on the base of the Amorwedge^{TM} block ($w1$) in rows 5 (**a**), 10 (**b**), 15 (**c**), 25 (**d**), 30 (**e**) and 35 (**f**) for unit flow rates (q): (i) 0 $m^2\ s^{-1}$, (ii) 0.04 $m^2\ s^{-1}$, (iii) 0.08 $m^2\ s^{-1}$, (iv) 0.10 $m^2\ s^{-1}$, (v) 0.12 $m^2\ s^{-1}$, (vi) 0.16 $m^2\ s^{-1}$, (vii) 0.20 $m^2\ s^{-1}$, (viii) 0.24 $m^2\ s^{-1}$ and (ix) 0 $m^2\ s^{-1}$.

It was observed in the test trials that the time required to reach the stationary state at a given flow rate amounted to very few seconds. However, some measurements were observed where this did not occur, such as in row 30 (Figure 20e) and with some of the flow rates in row 5 (Figure 20a) and 35 (Figure 20f). In general, it was also observed that the return to zero pressure did not occur when the flow discharge decreased to zero, an aspect that may be due to a hysteresis phenomenon. In most of the situations analyzed (Figure 20a–d), this phenomenon was modest; however, it was very prominent in rows 30 (Figure 20e) and 35 (Figure 20f). Nonetheless, it should be noted that the tests were repeated at least twice for each row, obtaining very similar results.

Comparison of the Armorwedge^{TM} ($w1$) and ACUÑA ($w2$) Blocks

Both types of blocks were placed on a drainage layer. The results for the ACUÑA and Armorwedge^{TM} block are shown in Figure 21. Positive uplift was registered with a lower flow rate discharge for the Armorwedge^{TM} block (0.12 m^2 s^{-1}) than for the ACUÑA block (0.16 m^2 s^{-1}).

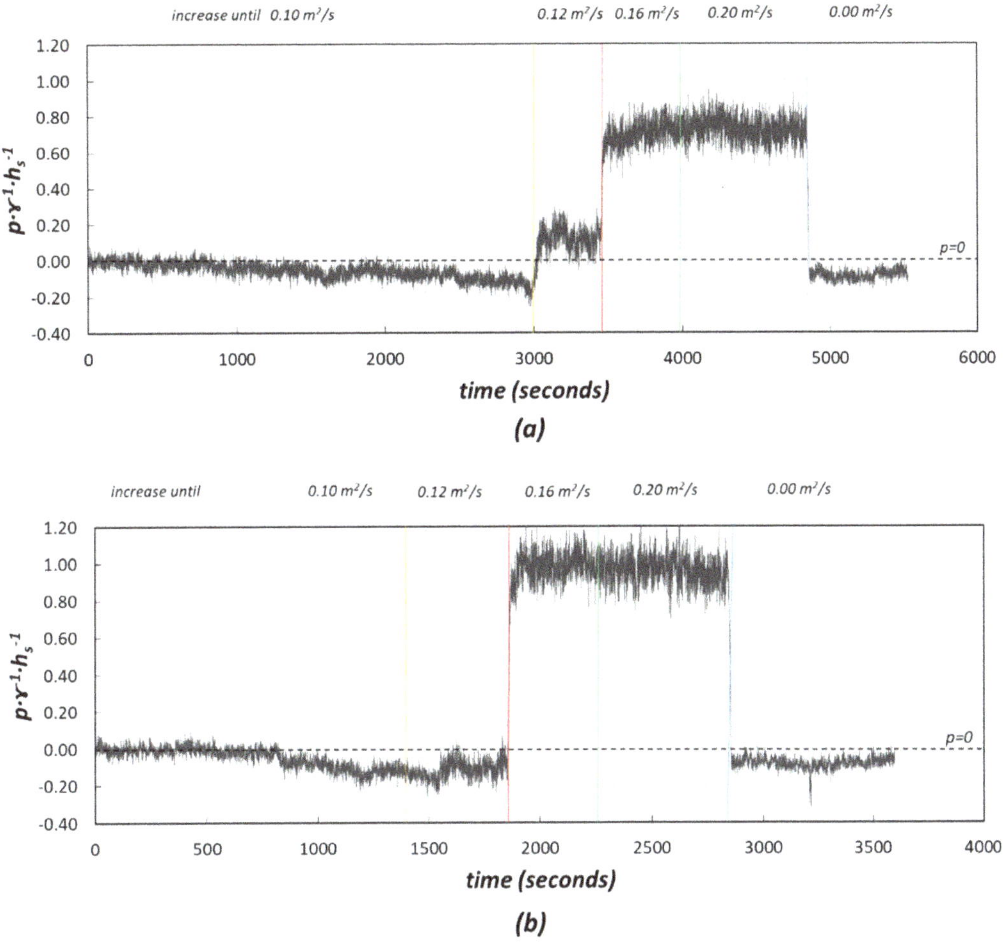

Figure 21. Hydrodynamic pressures on the base of the Armorwedge^{TM} block ($w1$) (**a**) and ACUÑA block ($w2$) (**b**) in row 25 for unit flow rates (q). Pressures range from 0.04 to 0.24 m^2 s^{-1} and a decrease to zero discharge.

The drainage layer was not saturated for the flow rate of 0.12 m^2 s^{-1}, whereas it was for the Armorwedge^{TM} block. This might be due either to a lesser drainage flow from the chute towards the drainage layer or to a greater capacity of leakage reintegration from the drainage layer to the chute as a consequence of increased negative suction pressure through the vents. This effect was significantly greater when the joints between blocks were sealed (Figure 22). However, when the drainage layer became saturated, the uplift on the base of the ACUÑA block was greater than that on the Armorwedge^{TM} block for the same discharge value. It is possible that the air vents of the Armorwedge^{TM} block, being in the lower part of the block riser, facilitated the outlet of the water contained in the drainage layer better than those of the ACUÑA block and, therefore, the relief of uplift pressures.

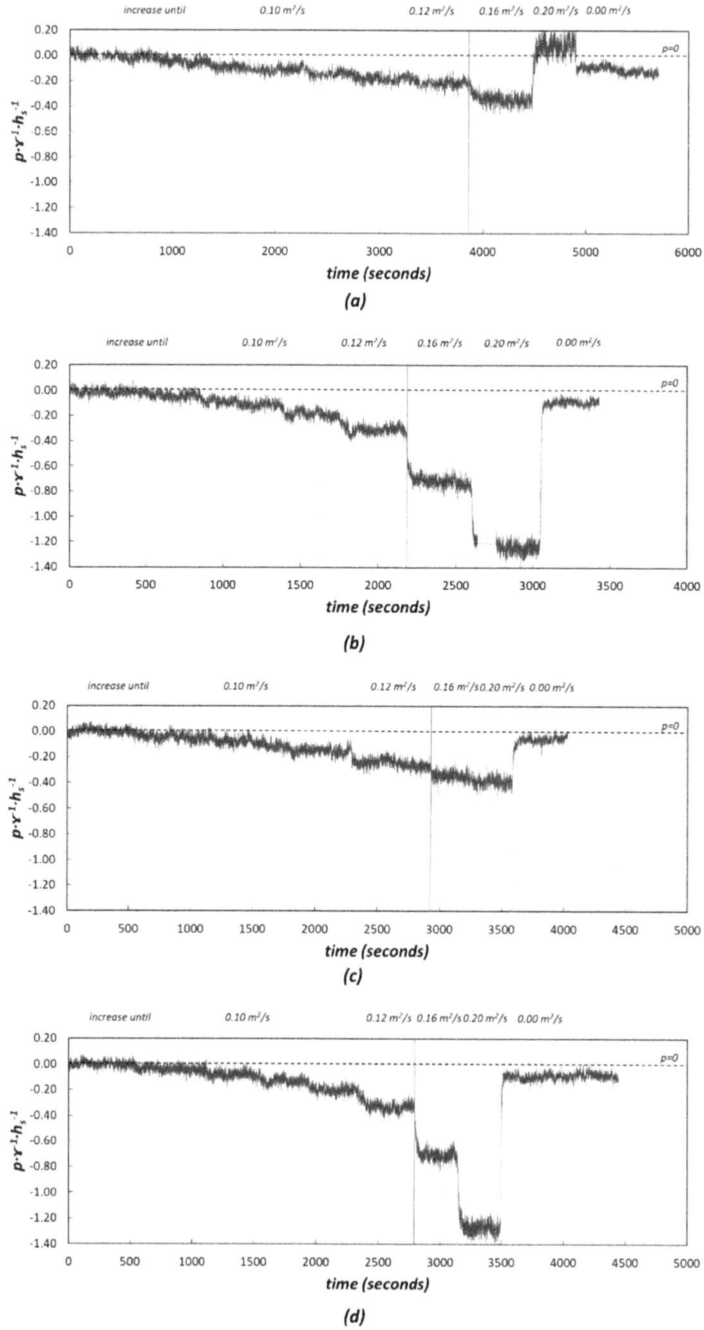

Figure 22. Hydrodynamic pressures on the base in row 25 for unit flow rates (q) from 0.04 to 0.24 $m^2\ s^{-1}$ and return to non-discharge. Longitudinal joints sealing: (**a**) Armorwedge^{TM} block ($w1$) and (**b**) ACUÑA block ($w2$). Complete sealing of the joints: (**c**) Armorwedge^{TM} block and (**d**) ACUÑA block.

3.2. Hydrodynamic Forces on the WSBs

Hydrodynamic forces play a critical role in the stability of WSBs. They can be classified as the following: forces due to the impact of the main discharge flow on the tread of the block, suction forces developed as a consequence of negative pressures generated in the block tread immediately downstream of each step, uplift on the base due to the saturation of the drainage layer ([8,11,50]), and suction force on the block base if the drainage layer is not saturated. The magnitude of the forces varies with the discharge and block location along the chute. Assuming the same block weight, the difference is negligible; the type of WSB with a more favorable hydrodynamic resultant force will be more stable. The two main hydrodynamic forces acting on the blocks with a saturated drainage layer are the force due to the jet flow impact on the block tread and the uplift force on the base of the block [13]. A pressure field on the block tread was defined based on the pressure measurements (Figures 14 and 15). Uniform distribution with the uplift pressure mean values registered was considered appropriate for the base of the blocks. The difference between total forces on both faces was the resultant hydrodynamic force on the WSB. A comparison of the two types of WSBs, Armorwenge^{TM} and ACUÑA, was carried out in three different scenarios considering the sealing of the joints between blocks: no sealing, longitudinal sealing and complete sealing (longitudinal and transversal), as described in Section 3.1.3.

Figure 23 shows the resultant hydrodynamic force per unit width on Armorwedge^{TM} and ACUÑA WSBs in rows 10 and 25 as a function of $h_c\ h_s{}^{-1}$. The positive values correspond to stabilizing forces. It is relevant to note that the blocks were inherently stable due to the hydrodynamic force, even without considering the contribution of the weight of the block. The weight of the block per unit width was 272 N/m. The hydrodynamic resultant force ranged between 13% and 59% of the weight of the block in cases without sealing of the joints and between 11% and 99% with sealed joints. In row 10, in the upper part of the chute, the stabilizing force systematically increased with discharge flow rate, and the ACUÑA block was more stable than Armorwedge^{TM}. This happened independently of the condition of the joint, sealed or unsealed. In row 25, in the lower part of the chute with fully developed velocity and with unsealed joints, the stabilizing force was reduced with a discharge flow rate equal to or greater than 0.12 $m^2\ s^{-1}$ ($h_c\ h_s{}^{-1}$ = 2.84) on the Armorwedge^{TM} block. The limit value was slightly higher (0.16 $m^2\ s^{-1}$ or $h_c\ h_s{}^{-1}$ = 3.44) for the ACUÑA block. The cause was undoubtedly the uplift due to the saturation of the drainage layer. The Armorwedge block was more stable than the ACUÑA block with the saturated drainage layer (Figure 23a). However, in row 25 in the two scenarios with sealed joints (Figure 23b,c), the ACUÑA block systematically increased its stability with the flow rate discharge. Additionally, for the flow rates in the skimming flow regime, the ACUÑA block was more stable than the Armorwedge^{TM} block.

Although, in all cases the blocks were inherently stable (even more so considering that their weight and the force generated by the interlocking between blocks were not taken into account in the presented calculation), the discussion concerns whether it is realistic for joint sealing to be envisaged in design practice or not. This action is considered viable since, as has been demonstrated, the sealing of the joints (mainly the longitudinal ones and in the upper part of the flume, as presented in Section 3.4) can be very useful for better performance of the technology. Nonetheless, the use of precast blocks that already incorporate waterproofing strips in the contacts between the joints could also be considered.

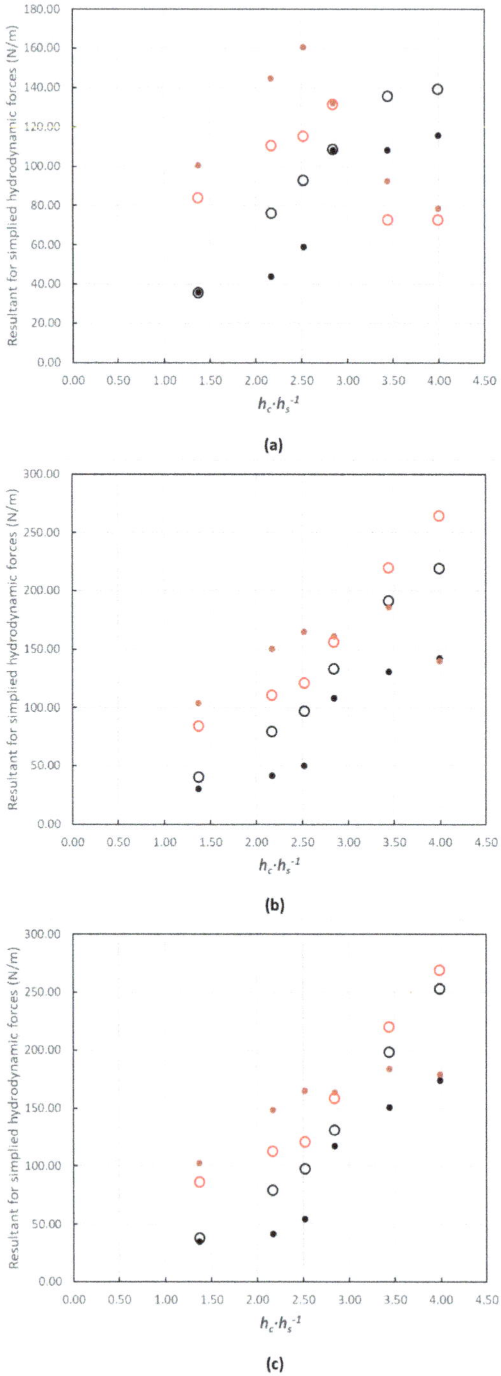

Figure 23. Resultant of the hydrodynamic forces on the block tread and base for the Armorwedge™ ($w1$) and ACUÑA ($w2$) block for unit flow rates (q) from 0.04 ($h_c\ h_s^{-1}$ = 1.37) to 0.24 $m^2\ s^{-1}$ ($h_c\ h_s^{-1}$ = 3.99): (**a**) joints unsealed; (**b**) longitudinal joints sealed; (**c**) longitudinal and transversal joints sealed.

3.3. Drainage Flow

The registered drainage flow rates (Figure 24) were compared with the results of the tests performed by Relvas and Pinheiro [16,18]. Their results show a slight reduction in the drainage flow as the inlet discharge of the tests increases. In our experiments, the drainage flow rate shows a clear difference depending on the drainage configuration: free drainage (*d*1) or with a granular drainage layer (*d*2).

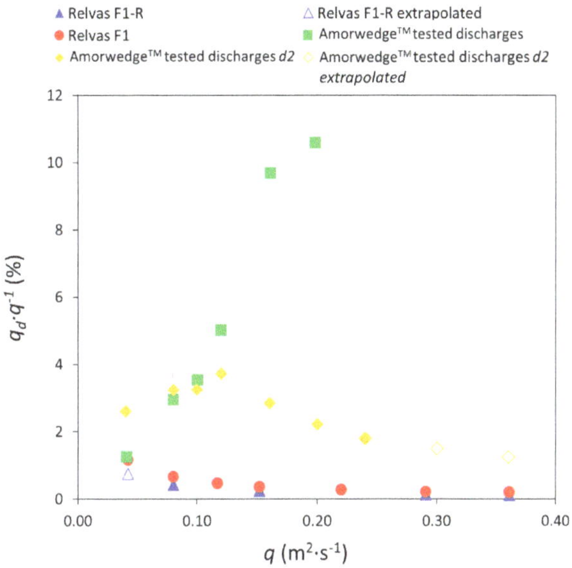

Figure 24. The unit drainage flow (q_d) expressed as a percentage of the inlet flow rate (q).

Configuration *d*1 prevents the saturation of the drainage area and allows the complete evacuation of the leakage flow through the joints. In this situation, unlike the results reported by Relvas and Pinheiro, there is an increase in the percentage of drainage flow as the main discharge flow becomes higher, reaching values up to 10% for the higher flow rates tested (0.20 $m^2 s^{-1}$) (Figure 1). However, the tests made with the blocks placed over a granular drainage layer (*d*2) showed a different behavior of the drainage unit flow, with evolution more similar to that described by Relvas and Pinheiro. Nevertheless, the values of the drainage flow rates (q_d/q_i) of the configuration *d*2 present a significant difference compared with those observed by Relvas and Pinheiro. This might be due to the different particle sizes of the materials of the drainage layers. Relvas reported a 0.20 m thick drainage layer that integrates two sublayers: a 0.15 m thick upper layer in contact with the concrete blocks and gravel varying between 4/6 (D_{50} of 5 mm, D_{10} of 2.6 mm, *Cu* of 2.3) and10/20 (D_{50} of 14.9 mm, D_{10} of 10.7 mm, *Cu* of 1.5) and a 0.05 m thick bottom layer of sand (D_{50} of 0.8 mm, D_{10} of 0.3 mm, *Cu* of 2.9) [17].

The different drainage flow rates between configurations *d*1 and *d*2 might be due to the degree of saturation underneath the WSBs. Thus, configuration *d*2 can be completely saturated when the maximum seepage capacity of the drainage layer is reached. At that point, once the drainage layer is saturated, the higher the discharge flow, the lower the drainage flow rate. In this situation, when the negative pressure is developed in the vents at the base of the riser of each block, a fraction of the drainage flow is expected to be sucked upward towards the main flow area.

3.4. Effect of the Joints among WSBs on the Drainage Flow

The set of tests with and without sealing of the joints allowed us to determine the origin of the drainage flow. A clear predominance of the leakage flow through the longitudinal joints was observed. The longitudinal joints were the origin of 55–80% of the total drainage flow; the higher percentage corresponds to the lower flow rate. The rest of the leakage was produced by both of the horizontal transverse joints between the blocks and the air vents. These results suggest that a reduction in the length of longitudinal joints could significantly reduce the leakage flows towards the drainage layer. The quantity of total drainage flow was quite similar for both types of WSB, Armorwegde™ and the proposed ACUÑA. This is logical, taking into account the significant predominance of leakage through the longitudinal joints.

The registered leakage flow rates through the air vents of the ACUÑA block were negligible for flow rates lower than 0.1 $m^2\ s^{-1}$. Such drainage flow rates increased up to 1–2% of the total leakage for higher inlet flow rates, reaching the maximum value for the inlet discharge of 0.24 $m^2\ s^{-1}$. Thus, the tests conducted with the ACUÑA block showed a reduction in the total drainage flow rate, compared with the Armorwedge™ block, from 12–15% for inlet flow rates between 0.04 and 0.24 $m^2\ s^{-1}$, respectively. The reduction in the drainage flow rate of the ACUÑA block might explain the relief of the uplift pressures (Figures 21 and 22) compared with those registered for the Armorwedge™ WSB.

Additional tests were performed to find out how the leakage flow rate was distributed along the chute for the ACUÑA WSB (Figure 25). Most of the drainage flow originated in the first part of the chute (the first eight rows, see folder "07_*w*2_*d*2_2019_no_sealing" in supplementary materials), so the sealing of the first quarter of the chute may significantly reduce the flow towards the drainage layer (see folder "08_*w*2_*d*2_2019_sealing_8rows" in supplementary materials") for the tested flow rates (up to 0.12 $m^2\ s^{-1}$). In fact, it was observed that the effect of the sealing has a greater impact on the upper area of the chute. The sealing of the first four rows (Figure 24) led to very similar results to those obtained with the sealing of the first eight rows. The low velocity of the main flow in the upstream area of the chute probably favors the leakage through the joints among the blocks.

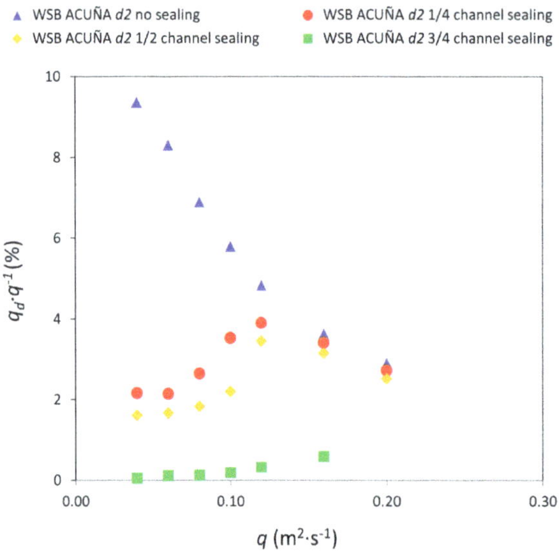

Figure 25. The unit flow rate through the drainage layer (q_d) expressed as a percentage of the flow rate (q) for different q and joint-sealing situations.

4. Conclusions

A reference WSB, the Armorwedge^{TM} block, and a new WSB, the ACUÑA block, were tested. The tests were performed in two different drainage conditions: free drainage, evacuating all the leakage flow towards the drainage layer without reaching saturation conditions and drainage through a granular layer with limited hydraulic capacity.

The main conclusions are as follows:

- Hydrodynamic pressures on the blocks tread were similar for the Armorwedge^{TM} and ACUÑA blocks, although a slightly higher pressure was observed on the ACUÑA block for the highest discharge flows in the lower part of the chute. Although a limited effect, this is favorable for the stability of the block.
- Pressure records in the riser of the ACUÑA block were negative or close to zero, with the greatest suction located in the upper third of the riser. This fact was also previously observed by numerical modeling and led to the new WSB design, ACUÑA, with air vents in the upper part of the riser.
- Negative suction pressures were registered at the base of the two types of blocks when the drainage layer was not saturated. This is favorable for the stability of the block. The suction at the base was higher when the longitudinal joints between blocks were sealed. The effect of sealing just the upper part of the chute was remarkable. The leakage towards the drainage layer was significantly reduced, delaying or avoiding its saturation and, hence, the uplift force.
- The drainage flow rate increased significantly with the inlet discharge flow when the drainage layer was not saturated; however, it (expressed as a fraction of the inlet flow) decreased with inlet flow if the drainage layer was saturated.
- It should be noted that in some cases, positive pressures, although low, were detected in the lower part of the riser. In these cases, the air vents presumably allowed the water to enter the drainage layer if air vents were located at the base of the riser, as was the case for the Armorwedge^{TM} block.
- In the upper part of the channel, the hydrodynamic stabilizing force increased systematically with the discharge flow. The ACUÑA block was more stable than the Armorwedge^{TM} block for all the tested cases. In the lower part of the channel, the stabilizing force was reduced with the discharge flow due to the saturation of the drainage layer and uplift pressures appearing at the base of the block. In this situation, the Armorwedge^{TM} block was more stable than the ACUÑA block.
- When the joints between blocks were sealed, and the drainage layer was unsaturated, the stabilizing forces increased with the discharge flow, and the ACUÑA block was more stable than the Armorwedge^{TM} block for all cases with the skimming flow regime.
- In both WSBs, the longitudinal joints between blocks were the source of the highest percentage of the total leakage flow. In addition, these leaks came mostly from the upper area of the flume.
- Although joint sealing is not a usual practice, it is advisable to consider the benefits and implement a cost-effective way for sealing the joints in new WSB dam protection against overtopping or spillways, especially in the upper sections.

5. Patents

The main result of the research is the development of a new design of WSB (Figure 26) termed ACUÑA (patented in Spain in May 2017 with the code ES2595852). The main difference of the ACUÑA model as compared with the previously developed WSBs is the position of the air vents located in the upper area of the riser. Additionally, the edge between the block tread and the riser was chamfered to facilitate the development of negative pressures in that area. This new layout reduces the leakage flow to the granular drainage layer. In addition, the ACUÑA block has a transverse orifice with the objective of transmitting the negative pressures to the lateral faces. This orifice presents other constructive advantages; it passes through the center of gravity to ease the transportation to the construction site and also allows the tying of each row of blocks with a cable to avoid failure of an isolated block.

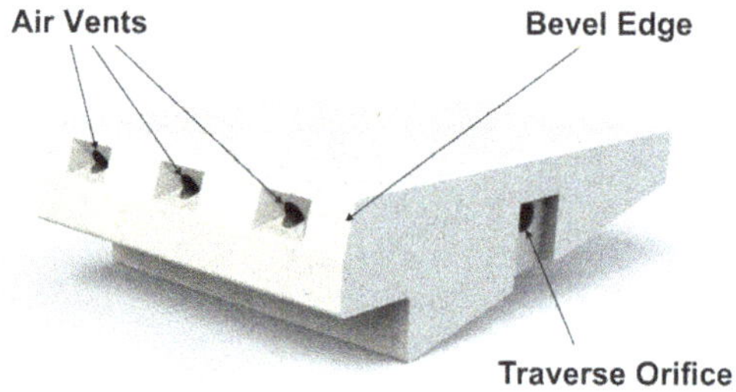

Figure 26. ACUÑA block (patent number ES2595852).

Supplementary Materials: The following are available online at https://zenodo.org/record/4957363#.YMi660wRVPY, Figure S1: (**a**) Aerial view of Barriga dam. Burgos, Spain (Source: Regional Authority of Castilla y León, JCYL) (**b**) View from the top of the dam (**c**) Spillway dicharge (Source: José Manuel Ruiz, JCYL, May 2008) (**d**) Spillway (**e**) and (**f**) Details of the Amorwedge^{TM} WSB chute spillway of the Barriga dam; Figure S2: Experimental setup. (**a**) and (**b**) Schemes. (**c**) Leakage flow collection system (**d**), Stilling basin, triangular thin-plate weir for leakage flow measurement and rectangular thin-plate weir for discharge flow measurement (**e**) ACUÑA WSBs placed over a fixed metallic grid (free drainage condition, *d*1) (**f**) ACUÑA WSBs placed over a layer of homogeneous gravel (granular drainage condition, *d*2); Figure S3: Inception point location; Figure S4: Uniform flow depth location; Figure S5: Mean pressures and standard deviation on the tread of Amorwedge^{TM} block at different rows of the chute for skimming flow and unit flow: (**a**) 0.12 m^2 s^{-1} (**b**) 0.16 m^2 s^{-1} (**c**) 0.20 m^2 s^{-1} (**d**) 0.24 m^2 s^{-1}; Figure S6: Average and standard deviation of the hydrodynamic pressures registered on the block riser of Amorwedge^{TM} block at row 25 of the chute for a different unit discharges simulate free drainage conditions (*d*1): (**a**) Average results compared (**b**) 0.04 m^2 s^{-1} (**c**) 0.08 m^2 s^{-1} (**d**) 0.10 m^2 s^{-1} (**e**) 0.12 m^2 s^{-1} (**f**) 0.16 m^2 s^{-1} (**g**) 0.20 m^2 s^{-1}; Figure S7: Average and standard deviation of the hydrodynamic pressures registered on the block riser of Amorwedge^{TM} block at row 25 of the chute for a different unit discharges simulate the blocks laid over an impervious soil with an intermediate and permeable bedding granular layer (*d*2): (**a**) Average results compared (**b**) 0.04 m^2 s^{-1} (**c**) 0.08 m^2 s^{-1} (**d**) 0.10 m^2 s^{-1} (**e**) 0.12 m^2 s^{-1} (**f**) 0.16 m^2 s^{-1} (**g**) 0.20 m^2 s^{-1}; Figure S8:Average and standard deviation of the hydrodynamic pressures registered on the block riser of Amorwedge^{TM} block at row 10 of the chute for a different unit discharges simulate free drainage conditions (*d*1): (**a**) Average results compared (**b**) 0.04 m^2 s^{-1} (**c**) 0.08 m^2 s^{-1} (**d**) 0.10 m^2 s^{-1} (**e**) 0.12 m^2 s^{-1} (**f**) 0.16 m^2 s^{-1} (**g**) 0.20 m^2 s^{-1}; Figure S9: Average and standard deviation of the hydrodynamic pressures registered on the block riser of Amorwedge^{TM} block at row 10 of the chute for a different unit discharges simulate the blocks laid over an impervious soil with an intermediate and permeable bedding granular layer (*d*2): (**a**) Average results compared (**b**) 0.04 m^2 s^{-1} (**c**) 0.08 m^2 s^{-1} (**d**) 0.10 m^2 s^{-1} (**e**) 0.12 m^2 s^{-1} (**f**) 0.16 m^2 s^{-1} (**g**) 0.20 m^2 s^{-1}; 00_Experimental_set_up (3 videos); 01_No_saturation_d2 (1 photo and 1 video); 02_Suction (2 videos); 03_w1_d1_2017 (1 video); 04_w2_d1_2017 (2 videos); 05_w1_d2_2018 (2 photos and 1 video); 06_w2_d2_2018 (2 videos); 07_w2_d2_2019_no_sealing (3 videos); and 08_w2_d2_2019_sealing_8rows (2 photos and 2 videos).

Author Contributions: Conceptualization, F.J.C., M.Á.T. and R.M.; methodology, F.J.C.; formal analysis, F.J.C., M.Á.T. and R.M.; investigation, F.J.C.; data curation, F.J.C.; visualization, F.J.C.; numerical modelling, J.S.M.; writing—original draft preparation, F.J.C.; writing—review and editing, R.M. and M.Á.T.; supervision, R.M. and M.Á.T. All authors have read and agreed to the published version of the manuscript.

Funding: This research was funded by the Spanish Ministry of Science and Innovation (Ministerio de Ciencia e Innovación, MICINN) through the projects ACUÑA (grant number IPT-2011-0997-020000), DIABLO (grant number RTC-2014-2081-5) and PABLO (grant number RTC-2017-6196-5). In addition, the ACUÑA project was partially financed by the European Regional Development Fund (ERDF) of the European Commission. Finally, the project InnovaDAM (grant number STARTUP RIS3-2019/L1-450), awarded to ACIS2in and funded by the Regional Authority of Madrid (CAM) and partially funded by the European Regional Development Fund (ERDF) of the European Commission, served to disseminate the technology and the works carried out in the previous projects to interested public and private entities in Spain during 2020.

Institutional Review Board Statement: Not applicable.

Informed Consent Statement: Not applicable.

Data Availability Statement: MDPI Research Data Policies.

Acknowledgments: The authors would like to acknowledge the collaboration of Luis Ruano† from PREHORQUISA, without whose involvement these projects would not have been possible, and the Red de Aulas CIMNE. The authors would also like to acknowledge Cristian Ponce-Farfán, Ricardo Monteiro-Alves, Javier Peraita and León Morera from the Dam Safety Research Group (SERPA) of UPM; Lucía Turrero, José Luis García-Ramos and Luis Balairón from CEDEX; Fernando Salazar from CIMNE; and Luis Jaime Caballero for the support to this research.

Conflicts of Interest: The authors declare no conflict of interest.

Nomenclature

C_u	uniformity coefficient
$d1$	free drainage condition
$d2$	the granular layer drainage condition
D_{10}	size of which 10% of the particles, in weight, are finer (m)
D_{50}	size of which 50% of the particles, in weight, are finer (m)
F^*	roughness Froude number
h_c	critical depth
h_s	height of the block riser
L_i	longitudinal distance between the critical depth position on the crest of the dam and the horizontal face where the inception point is located
L_u	longitudinal position from the dam or flume crest of the beginning of the quasi-uniform region
l	the partially exposed length of the top surface of the block
l_s	the total exposed length of the top surface of the block
L'	partial length of the base of the block
L'_b	the total length of the base of the block
p	average hydrodynamic pressure
q	flow-rate inlet in the chute
q_d	drainage flow-rate
$w1$	Armorwedge^{TM} block
$w2$	ACUÑA block
x	coordinate horizontal to the crest of the dam
y	coordinate perpendicular to the crest of the dam
γ	water volumic weight
θ	the angle of the slope of the chute

References

1. FEMA. *Technical Manual: Overtopping Protection for Dams*; U.S. Department of Homeland Security: Washington, DC, USA, 2014.
2. Grinchuk, A.S.; Pravdivets, Y.P.; Shekhtman, N.V. Test of earth slope revetments permitting flow of water at large specific discharges. *Hydrotech. Constr.* **1977**, *11*, 22–26. [CrossRef]
3. Hewlett, H.; Baker, R.; May, R.; Pravdivets, Y.P. *Design of Stepped-Block Spillways*; Construction Industry Research and Information Association: London, UK, 1997.
4. Pravdivets, Y.P.; Slissky, S.M. Passing floodwaters over embankment dams. *Int. Water Power Dam Constr.* **1981**, *33*, 30–32.

5. Baker, R.; Gardiner, K. Construction and performance of a wedge block spillway at Brushes Clough reservoir. In Proceedings of the 8th Conference of the British Dam Society on Reservoir Safety and the Environment, Exeter, UK, 14–17 September 1994; pp. 214–223.
6. Baker, R.; Pravdivets, Y.; Hewlett, H. Design considerations for the use of wedge-shaped precast concrete blocks for dam spillways. *Proc. Inst. Civ. Eng. Water Marit. Eng.* **1994**, *106*, 317–323. [CrossRef]
7. Baker, R.; Gardiner, K.D. Building blocks. *Int. Water Power Dam Constr.* **1995**, *47*, 2.
8. Bramley, M.; May, R.; Baker, R. *Performance of Wedge-Shaped Blocks in High Velocity Flow—CIRIA Research Project 407 Stage 1 Report*; CIRIA: London, UK, 1989; 163p.
9. Bramley, M.; May, R.; Baker, R. *Performance of Wedge-Shaped Blocks in High Velocity Flow—CIRIA Research Project 407 Stage 2 Laboratory Study*; CIRIA: London, UK, 1991; 112p.
10. Cloppper, P.E. *Hydraulic Stability of Articulated Concrete Block Revetment Systems during Overtopping Flow*; Federal Highway Administration: McLean, VA, USA, 1989.
11. Slovensky, G.G., Jr. Near-Prototype Testing of Wedge-Block Overtopping Protection. Master's Thesis, Colorado State University, Fort Collins, CO, USA, 1993.
12. Gaston, M.L. Air Entrainment and Energy Dissipation on a Stepped Block Spillway. Master's Thesis, Colorado State University, Fort Collins, CO, USA, 1995.
13. Frizell, K.H. Protecting embankment dams with concrete stepped overlays. *Hydro Rev.* **1997**, *16*, 36–45.
14. Frizell, K.H. *Armorwedge^{TM} Analysis Report: Block Size Scaling and Bedding Information*, 2007; (Confidencial).
15. Thornton, C.I.; Robeson, M.D.; Varyu, D.R. *Armorwedge^{TM} Data Report 2006 Testing for Armortec Erosion Control Solutions*; Unpublished manuscript (Confidential).
16. Relvas, A.T.; Pinheiro, A.N. Closure of "Inception Point and Air Concentration in Flows on Stepped Chutes Lined with Wedge-Shaped Concrete Blocks". *J. Hydraul. Eng.* **2010**, *136*, 86–88. [CrossRef]
17. Relvas, A.T. Descarregadores de Cheias de Blocos de Betão Prefabricados em Forma de Cunha sobre Barragens de Aterro. Ph.D. Thesis, Instituto Superior Técnico, Universidade Tecnica de Lisboa, Lisbon, Portugal, 2008. (In Portuguese).
18. Relvas, A.T.; Pinheiro, A.N. Inception Point and Air Concentration in Flows on Stepped Chutes Lined with Wedge-Shaped Concrete Blocks. *J. Hydraul. Eng.* **2008**, *134*, 1042–1051. [CrossRef]
19. Relvas, A.T.; Pinheiro, A.N. Stepped chutes lined with wedge-shaped concrete blocks: Hydrodynamic pressures on blocks and stability analysis. *Can. J. Civ. Eng.* **2011**, *38*, 338–349. [CrossRef]
20. Relvas, A.T.; Pinheiro, A.N. Velocity Distribution and Energy Dissipation along Stepped Chutes Lined with Wedge-Shaped Concrete Blocks. *J. Hydraul. Eng.* **2011**, *137*, 423–431. [CrossRef]
21. Morán, R. Wedge-shaped Blocks: A Historical Review. In Proceedings of the 2nd International Seminar on Dam Protections against Overtopping Protections 2016, Colorado State University (CSU), Fort Collins, CO, USA, 7–9 September 2016. Available online: https://mountainscholar.org/handle/10217/179790 (accessed on 16 February 2017).
22. Morán, R.; Toledo, M.A. Design and construction of the Barriga dam spillway through an improved wedge-shaped block technology. *Can. J. Civ. Eng.* **2014**, *41*, 924–927. [CrossRef]
23. Pether, R.; Marsh, P.; Cartwright, P. An innovative new spillway for bruton flood storage reservoir. *Dams Reserv.* **2009**, *19*, 67–72. [CrossRef]
24. Booth, N. Technical visit: Grane Valley reservoir spillway improvements. *Dams Reserv.* **2012**, *22*, 113–115. [CrossRef]
25. Thomson, D. United Efforts at Grane Valley Reservoir. Water Power and Dams Construction. 2014. Available online: http://www.waterpowermagazine.com/features/featureunited-efforts-at-grane-valley-reservoir-4273851/ (accessed on 26 May 2014).
26. Morton, D.; Binnie, M.; Christopher; Crowson, D.; Herrington, J.; Marsh, P.; Seaton, M. Norton Fitzwarren Flood Alleviation Dam. Available online: http://www.hyderconsulting.com/insights/Documents/ (accessed on 6 April 2016).
27. Caballero, F.J.; Toledo, M.A.; San Mauro, J.; Salazar, F. Physical and numerical modeling for understanding the hydraulic behavior of wedge-shaped-blocks (WSB) spillways. In *Dam Protections against Overtopping and Accidental Leakage*; CRC Press: Boca Raton, FL, USA, 2015; pp. 193–206. ISBN 978-1-138-02808-1.
28. Caballero, F.J.; San Mauro, J.; Toledo, M.A.; Moran, R.; Salazar, F. Advances in the understanding of the hydraulic behavior of wedge-shape block spillways. In Proceedings of the 2nd International Seminar on Dam Protections against Overtopping, Protections 2016, Colorado State University (CSU), Fort Collins, CO, USA, 7–9 September 2016. Available online: https://mountainscholar.org/handle/10217/179781 (accessed on 16 February 2017).
29. San Mauro, J.; Larese, A.; Salazar, F.; Irazábal, J.; Morán, R.; Toledo, M.Á. Hydraulic and stability analysis of the supporting layer of wedge-shaped blocks. In Proceedings of the 2nd International Seminar on Dam Protections against Overtopping and Accidental Leakage, Protections 2016, Colorado State University (CSU), Fort Collins, CO, USA, 7–9 September 2016. Available online: https://mountainscholar.org/handle/10217/179782 (accessed on 16 February 2017).
30. Caballero, F.J.; San Mauro, J.; Toledo, M.A.; Morán, R.; Salazar, F.; Moreno, E. A new evolution on the wedge-shaped block for overtopping protection of embankment dams. In Proceedings of the 3nd International Seminar on Dam Protections against Overtopping, Protections 2018, Grange-over-Sands, UK, 6–8 June 2018. Available online: http://eprints.hrwallingford.co.uk/1519/ (accessed on 2 April 2020).

31. Mauro, J.S.; Toledo, M.; Salazar, F.; Caballero, F.F.J. A methodology for the design of dam spillways with wedge shaped blocks based on numerical modeling. *Rev. Int. Métodos Numér. Cálc. Diseño Ing.* **2019**, *35*. Available online: https://www.scipedia.com/public/San_Mauro_et_al_2018a (accessed on 8 February 2019). [CrossRef]
32. Blöschl, G.; Hall, J.; Parajka, J.; Perdigão, R.A.P.; Merz, B.; Arheimer, B.; Aronica, G.T.; Bilibashi, A.; Bonacci, O.; Borga, M.; et al. Changing climate shifts timing of European floods. *Science* **2017**, *357*, 588–590. [CrossRef] [PubMed]
33. Blöschl, G.; Hall, J.; Viglione, A.; Perdigão, R.A.P.; Parajka, J.; Merz, B.; Lun, D.; Arheimer, B.; Aronica, G.T.; Bilibashi, A.; et al. Changing climate both increases and decreases European river floods. *Nature* **2019**, *573*, 108–111. [CrossRef] [PubMed]
34. Gabriel-Martín, I.; Sordo-Ward, Á.; Garrote, L. Influence of initial reservoir level on the allocation of seasonal maximum conservation levels/Influencia del nivel inicial en la definición de resguardos estacionales en presas. *Ing. Agua* **2018**, *22*, 225–238. Available online: https://polipapers.upv.es/index.php/ia/article/view/9526 (accessed on 8 February 2019). (In Spanish). [CrossRef]
35. Gabriel-Martin, I.; Sordo-Ward, A.; Garrote, L.; Granados, I. Hydrological Risk Analysis of Dams: The Influence of Initial Reservoir Level Conditions. *Water* **2019**, *11*, 461. [CrossRef]
36. Garijo, C.; Mediero, L.; Garrote, L. Usefulness of AEMET generated climate projections for climate change impact studies on floods at national-scale (Spain)/Utilidad de las proyecciones climáticas generadas por AEMET para estudios de impacto del cambio climático sobre avenidas a escala nacional. *Ing. Agua* **2018**, *22*, 153–166. Available online: https://polipapers.upv.es/index.php/IA/article/view/9312 (accessed on 8 February 2019). (In Spanish) [CrossRef]
37. Garijo, C.; Mediero, L. Assessment of Changes in Annual Maximum Precipitations in the Iberian Peninsula under Climate Change. *Water* **2019**, *11*, 2375. [CrossRef]
38. Quintero, F.; Mantilla, R.; Anderson, C.; Claman, D.; Krajewski, W. Assessment of Changes in Flood Frequency Due to the Effects of Climate Change: Implications for Engineering Design. *Hydrology* **2018**, *5*, 19. [CrossRef]
39. Soriano, E.; Mediero, L.; Garijo, C. Quantification of Expected Changes in Peak Flow Quantiles in Climate Change by Combining Continuous Hydrological Modelling with the Modified Curve Number Method. *Water Resour. Manag.* **2020**, *34*, 4381–4397. [CrossRef]
40. Chamani, M.R.; Rajaratnam, N. Onset of skimming flow on stepped spillways. *J. Hydraul. Eng.* **1999**, *125*, 969–971. [CrossRef]
41. André, S. High Velocity Aerated Flow on Stepped Chutes with Macro-Roughness Elements. Ph.D. Thesis, Ecole Polytechnique Fédérale de Lausanne (EPFL), Lausanne, Switzerland, 2004.
42. Boes, R.M.; Hager, W.H. Two-phase flow characteristics of stepped spillways. *J. Hydraul. Eng.* **2003**, *129*, 661–670. [CrossRef]
43. Chanson, H. State of the art of the hydraulic design of stepped chute spillways. *Int. J. Hydropower Dams* **1994**, *1*, 33–42.
44. Matos, J. Emulsionamento de Ar e Dissipação de Energia do Escoamento em Descarregadores em Degraus. Ph.D. Thesis, Technical University of Lisbon, Lisbon, Portugal, 1999.
45. Yasuda, Y.; Ohtsu, I. Flow resistance of skimming flow in stepped channels. In Proceedings of the 28th IAHR Congress, Graz, Austria, 22–27 August 1999; Volume B14.
46. Mateos, I.; Elviro, G. Initiation of aeration in stepped spillways. In Proceedings of the 27th IAHR Congress, Theme D, ASCE, San Francisco, CA, USA, 10–15 August 1997; pp. 589–594.
47. Chanson, H. Hydraulic design of stepped spillways and downstream energy dissipators. *Dam Eng.* **2001**, *11*, 205–242.
48. Boes, R.; Minor, H.E. Guidelines for the Hydraulic Design of Stepped Spillways. In Proceedings of the International Workshop on Hydraulics of Stepped Spillways, Zurich, Switzerland, 22–24 March 2000; Minor, H.-E., Hager, W., Eds.; Balkema: Boca Raton, FL, USA, 2000; pp. 163–170.
49. Christodoulou, G. Design of stepped spillways for optimal energy dissipation. *Int. J. Hydropower Dams* **1999**, *5*, 90–93.
50. Frizell, K.H.; Smith, D.H.; Ruff, J.F. Stepped overlays proven for use in protecting overtopped embankment dams. In Proceedings of the ASDSO Annual Conference on Dam Safety, Boston, MA, USA, 9–13 September 1994; pp. 11–14.

water

MDPI

Article

Energy Dissipation in Stilling Basins with Side Jets from Highly Convergent Chutes

Rafael Moran [1,2,*], Miguel Ángel Toledo [1], Javier Peraita [1,3] and Raffaella Pellegrino [1]

1 Department of Civil Engineering: Hydraulics, Energy and Environment, E.T.S. de Ingenieros de Caminos, Canales y Puertos, Universidad Politécnica de Madrid, Profesor Aranguren s/n, 28040 Madrid, Spain; miguelangel.toledo@upm.es (M.Á.T.); j.peraita@upm.es (J.P.); r.pellegrino@upm.es (R.P.)
2 Centre Internacional de Metodes Numerics en Enginyeria, Campus Norte UPC, Gran Capitán s/n, 08034 Barcelona, Spain
3 ACIS Innovación más Ingeniería, Planeta Urano 13, 28983 Parla, Spain
* Correspondence: r.moran@upm.es; Tel.: +34-649-97-4801

Abstract: Spillways with Highly Converging Chutes (HCCs) are a non-conventional alternative that can be applied to achieve a higher outflow capacity when the weir length exceeds the width of the valley at the toe of gravity or arch dams. This kind of spillway has been used in the past, but no general studies have yet been published. This article summarizes experimental research work aiming to increase the knowledge of the effect of some design parameters of HCCs on the energy dissipation in the stilling basin at the toe of the dam. As a comparison reference, we use the Type I stilling basins, widely known by the technical dam engineering community. The obtained results show that spillways with HCCs are a promising alternative to traditional designs, combining the ability to increase the weir length with a high capacity to dissipate energy through the impingement effect of the frontal and the side jets inside the stilling basin.

Keywords: spillway; dam; stilling basin; bucket; chute; flood; weir; safety; protection; dam protection

Citation: Moran, R.; Toledo, M.Á.; Peraita, J.; Pellegrino, R. Energy Dissipation in Stilling Basins with Side Jets from Highly Convergent Chutes. *Water* **2021**, *13*, 1343. https://doi.org/10.3390/w13101343

Academic Editor: Helena M. Ramos

Received: 6 April 2021
Accepted: 10 May 2021
Published: 12 May 2021

1. Introduction

Increasing flooding due to climate change along with more restrictive modern criteria for dam safety have necessitated comprehensive reviews of the design of the spillways of existing dams. Thus, recent research [1–8] has been conducted to deepen our knowledge of methods for increasing the discharge capacity of spillways. In addition, new technical guidelines for dam protection against overtopping [9–11] have been published to ensure dam safety during extreme floods, when the spillway is not able to pass the outlet discharge and the dam crest is overtopped.

The use of highly convergent chutes (HCCs) can be a cost-effective alternative to increase the length of the spillway weir (i.e., increase the spillway capacity) in concrete dams. This type of spillway is characterized by a concrete chute downstream that changes the direction of the inlet flow and leads it to a central area of the river valley (Figure 1a). Thus, the wall of this chute is highly convergent (Figure 1b) in comparison with the usual design criteria for the maximum convergence of the sidewalls of spillway chutes [12].

HCCs have been successfully adopted in several cases such as the Hammam Grouz Dam (Algeria); Vatnsfell (Iceland); Asari, Atagi, Chubetsu, Haidsuka, Kodama, Fukutomi, Fukashiro, Nunome, Ishigoya, Masudagawa, Onbe, Oonagami, Origawa, Sasakura, Shikogawa, Shinmiyagawa, Tomada, and Yasutomi Dams (Japan); Pagade Dam (Portugal); and Baíña, Doña Ana, Rambla del Moro, Torre de Abraham, Vilasouto, and Zapardiel Dams (Spain) [13,14]. There is also a case of HCCs being used to protect the downstream toe of the dam against potential scour caused by overtopping flow (Tygart Dam, USA) [15]. A similar design of converging walls was applied to prevent scour at the abutment groin areas for the overtopping protection of embankment dams formed by lifts of roller-compacted concrete (Spring Creek Dam, USA) [5,11,16–18].

Even though HCCs have been successfully applied, design criteria and knowledge are lacking on the performance of the energy dissipation in comparison with other alternatives such as stilling basins or flip buckets. Thus, in most cases, the experience of the engineer, with the support of either physical or numerical modeling (or both), has been the only resource used to construct a suitable design. This context motivated the beginning of a new research effort, aimed to determine the effect of HCCs in the energy dissipation of the Type I stilling basin, as originally defined by the Bureau of Reclamation (BOR) [19]. This article summarizes part of our experimental research.

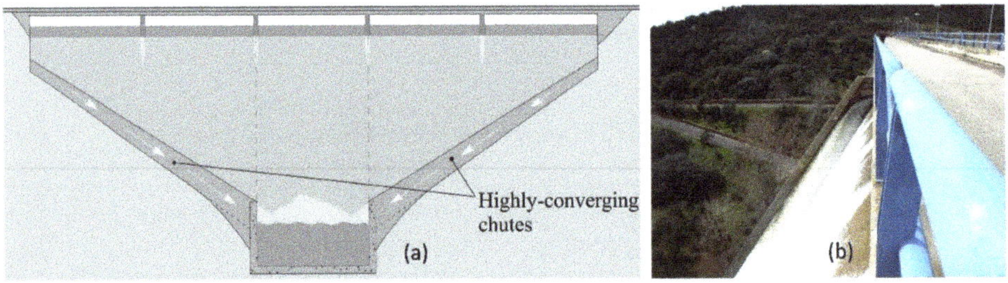

Figure 1. (**a**) Scheme of the HCCs of a gravity dam from downstream [13]. (**b**) View of one of the HCCs of Torre de Abraham Dam (Spain) in operation (courtesy of José R. González).

2. Aim and Scope

As mentioned above, the general aim of this study was to conduct experimental research on the energy dissipation of HCCs in Type I stilling basins to enable cost-effective design when adapting the stilling basins of existing dams to the operation of new flow inlets from newly added HCCs for purposes such as [13]:

- Increasing the spillway capacity.
- Protecting the downstream toe of the dam abutments against potential scour caused by overtopping.
- Enhancing the reservoir storage by increasing the full supply level, preserving the existing freeboards and the former capacity of the spillway.

In addition, the findings of the study are helpful for designing HCCs in new dam projects, where the length of the spillway weir may exceed the width of the floodplain of the river downstream, and the preferred option for the energy dissipation is a stilling basin.

The Type I basin mainly achieves energy dissipation through the formation of a hydraulic jump inside it when the inlet discharge (hereinafter referred to as frontal discharge) from the chute enters the basin following its longitudinal direction [19–22]. However, when side jets from HCCs impinge the flow from the frontal discharge, their collision causes a highly turbulent and complex pattern of flow within the basin [23]. This fact leads to a new energy dissipation process that differs drastically from the one provoked by a hydraulic jump.

Our secondary goals were:

1. To establish objective criteria for the suitable hydraulic conditions of the flow downstream from the stilling basin to enable the appropriate restitution of the flow to the riverbed according to widely accepted technical guidelines.
2. To perform a conceptual study of the effect of key HCCs design parameters on the energy dissipation at the stilling basin.
3. To assess the design adaptation needs of existing stilling basins depending on the increase in the outlet discharge.
4. To obtain experimental data to calibrate and validate numerical models for future research.

3. Methodology

This research was based on a set of tests performed at the hydraulic laboratory of the engineering college of Caminos, Canales y Puertos of the Universidad Politécnica de Madrid (Spain). The methodology of the research is summarized below through a brief description of the experimental facility and the procedures followed during the tests.

3.1. Experimental Facility

The experimental facility was used to simulate a typical gravity dam profile where different configurations of the spillway could be set. The physical model was obtained from previous studies on this topic [13], and was specifically adapted for the current investigation. The facility had the following features:

- Maximum length of the weir (W_w): 5 m
- Height of the weir over the bottom of the stilling basin (H_w): 1.5 m
- Vertical upstream slope of the cross-section of the gravity dam
- Downstream slope of the cross-section of the gravity dam: 0.8 H:V
- Width of the stilling basin (W_b): 1 m
- Maximum length of the stilling basin (L_b): 3.7 m
- Maximum height of the sidewalls of the stilling basin (H_b): 0.5 m

The model was set up to allow changing the length of the weir by adding (or removing) 0.5 m-wide steel plates (up to a maximum of 4 plates on each side of the central meter), which could be fixed to the upstream face of the dam body by screws (Figure 2). Thus, the weir length could be set to 1 m (as shown in Figure 2) for the case of frontal discharge only; and 2, 3, 4, to a maximum of 5 m for the tests where side jets (Q_s) and frontal (Q_f) discharges were combined. In every test, the symmetry of the inlet flow at the stilling basin remained unchanged. Two vertical steel plates (inlet plates in Figure 2), transversely placed to the longitudinal axis of the gravity dam, were placed on both sides of the upstream area of the weir to maintain appropriate inlet conditions over the weir, i.e., to ensure an orthogonal direction of the inlet flow with respect to the longitudinal axis of the weir in all tests.

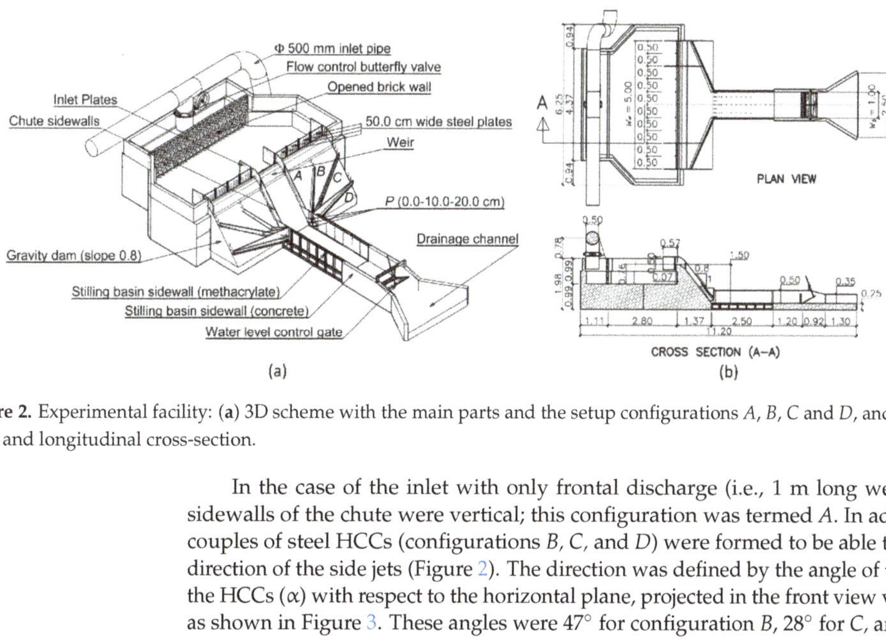

Figure 2. Experimental facility: (**a**) 3D scheme with the main parts and the setup configurations *A*, *B*, *C* and *D*, and (**b**) plan view and longitudinal cross-section.

In the case of the inlet with only frontal discharge (i.e., 1 m long weir), the steel sidewalls of the chute were vertical; this configuration was termed *A*. In addition, three couples of steel HCCs (configurations *B*, *C*, and *D*) were formed to be able to change the direction of the side jets (Figure 2). The direction was defined by the angle of the bottom of the HCCs (α) with respect to the horizontal plane, projected in the front view vertical plane, as shown in Figure 3. These angles were 47° for configuration *B*, 28° for *C*, and 10° for *D*.

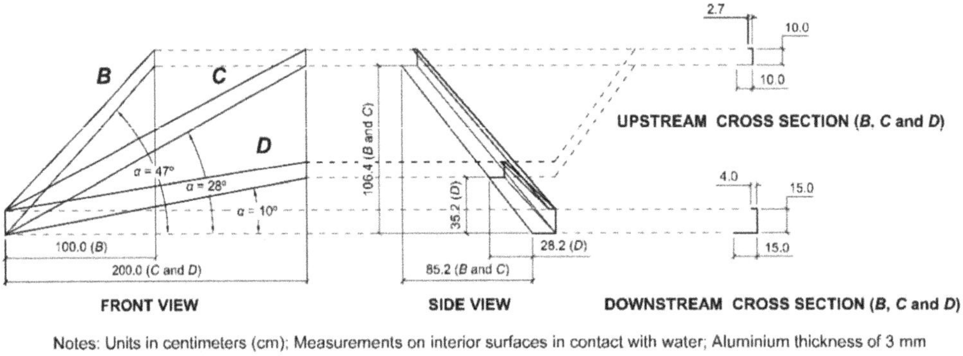

Figure 3. Drawings of the converging chutes for setup configurations *B*, *C*, and *D*.

For each HCCs configuration, the chutes were attached to the downstream slope of the gravity dam by steel supports. As mentioned above, the bottom of the HCCs had a constant slope with the horizontal plane. The width of the bottom of the HCCs changed linearly from 10 cm in the upstream cross-section to 15 cm at the downstream end, which was located next to the sidewall of the stilling basin. The sidewall of the converging chute was vertical, 10 cm high in the upstream cross-section and 15 cm at the downstream end (Figure 3). In every cross-section, the top of the vertical sidewall was closed by a 5 cm wide deflector, parallel to the bottom of the converging chute, to both prevent spilling and allow the water to flip, forming a rolling wave as shown in Figure 1b. In addition, the height of the step from the bottom of the basin to the bottom of the downstream section of the HCCs (*P*) could be modified during the tests.

The stilling basin was formed by a 3.7 m-long, 1 m-wide rectangular channel located at the toe of the dam in front of the central meter of the weir. The bottom of the channel had a horizontal slope. From upstream to downstream, the first 2.5 m of the channel was composed of transparent methacrylate that was reinforced externally with steel frames. The bottom of this part of the channel had 49 threaded holes used to install dynamic pressure devices underneath. Likewise, the last 1.2 m of the channel was formed by a brick wall covered by cement mortar. A gate for the control of the downstream water elevation was installed at the end of this area. Downstream of the stilling basin, the outlet flow dropped into the returning channel of the laboratory, which sent the outlet flow back to the main tank of the laboratory. A calibrated thin-plate weir was installed inside the returning channel to measure the discharge during the tests.

The required discharge in each test was achieved using a variable-speed pump drive, which was automatically controlled by the flowmeter installed at the inlet pipe. The maximum available discharge of the pumping system was 0.25 $m^3\ s^{-1}$.

3.2. Instrumentation

The experimental facility was equipped with the following items:

- An electromagnetic flowmeter installed at the inlet pipe of the facility. The measured discharge was used as the variable controlling the inlet flow provided by the pumping system.
- An ultrasonic limnimeter for measuring the water level at the inlet tank, placed upstream of the gravity dam.
- A total of 21 dynamic pressure gauges located at the bottom of the stilling basin (Figure 4). The measurements analyzed in this study were those obtained from the 11 sensors located along the longitudinal axis of the basin.
- A set of three limnimeters attached to a movable carriage located at the stilling basin, 3.2 m downstream of the dam toe.

- An electronic limnimeter located upstream of the rectangular thin-plate weir at the returning channel, downstream of the model. This sensor provided an indirect measurement of the testing discharge through the calibrated rating curve of the weir.
- Photo and video cameras. The images showed the elapsed time of each test provided by the large chronometer located next to the right sidewall of the experimental facility.
- A self-programmed data acquisition system to gather real-time measurements of the devices. The system included acquisition cards controlled by a laptop computer.

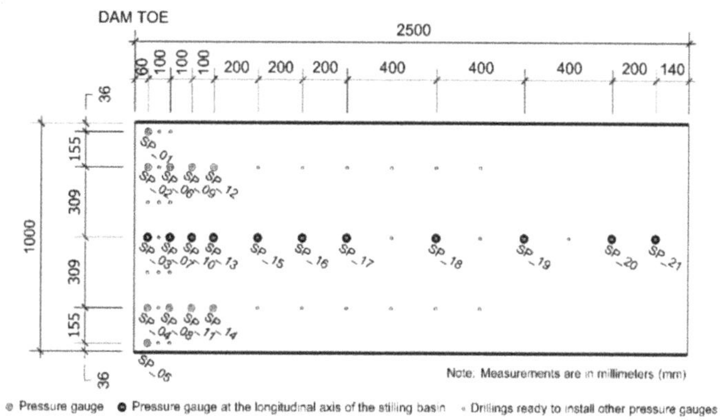

Figure 4. Plan view of the location of the dynamic pressure devices at the bottom of the stilling basin. The flow traveled from left to right, being the left border of the dam toe.

3.3. *Research Approach*

The research work was divided into three phases. The first phase included the general approach, the selection of the parameters of the study, and the definition of the scope. After this phase, we concluded that it was necessary to establish a comparison framework to objectively evaluate the energy dissipation criteria. This work was conducted in phase two. Finally, the core of the experimental research on the performance of stilling basins with HCCs was performed in phase three. This work is briefly described below.

3.3.1. Phase 1: Scope and Selection of Parameters

The parameters H_b, H_w, L_b, and W_b; the geometry of the HCCs corresponding to each configuration (*B*, *C*, and *D*); and the geometry of the gravity dam, including the shape of the weir, remained unchanged in every test. The size of the experimental facility limited the maximum suitable discharge to 0.15 $m^3\ s^{-1}$. The parameters that were modified during the tests were *P*, the ratio (K_Q) between the discharge from the side jets (Q_s) and the total discharge (*Q*), and the value of the angle of the bottom of the HCCs with respect the horizontal plane (α). The reference case for the comparison of each test was the one corresponding to frontal discharge only (i.e., $Q_f = Q$), with energy dissipation achieved by a classical hydraulic jump.

3.3.2. Phase 2: Determination of the Acceptance Criteria for the Energy Dissipation

The index selected to establish the flow condition at the exit of the stilling basin was the coefficient of variation (C_V) of the dynamic pressure measured at the pressure gauges located along the longitudinal axis of the stilling basin. The objective of this selection was to characterize the degree of turbulence at the end of the basin. For this purpose, we decided to use C_V because it could be obtained from the measurements of the dynamic pressure (p). To apply a criterion based on the C_V index, we needed to determine the reference values of the index according to widely accepted technical guidelines. As such, a set of tests with only frontal discharge (i.e., without discharge from side jets) was performed at the experimental facility. The tailwater elevation downstream of the stilling basin was imposed by the control gate for different discharges between 0.010 and 0.150 m^3 s^{-1} until a stable hydraulic jump was generated with its upstream end located at the toe of the dam. The water depth downstream the jump was termed d_{f2}. Applying the hydraulic jump equation [24] between the inlet (subscript 1) and outlet (subscript 2) sections of the hydraulic jump, the water depth of the flow entering the hydraulic jump (d_{f1}) could be calculated as follows (Equation (1)):

$$d_{f1} = \frac{d_{f2}}{2} \cdot \left(\sqrt{1 + 8 \cdot F_{f2}{}^2} - 1 \right), \quad (1)$$

where F_{f2} is the Froude number of the flow downstream of the hydraulic jump, which can be obtained through:

$$F_{f2} = \frac{V_{f2}}{\sqrt{g \cdot d_{f2}}}. \quad (2)$$

The average velocity (V_{f2}) corresponding to the inlet cross-section of the basin is related to the discharge (Q) according to Equation (3):

$$V_{f2} = \frac{Q}{W_b \cdot d_{f2}}. \quad (3)$$

As the value of d_{f2} and Q were known through the measurements for the tests (W_b is constant and equal to 1 m), V_{f2} was obtained using Equation (3). Then, substituting V_{f2} in Equation (2), F_{f2} was calculated (Equation (2)). Finally, d_{f1} and V_{f1} were determined applying Equations (1) and (3), respectively (changing subindex 2 for 1 in Equation (3)).

The process described above was aimed at indirectly determining the inlet conditions (i.e., d_{f1} and V_{f1}). This was considered necessary due to the low and fluctuating values of d_{f1} that were reached during the tests. This finding prevented the accurate measurement of the values of d_{f1}, so we used the more reliable measurements of the water depth registered downstream of the hydraulic jump.

Once the values of d_{f1}, V_{f1}, and F_{f1} were known, the length of the basin with only frontal discharge (L_f) was obtained following the recommendations of BOR [19] according to Figure 5.

Then, after the hydraulic jump was developed and stable, in steady-state conditions, dynamic pressures (p) were registered at the pressure gauges located along the longitudinal axis of the stilling basin over 20 s. Thus, it was possible to obtain the C_V (Equation (4)) of each of the series of records corresponding to each device.

$$C_V = \frac{\sigma_p}{\overline{p}} \cdot 100 \quad (4)$$

where σ_p is the standard deviation of the pressure values, and $\overline{p}$ is the mean of the pressure values.

A linear interpolation between two consecutive C_V values in adjacent devices was used to calculate the C_V corresponding to the length of the stilling basin recommended by the BOR (C_{Vf}). This C_{Vf} was adopted as the reference index for each tested discharge (Q, coincident with Q_f in phase 2). Next, the sets of pairs (Q, L_f) and (Q, C_{Vf}) could be obtained and applied as an acceptance criterion of the energy dissipation of the basins with inlet flows from HCCs. In this phase, the maximum discharge was limited by the length of the jump. Thus, for discharges above 0.147 m^3 s^{-1}, this length was so high that the jump was out of the area of the basin equipped with pressure sensors and it was not possible to obtain the C_{Vf}. The total of 49 tests were conducted in phase 2.

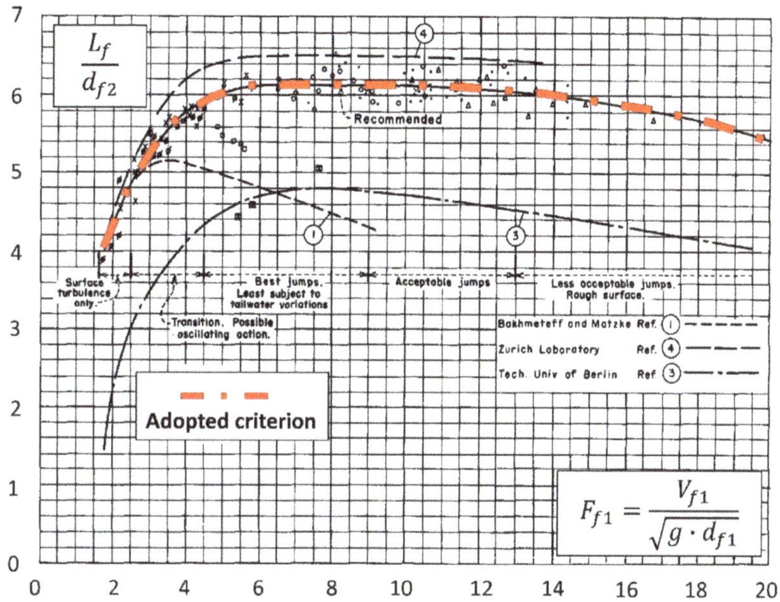

Figure 5. Length of the Type I basin recommended by the BOR (Adapted from Ref. [19]).

3.3.3. Phase 3: Core of the Experimental Research

Once the acceptance criteria were determined in phase 2, the test program was set up according to the scope of the research determined in phase 1. A total of 649 tests were performed in phase 3, as summarized in Table 1.

Table 1. Number of tests performed in phase 3.

Configuration	α (°)	K_Q (–)	Number of Tests
B	47	0.50 and 0.67	132
C	28	0.50, 0.67, 0.75, and 0.80	256
D	10	0.50, 0.67, 0.75, and 0.80	261

Tests were performed for the following target discharges (Q_T): 0.010, 0.025, 0.050, 0.100, and 0.150 m^3 s^{-1}. The target discharge was defined as the flowrate that was required by the pumping system for the inlet discharge at the facility. However, the variability in the discharge values registered by the flowmeter necessitated a more reliable measurement of the flow discharge (Q). The values of Q_T and Q were similar for each test.

Thus, the series of Q_T values indicated above were applied for every geometrical configuration (i.e., at every couple of values of α and K_Q) when possible. In some cases, for the highest value of Q_T (0.150 m^3 s^{-1}), the maximum sidewall height was not high enough to obtain the needed water depth to produce a suitable dissipation of energy. Conversely, the tests with a lower value of Q_T (0.010 m^3 s^{-1}) were considered useless after the review of the registered pressure data. Therefore, the tests corresponding to this target discharge were not considered in the discussion.

It was also necessary to adopt criteria to establish the tailwater depth (d^*) in each test. The variability in Q prevented setting up preset tailwater depths for all tests. Thus, the lowest water depth in each test was determined through our direct observations when the water flow upstream the control gate was considered as subcritical. When possible, three higher water depths were applied in each test for a total of 4. The objective was to test higher tailwater levels to achieve suitable energy dissipation with partially or totally submerged conditions within the basin. Then, after a certain tailwater depth was fixed for each test, the dynamic pressure was registered in the pressure gauges over 20 s. Finally, during data post-processing, the required length of the basin (L^*) for each d^* was determined by linearly interpolating the C_{Vf} (corresponding to the acceptance criteria for each Q) between the C_V registered in the consecutive pressure gauges located along the longitudinal axis of the basin (i.e., gauges *SP_03*, *SP_07*, *SP_10*, *SP_13*, *SP_15*, *SP_16*, *SP_17*, *SP_18*, *SP_19*, and *SP_20*). An example of this is shown in Figure 6.

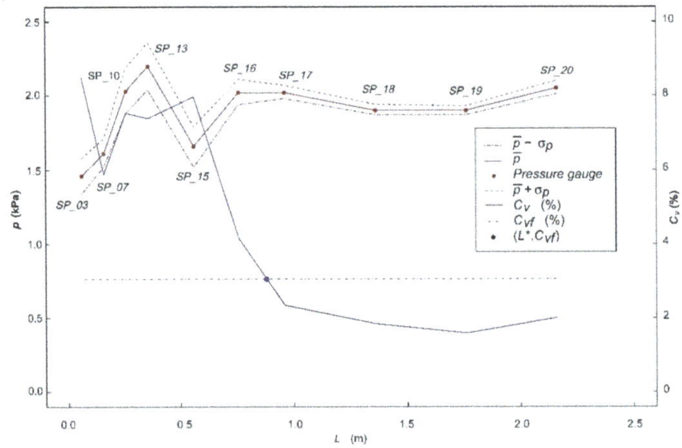

Figure 6. Example of the determination of the required length of the HCCs basin (L^*) in phase 3 tests. The figure shows the results obtained from the measurements of the dynamic pressure gauges located in the longitudinal axis of the basin: mean dynamic pressure ($\overline{p}$), standard deviation (σ_p), and coefficient of variation (C_V).

Thus, a relevant result for each test was a pair of values (d^*, L^*), which met the acceptance criteria determined in phase 2, that is, the pair of values needed to dissipate the energy of the same discharge (Q) in similar conditions (i.e., the same coefficient of variation of dynamic pressures along the axis of the basin) as in Type I stilling basins. In this way, we compared the pairs (d^*, L^*) and (d_{f2}, L_f) to evaluate the required size depending on the inlet conditions.

4. Results and Discussion

This section is divided into three parts: (1) a discussion of the results obtained in phase 2, including the determination of the acceptance criteria for the dissipation of energy; (2) the influence of each of the parameters considered in the operation of the stilling basin with an inlet from HCCs (i.e., P, α, and K_Q) are discussed; and (3) general conclusions about the flow pattern observed during the tests are provided.

4.1. Energy Dissipation Criteria

As shown in the example in Figure 6, the C_V decreases with increasing distance from the dam toe. As expected, $\overline{p}$ also increases with L given that the water depth is higher when subcritical flow conditions are achieved. The C_V value that strictly fulfills the acceptance criteria for each particular discharge (Q) can be obtained through a linear interpolation between the immediately higher and lower values of C_{Vf} in Figure 6 as the intersection between the horizontal line (C_{Vf}) and the C_V corresponding to each test.

A summary of the results of the tests conducted in phase 2 is presented in Figure 7.

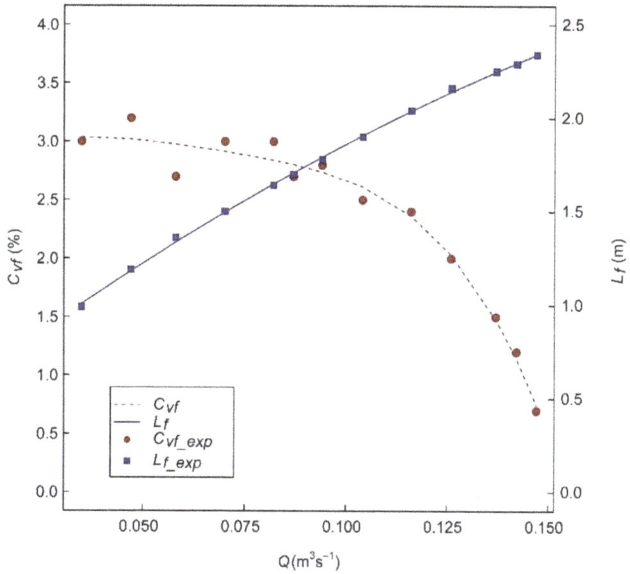

Figure 7. Evolution of the C_{Vf} and Lf for different discharges tested in phase 2.

The evolution of C_{Vf} with Q (coincident with Q_f in this phase) shows that the interval of the registered values ranges between 0.5% and 3.0%. This evolution shows a gradual decrease as Q increases in the range of flow rates used in the tests. The C_{Vf} values were not reliable for Q below 0.034 m^3 s^{-1}. As described in Section 3.3, for discharges above 0.147 m^3 s^{-1}, the length of the hydraulic jump exceeded the area of the basin where dynamic pressures could be registered. Therefore, the C_{Vf} values that could be experimentally obtained ranged from 0.034 and 0.147 m^3 s^{-1}. Within this interval, the C_{Vf} could be fitted to a fourth-grade polynomic expression (Equation (5)) with a coefficient of determination of 0.96.

$$C_{Vf} = -3.493503 \times 10^4 \cdot Q^4 + 9.13858 \times 10^3 \cdot Q^3 - 9.2940 \cdot 10^2 \times Q^2 + 38.20 \times Q + 2.50 \quad (5)$$

For discharges below 0.034 m^3 s^{-1}, the C_{Vf} considered for the acceptance criteria was constant, with a value of 3.05% (i.e., the minimum available discharge in this phase). As the maximum discharge used in phase 3 was 0.162 m^3 s^{-1}, which is close to 0.147 m^3 s^{-1}, a constant C_{Vf} value of 0.5 % was adopted for discharges higher than 0.147 m^3 s^{-1}. The criteria to assess the energy dissipation can be modified depending on the conditions (such as geological or environmental conditions, the potential scour, or others) of the downstream riverbed. However, as indicated in the methodology, in this case, the intention was to establish a consistent approach with the widely accepted criteria for Type I basins.

The results of the length of the hydraulic jump with frontal discharge (L_f) for different discharges are shown in Figure 7. The results were adjusted to a simple linear regression (Equation (6)):

$$L_f = -26.5 \times Q^2 + 16.7 \times Q + 0.45 \quad (6)$$

Th results displayed in Figure 8 are the combinations of L^* and d^* of every test performed in this study. Thus, every symbol represents one of the tests performed in phases 2 and 3. The red dots are the combinations of L^* and d^* that fulfilled the energy dissipation criteria obtained in phase 2.

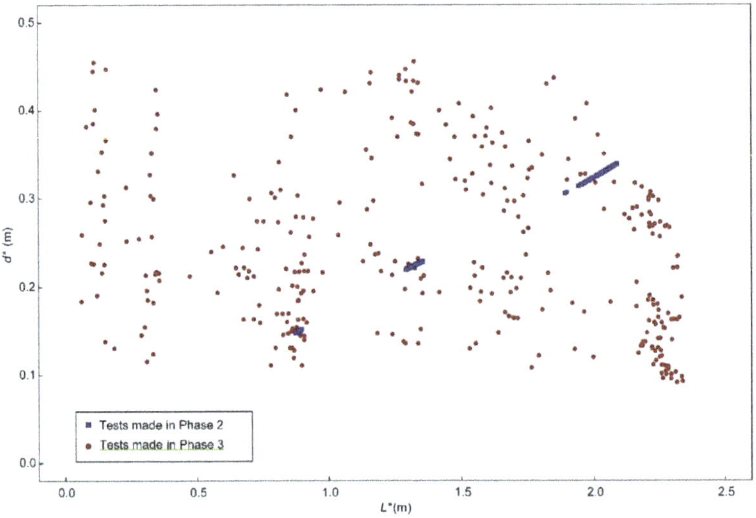

Figure 8. Combinations of d^* (m) and L^* (m) of the phase 2 tests conducted to formulate the energy dissipation criteria with only frontal flow, and the phase 3 tests that fulfilled such criteria with side jet flow.

4.2. *Influence of the Parameters on the Performance of the Stilling Basin*

The discussion begins with the analysis of the parameter P, as we observed signs that this parameter may be the easiest to interpret among those considered in the research. After the effect of P is discussed, parameters α and K_Q are jointly analyzed, specifically for the cases where P was more effective in terms of energy dissipation.

4.2.1. Analysis of the Height of the Bottom of the HCCs over the Bottom of the Basin (P)

The results in Table 2 show the percentage of tests that fulfilled the following conditions: $d^* < d_{f2}$ (termed as condition I), $L^* < L_f$, (condition II), and both I and II, for different values of P and Q_T. The latter condition (i.e., I and II) represents the percentage of cases where the size of the basin needed to dissipate the energy with side inlets from HCCs is smaller than the size with only frontal flow (i.e., with a dissipation achieved by a hydraulic jump). For this analysis, the freeboard needed for the sidewall height was neglected, so that the comparison focused only on the water depth and the length of the basin (note that the width of the basin was constant for every test) required to fulfill the acceptance criteria for the dissipation of energy.

Table 2. Number of tests that met the energy dissipation criteria for different basin dimension conditions for different values of P.

Condition	I ($d^* < d_{f2}$)			II ($L^* < L_f$)			I and II		
P (m)	0.0	0.1	0.2	0.0	0.1	0.2	0.0	0.1	0.2
Q_T ($m^3\ s^{-1}$)	Number of Tests Fulfilling the Condition/Total Tests for Each P and Q_T (%)								
0.010 [1]	21/49 (42.9)	15/34 (44.1)	10/29 (34.5)	8/49 (16.3)	8/34 (23.5)	6/29 (20.7)	0/49 (0.0)	0/34 (0.0)	0/29 (0.0)
0.025	17/33 (51.5)	23/39 (59.0)	18/31 (58.1)	26/33 (78.8)	13/39 (33.3)	9/31 (29.0)	13/33 (39.4)	1/39 (2.6)	0/31 (0.0)
0.050	25/36 (69.4)	28/40 (70.0)	16/34 (47.1)	27/36 (75.0)	16/40 (40.0)	16/34 (47.1)	16/36 (44.4)	4/40 (10.0)	0/34 (0.0)
0.100	11/23 (47.8)	25/44 (56.8)	22/49 (44.9)	23/23 (100)	27/44 (61.4)	30/49 (61.2)	11/23 (47.8)	8/44 (18.2)	4/49 (8.2)
0.150 [2]	0/0 (0.0)	0/4 (0.0)	0/1 (0.0)	0/0 (0.0)	4/4 (100.0)	1/1 (100.0)	0/0 (0.0)	0/4 (0.0)	0/1 (0.0)
All discharges	74/141 (52.5)	91/161 (56.5)	66/144 (45.8)	84/141 (59.6)	68/161 (42.2)	61/144 (42.4)	40/141 (28.4)	13/161 (8.1)	4/144 (2.8)

[1] This target discharge was not be considered representative for the discussion as mentioned in Section 3.3.3.
[2] Most of the tests performed for this target discharge did not meet the acceptance criteria due to the size limits of the facility so they were not considered statistically representative.

Table 3 presents the values of the rates between the mean value of d_{f2} with respect to the mean value of d^* ($\overline{d_{f2}}/\overline{d^*}$), and the mean value of L_f with respect the mean value of L^* ($\overline{L_f}/\overline{L^*}$).

Table 3. Values of the rates between the mean value of d_{f2} with respect to the mean value of d^* ($\overline{d_{f2}}/\overline{d^*}$), and the mean value of L_f with respect the mean value of L^* ($\overline{L_f}/\overline{L^*}$) for every value of P.

P (m)	0.0		0.1		0.2	
Q_T ($m^3\ s^{-1}$)	$\overline{d_{f2}}/\overline{d^*}$	$\overline{L_f}/\overline{L^*}$	$\overline{d_{f2}}/\overline{d^*}$	$\overline{L_f}/\overline{L^*}$	$\overline{d_{f2}}/\overline{d^*}$	$\overline{L_f}/\overline{L^*}$
0.010 [1]	0.85	0.51	0.83	0.43	0.76	0.49
0.025	0.93	1.30	0.96	0.64	0.92	0.58
0.050	1.09	1.35	1.08	0.89	0.90	0.98
0.100	0.97	1.54	1.02	1.24	0.94	1.18
0.150 [2]	-	-	0.34	1.05	0.95	1.01

[1] This target discharge was not considered representative for the discussion as mentioned in Section 3.3.3. [2] Most of the tests performed for this target discharge did not meet the acceptance criteria due to the size limits of the facility so they were not considered statistically representative.

The results presented in Table 2 show that the frequency of fulfillment of conditions I and II (and both simultaneously) increased when P was lower. This finding was evident for the representative target discharges (i.e., 0.025, 0.050, and 0.100 $m^3\ s^{-1}$) among the tests conducted. Thus, for representative values of Q_T, and when P equals zero, 56.8% to 70% of the tests complied with $d^* < d_{f2}$. Likewise, 75% to 100% of the tests complied with $L^* < L_f$, and 39.4% to 47.8% fulfilled both conditions. Therefore, if P is zero, i.e., if the bottom of the HCCs enters the basin at the same elevation of the sill of the basin, it is more likely that a smaller basin would be required. In other words, HCCs can be used to increase the capacity of existing spillways without any change in the original basin, especially when P is zero. In most of the tests performed with P equaling 0.20 m, both d^* and L^* were higher than when P was zero. This finding led to infer that the higher the depth of entry to the basin, the greater the dissipation of energy. This was observed especially when the basin operated in submergence conditions (i.e., when the tailwater level was high enough that the side jets impinged into the water mass).

Table 3 provides results that are useful for quantifying the changes in mean values of d^* and L^* in comparison with d_{f2} and L_f, respectively. In the table, values higher than one indicate that the tailwater depths and lengths of the basin (when HCCs are in operation) are lower than those corresponding to a classical hydraulic jump with only frontal flow for the same Q_T. Focusing on representative target discharges (i.e., 0.025, 0.050, and 0.100 m^3 s^{-1}), and considering the $\overline{d_{f2}/d^*}$ rate, the ranges of values are 0.93 to 1.09 for P = 0 m, 0.96 to 1.08 for P = 0.1 m, and 0.90 to 0.94 for P = 0.2 m. The $\overline{L_f/L^*}$ rate has a wider range of variation: 1.30 to 1.54 for P = 0 m, 0.64 to 1.24 for P = 0.1 m, and 0.58 to 1.18 for P = 0.2 m. Thus, the results presented in Table 3 indicate that the differences between the height of the basin (i.e., d^* and d_{f2}) for HCCs or hydraulic jump are not significant. Conversely, the effect of the operation of HCCs can achieve reductions of up to 54% of the required length of the basin, which can be meaningful from an economical point of view in actual cases. However, in some of the tests with P values of 0.10 or 0.20 m, reductions in d^* and L^* were also achieved; so, in some cases, it may be possible to design HCCs with inlets raised from the bottom of the basin that provide good performance.

4.2.2. Influence of the Angle between the Bottom of the HCCs with Respect to the Horizontal Plane (α) and the Discharge Rate of the Side Jets (K_Q)

As discussed in Section 4.2.1, the length of the basin (L^*) was proven to be the parameter with the most potential for optimization. Therefore, the discussion firstly focuses on the response of L^* to the tested values of α and K_Q. As shown in Figure 9, which depicts results of the tests for a P of zero, most of the highest reductions in L^* occurred for the lowest value of α (i.e., 10°). Such reductions are especially relevant for higher values of K_Q (i.e., 0.75 and 0.80). The possible reason for this result may be the higher energy dissipation achieved by the impingement of the side jets. Thus, the lowest value of α (10°) more directly impinges the side jets, which produces a quasi-frontal impact. In addition, the higher the values of K_Q, the higher the proportion of the discharge involved in the impact dissipation of the side jets in relation to the frontal discharge. Therefore, the results show that the combination of both effects has significative consequences for the reduction in L^*.

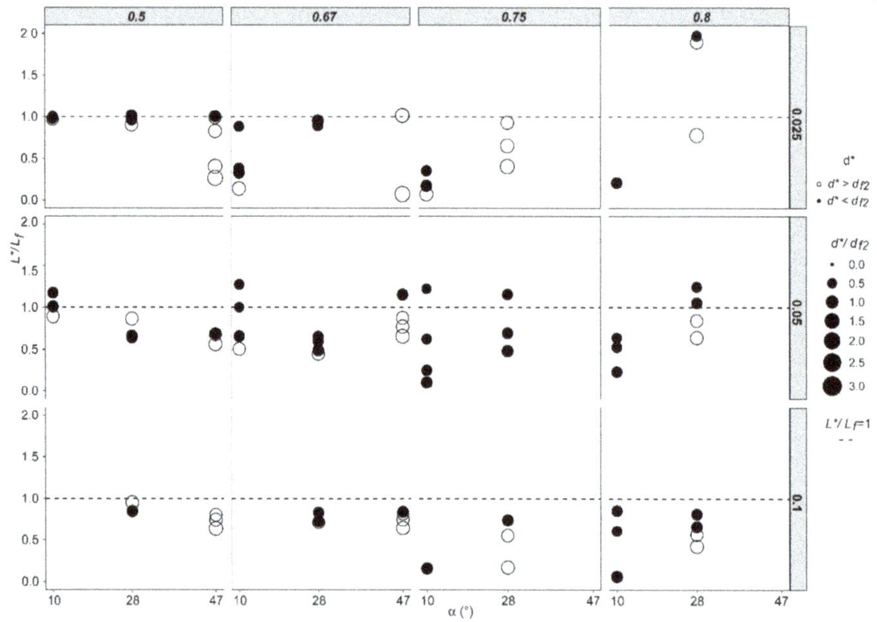

Figure 9. The L^*/L_f (−) rate with respect to α (°) for tests with a P of zero. The figure is organized in a matrix layout, with vertical columns representing K_Q (−) and horizontal rows Q_T (m^3 s^{-1}). Filled circle represent cases where $d^* < d_{f2}$; empty circles represent $d^* > d_{f2}$. The size of every circle represents the d^*/d_{f2} (−) rate.

Next, we discuss the effect of d^*. As shown in Figure 10, the angle α does not seem to have a critical effect on the required tailwater elevation. Thus, the values of d^* are not far from the value of the conjugated water elevation (d_{f2}), with a usual range of variation in d^*/d_{f2} between 0.50 to 1.50. In general, with some exceptions (see Q_T 0.025 m^3 s^{-1} and K_Q 0.8), the higher the α, the higher the d^*. However, this tendency is not evident, as noted in the following cases: Q_T 0.1 m^3 s^{-1} and K_Q 0.67, Q_T 0.1 m^3 s^{-1} and K_Q 0.75, and Q_T 0.05 m^3 s^{-1} and K_Q 0.5. However, no conclusive conclusions could be drawn about the effect of K_Q on the value of d^*.

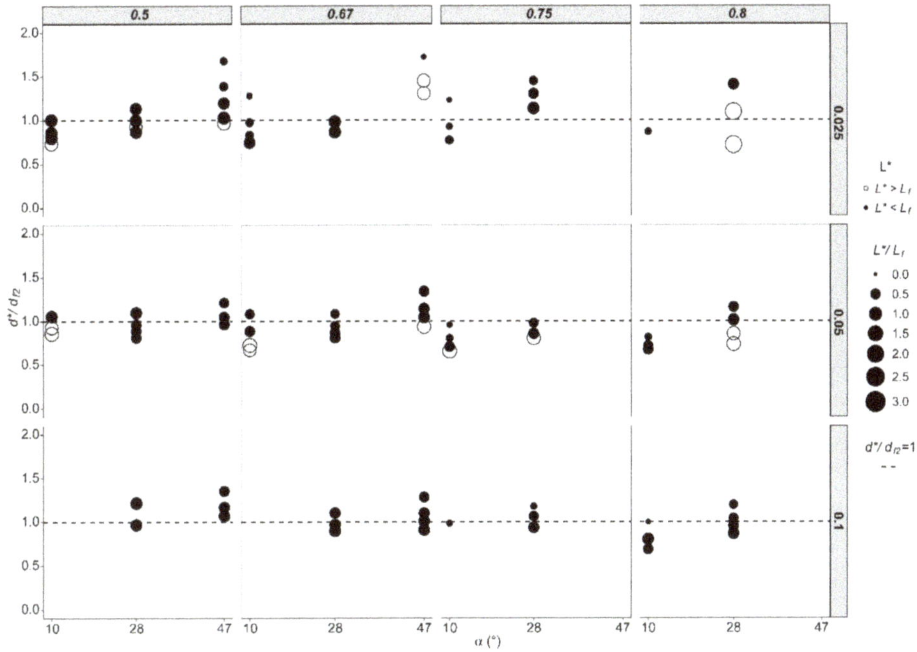

Figure 10. d^*/d_{f2} (−) with respect to α (°) for tests when P is zero. The figure is organized in a matrix layout, with vertical columns representing K_Q (−) and horizontal rows representing Q_T (m^3 s^{-1}). Filled circles represent cases where $L^* < L_f$; empty circles represent $L^* > L_f$. The size of every circle represents L^*/L_f (−).

Another remarkable conclusion from both Figures 9 and 10 is that in every test, there was always a pair of values of d^* and L^* that was less than the corresponding d_{f2} and L_f, values (every filled circle below the dotted line in each figure). That is, it is highly probable that a smaller basin size can be found that can dissipate the energy. In other words, it may be possible to adapt an existing stilling basin to dissipate the energy of a larger outflow discharge without varying the size of the basin. This conclusion can be useful in cases where an increase in the capacity of the spillway is needed in gravity (or arch-gravity) dams with a stilling basin as the energy dissipator.

4.3. *Effect of Submergence on Flow Distribution*

The flow pattern originating in the area where the front and side jets meet causes three-dimensional and highly turbulent functioning. Thus, the hydraulic behavior and the energy dissipation pattern are essentially different than those that occur in a hydraulic jump. As shown in Figure 11a1,a2 and Video S1, where low tailwater was used to meet the required dissipation criteria, the impingement of the jets causes a rise in the water level at the toe of the dam, approximately coinciding with the longitudinal axis of the basin.

Such concentration of the flow at the center of the basin generates lateral flow toward the sidewalls that spreads downstream (Figure 11a1).

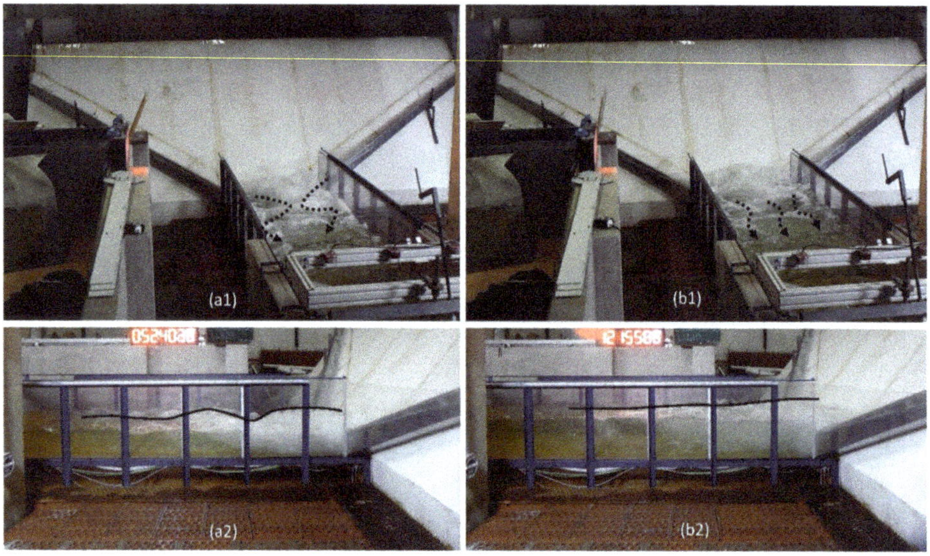

Figure 11. Pictures showing the hydraulic conditions in a stilling basin with low ((**a1**,**a2**), views from downstream and left side, respectively) and high submergence ((**b1**,**b2**), same positions as a1 and a2, respectively) due to different tailwater elevations in the test with $Q_T = 0.050$ m^3 s^{-1}, $\alpha = 28°$ (configuration *B*), $P = 0$, and $K_Q = 0.80$.

Conversely, as illustrated in Figure 11b1,b2 and Video S2, when the energy dissipation is achieved with higher tailwater elevations (i.e., higher submergence condition), the propagation of shockwaves is reduced, and the downstream flow distributes uniformly along the width of the basin. The effect of submergence described above for this test setup was observed in most of the cases. Therefore, we concluded that operation of the basin in higher submergence conditions can effectively achieve more uniform distribution of the flow downstream the basin.

In addition to the effect of submergence, there are other aspects that can be analyzed for a particular case through numerical modeling with specific software (*CALA*) developed during the research project [2,3,25]. This software was calibrated and validated by the experimental results presented in this paper, which may help the technical community apply the findings of this paper to real-sized cases.

5. Conclusions

The experimental research summarized in this article showed that the use of HCCs spillways with symmetric side jet inlets can effectively dissipate the flow energy in the stilling basins of gravity dams, with a likely reduction in the size of the corresponding Type I basin. This is particularly interesting for cases where the outflow capacity of an existing spillway needs to be increased and the stilling basin has already been built. It can also be applied in cases where it is intended to protect the downstream toe of the dam against erosion due to overtopping.

The methodology included a comprehensive study of the coefficient of variation of the dynamic pressures at the outlet area of Type I basins with a hydraulic jump dissipation pattern. In this study, we obtained an objective dissipation criterion to be met by the tests performed with side jet inlets from HCCs.

In general, the test setup with the inlet of the HCCs at the bottom of the basin (i.e., when *P* was 0) resulted in the minimum sizes of the HCCs basins, with average reductions in the length of the basin (L^*) from 23% to 35% in comparison with Type I basin. The results of the required tailwater depths (d^*) are closer to those needed by Type I basins (average reductions up to 8%, and average increases up to 7%). In any case, there were tests in which reductions in both L^* and d^* were achieved from the corresponding values (L_f and d_{f2}) of the Type I basins for a given flow discharge.

Among the tests with the inlet of the HCCs at the bottom of the basin, greater size reductions were achieved when the direction of the impingement of the symmetrical side jets was more frontal ($\alpha = 10°$) and most of the water flow in comparison with the frontal flow was from the HCCs side jets (K_Q was 0.8).

Finally, the observations of the flow pattern in the HCC basins showed that operation in submergence conditions usually caused a more uniform distribution of the flow along the whole width of the basin. Conversely, lower tailwater elevations led to the formation of shockwaves along the basin and flow concentrations in some areas of its cross-section.

Supplementary Materials: The following are available online at https://www.mdpi.com/article/10.3390/w13101343/s1. Video S1: VideoFig11a, Video S2: VideoFig11b.

Author Contributions: Conceptualization, M.Á.T. and R.M.; methodology, J.P.; formal analysis, R.M.; investigation, J.P.; data curation, R.P.; writing—original draft preparation, R.M.; writing—review and editing, M.Á.T.; visualization, R.P.; supervision, M.Á.T. All authors have read and agreed to the published version of the manuscript.

Funding: This research was funded by the Ministerio de Ciencia e Innovación of Spain within the CALA research project (grant number RTC-2016-4581-5).

Institutional Review Board Statement: Not applicable.

Informed Consent Statement: Not applicable.

Data Availability Statement: MDPI Research Data Policies.

Acknowledgments: The authors would like to acknowledge collaboration with the Red de Aulas CIMNE. The authors would also like to acknowledge Javier Caballero, Ricardo Monteiro-Alves, Víctor Flórez, Begoña Labalde, Jaime García, Javier San Mauro, Fernando Salazar, and Eugenio Oñate for supporting this research.

Conflicts of Interest: The authors declare no conflict of interest.

Nomenclature

C_V	coefficient of variation of the water pressure, expressed in percentage.
C_{Vf}	coefficient of variation of the pressure downstream the BOR's Type I stilling basin, expressed in percentage.
d_{f1}	water depth of the flow entering the jump in BOR's *Type I* basin.
d_{f2}	water depth downstream of the BOR's *Type I* basin.
d^*	water depth downstream of the basin after energy dissipation with side jets discharge.
F_{f1}	Froude number at the entrance of the hydraulic jump.
F_{f2}	Froude number downstream the hydraulic jump.
H_b	maximum height of the sidewalls of the stilling basin, equals to 0.5 m.
H_w	height of the weir over the bottom of the stilling basin, equals to 1.5 m.
K_Q	discharge rate of the side jets (i.e., Q_s/Q).
L	horizontal distance from the toe of the dam along the longitudinal axis of the basin.
L_b	maximum length of the stilling basin of the experimental facility: 3.7 m.
L_f	length of the basin recommended by the BOR criteria for Type I basins.
L^*	length of the basin required to a suitable energy dissipation with side jets discharge.
p	pressure.
$\overline{p}$	mean of the pressure values.
P	height of the step from the bottom of the basin to the bottom of the downstream section of the *HCCs*.

Q total discharge.
Q_f frontal discharge.
Q_s side discharge from the HCCs inlet.
Q_T target discharge.
V_{f1} velocity of the flow entering the jump in BOR's Type I basin.
V_{f2} velocity of the flow downstream of the BOR's Type I basin.
W_b width of the stilling basin, equals 1 m.
W_w maximum length of the weir, equals 5 m.
α angle of the bottom of the HCCs with respect the horizontal plane, projected in the front view vertical plane.
σ_p standard deviation of the pressure values.

References

1. San Mauro, J.; Morera, L.; Salazar, F.; Rossi, R.; Toledo, M.; Morán, R.; Martínez, B.; Caballero, F.; Oñate, E. Modelación Física Y Numérica De Aliviaderos Con Cajeros Altamente Convergentes. In Proceedings of the III Jornadas De Ingeniería Del Agua, Valencia, Spain, 23–24 October 2013.
2. Larese, A.; Salazar, F.; San Mauro, J.; Oñate, E.; Toledo, M.; Morán, R. Advanced Computational Methods for Dam Protections Against Overtopping. In Proceedings of the Protections 2018 (3rd International Conference on Protection against Overtopping), Grange-over-Sands, UK, 6–8 June 2018.
3. Mauro, S.; Salazar, F. Mejora De La Seguridad Hidrológica E Incremento De La Capacidad De Embalse De Presas De Fábrica Mediante Aliviaderos Con Cajeros Altamente Convergentes. 2019. Available online: https://www.scipedia.com/public/Peraita_et_al_2019a (accessed on 1 October 2020).
4. Lempérière, F.; Vigny, J.; Deroo, L. *New Methods and Criteria for Designing Spillways could Reduce Risks and Costs Significantly*; Hydropower & Dams: Wallington, UK, 2012; pp. 120–128.
5. Hunt, S.L. Design of Converging Stepped Spillways. Ph.D. Thesis, Colorado State Univeristy, Fort Collins, CO, USA, 2008.
6. Hunt, S.L.; Kadavy, K.C.; Abt, S.R.; Temple, D.M. Impact of Converging Chute Walls for Roller Compacted Concrete Stepped Spillways. *J. Hydraul. Eng.* **2008**, *134*, 1000–1003. [CrossRef]
7. Schleiss, A. From Labyrinth to Piano Key Weirs—A Historical Review. In *Proceedings of the Labyrinth and Piano Key Weirs*; Apple Academic Press: Palm Bay, FL, USA, 2011; pp. 3–15.
8. Macián-Pérez, J.F.; Vallés-Morán, F.J.; Sánchez-Gómez, S.; De-Rossi-Estrada, M.; García-Bartual, R. Experimental Characterization of the Hydraulic Jump Profile and Velocity Distribution in a Stilling Basin Physical Model. *Water* **2020**, *12*, 1758. [CrossRef]
9. NVE. Guidelines for Embankment Dams. Guideline 4/2012. Norwegian Water Resources and Energy Directorate. 2012. Available online: https://www.nve.no/damsikkerhet-og-kraftforsyningsberedskap/damsikkerhet/regelverk/veileder-for-fyllingsdammer/ (accessed on 1 October 2020).
10. Portland Cement Association. *Design Manual for RCC Spillways and Overtopping Protection*, 1st ed.; Portland Cement Association: Skokie, IL, USA, 2002; p. 100.
11. FEMA. *Technical Manual: Overtopping Protection for Dams*; U.S. Department of Homeland Security: Washington, DC, USA, 2014.
12. United States Bureau of Reclamation. *Design of Small Dams*; US Department of the Interior, Bureau of Reclamation: Washington, DC, USA, 1987.
13. Morera, L.; San Mauro, J.; Salazar, F.; Toledo, M.Á. *Highly-Converging Chutes as an Overtopping Protection for Concrete Dams: Physical and Numerical Modelling*; Taylor & Francis Group: London, UK, 2015.
14. Yang, J. Investigations at Vatnsfell. *Int. Water Power Dam. Constr.* **2007**, *59*, 9.
15. Zaitsoff, M. Overtopping modifications to Tygart Dam. In Proceedings of the National Dam Safety Technical Seminar: Overtopping Protection for Dams, Portland, OR, USA, 13–15 May 2003; pp. 20–21.
16. Talbot, J.; Robinson, K.; Kadavy, K. Hydraulic Model Study of a Roller Compacted Concrete Stepped Spillway with Converging Chute Walls. In Proceedings of the Association of State Dam Safety Officials Annual Conference, Pittsburg, PA, USA, 7–10 September 1997.
17. Hunt, S.L.; Temple, D.M.; Abt, S.R.; Kadavy, K.C.; Hanson, G. Converging Stepped Spillways: Simplified Momentum Analysis Approach. *J. Hydraul. Eng.* **2012**, *138*, 796–802. [CrossRef]
18. Hunt, S.L.; Kadavy, K.C.; Abt, S.R.; Temple, D.M. Impact of Converging Chute Walls for RCC Stepped Spillways. *Impacts Glob. Clim. Chang.* **2005**, 1–12. [CrossRef]
19. Peterka, A.J. *Hydraulic Design of Stilling Basins and Energy Dissipators*; Department of the Interior, Bureau of Reclamation: Washington, DC, USA, 1978.
20. Barjastehmaleki, S. Spillway Stilling Basins and Plunge Pools Lining Design. Ph.D. Thesis, University of Trieste, Trieste, Italy, 2016.
21. Babaali, H.; Shamsai, A.; Vosoughifar, H. Computational Modeling of the Hydraulic Jump in the Stilling Basin with Convergence Walls using CFD Codes. *Arab. J. Sci. Eng.* **2015**, *40*, 381–395. [CrossRef]
22. Vischer, D.; Hager, W. Energy Dissipators. *Oceanogr. Lit. Rev.* **1996**, *1*, 87.
23. Martín-Vide, J. The Design of Converging Overfall Spillways. *Int. J. Hydropower Dams* **1995**, *2*, 87.

24. Chanson, H. *Hydraulics of Open Channel Flow*; Elsevier: Amsterdam, The Netherlands, 2004.
25. San Mauro, J.; Salazar, F.; Morán, R.; Peraita, J.; Toledo, M.Á.; Conde, M.J.; Flórez, V.; Labalde, B.; Alcalde, F. Aliviaderos Con Cajeros Altamente Convergentes. ¿Posible Solución Para La Presa De Oroville? In Proceedings of the V Jornadas de Ingeniería del Agua, A Coruña, Spain, 24–26 October 2017.

water

MDPI

Article

Prediction of Concrete Dam Deformation through the Combination of Machine Learning Models

Patricia Alocén [1,2,*], Miguel Á. Fernández-Centeno [1,2] and Miguel Á. Toledo [1]

[1] E.T.S. de Ingenieros de Caminos, Canales y Puertos, Universidad Politécnica de Madrid (UPM), Profesor Aranguren s/n, 28040 Madrid, Spain; miguelangel.fernandez@upm.es (M.Á.F.-C.); miguelangel.toledo@upm.es (M.Á.T.)

[2] ACIS Innovation+Engineering S.L. (ACIS2in), Planeta Urano 13, P18 2°A, Parla, 28983 Madrid, Spain

* Correspondence: patricia.alocen@alumnos.upm.es

Abstract: Dam safety monitoring is of vital importance, due to the high number of fatalities and large economic damage that a failure might imply. This, along with the evolution of artificial intelligence, has led to machine learning techniques being increasingly applied in this field. Many researchers have successfully trained models to predict dam behavior, but errors vary depending on the method used, meaning that the optimal model is not always the same over time. The main goal of this paper is to improve model precision by combining different models. Our research focuses on the comparison of two successful integration strategies in other areas: Stacking and Blending. The methodology was applied to the prediction of radial movements of an arch-gravity dam and was divided into two parts. First, we compared the usual method of estimating model errors and their hyperparameters, i.e., Random Cross Validation and Blocked Cross Validation. This aspect is relevant not only for the importance of robust estimates, but also because it is the source of the data sets used to train meta-learners. The second and main research topic of this paper was the comparison of combination strategies, for which two different types of tests were performed. The results obtained suggest that Blocked CV outperforms the random approach in robustness and that Stacking provides better predictions than Blending. The generalized linear meta-learners trained by the Stacking strategy achieved higher accuracy than the individual models in most cases.

Keywords: stacking; blending; combination; meta-learner; experts; machine learning; Cross Validation; radial displacement

Citation: Alocén, P.; Fernández-Centeno, M.Á.; Toledo, M.Á. Prediction of Concrete Dam Deformation through the Combination of Machine Learning Models. *Water* **2022**, *14*, 1133. https://doi.org/10.3390/w14071133

Academic Editors: Paolo Mignosa, Francesco Gallerano and Giuseppe Pezzinga

Received: 25 February 2022
Accepted: 29 March 2022
Published: 1 April 2022

1. Introduction and Background

Monitoring the safety status and behavior of dams plays a crucial role in civil engineering, due to the high cost that dam failure can entail. Monitoring techniques that comprise the safety system of a dam and its follow-up have evolved over time with technological advances, including artificial intelligence.

In recent years, the development of predictive models with machine learning techniques has been widely applied to different practical problems. Specifically, in the field of dam safety, the area of machine learning is attracting growing attention because of the complexity of the dam system, involving materials of great heterogeneity. Machine learning models achieve high accuracy in the prediction of their behavior, and a comparison with the measured responses allows early detection of anomalous behavior that may reveal an internal failure of the infrastructure. It is therefore of vital importance to achieve the highest possible accuracy with the trained models. From this derives the main objective of this research, which is to increase the accuracy of the usual models through their combination.

Many researchers have already successfully applied these techniques, including Support Vector Machine (SVM) [1], Boosted Regression Trees (BRT) [2], Random Forest (RF) [3], and different types of Neural Networks (NN) [4].

Fernando Salazar et al., for example, obtained promising results by applying these techniques to real cases in the field of dam safety. However, he emphasizes the need for further generalization and validation [4]. They successfully used BRT in several of their research studies [2,5,6]. Furthermore, they demonstrated the effectiveness of the mentioned techniques compared to the usual statistical models, concluding that BRT was the best model over 14 target variables [7]. Support Vector Regression can also be used as an accurate model to predict displacement of dams [8], while J. Mata demonstrates that Neural Networks have great potential for assessing dam behavior [9]. Herrera et al. compared Machine Learning models of different nature and managed to accurately predict hourly urban water demand [10]. Kang et al. also obtained good results using the Machine Learning RBFN technique [11].

These models, called experts or first-level models, do not perform in the same way in all periods of the series. Therefore, the possibility of finding different optimal experts depending on patterns arise. By combining those experts, a second-level machine learning model, or meta-learner, can identify such patterns and maximize accuracy, which leads us to the main topic of this paper: combination of models through Stacking and Blending.

Stacked generalization was introduced by Wolpert in 1992, where the first-level models are trained and later combined during the training of the meta-learner. The inputs of the meta-learner are the predictions of each of the experts generated during the Cross Validation (CV) process [12]. If all the available predictions are used to train the meta-learner, we speak of Stacking, while if only 10 or 20% of the data is used, we speak of Blending [13]. The training set used in Blending is called the Hold-out set.

Before detailing the main topic of this paper, we consider it necessary to emphasize the importance of the Cross Validation process and robust estimation, since the inputs of the second-level model of Stacking and Blending are derived from this process.

The usual division of the data set between training and validation allows for the evaluation of the models in the latter subset, which has a reduced percentage of inputs. For a more reliable and accurate estimation, the concept of Cross Validation is introduced.

This evaluation method consists of dividing the training set into folds (usually 10), where a model is iteratively trained with all folds except one, which is used for testing. This is repeated until all folds have been used for testing. Hence, all the examples of the training set are used for training and testing at least once. The estimated error is the average of the errors committed across these test folds.

In all machine learning problems, CV plays a fundamental role since it is used to estimate not only the error that the model will make on future data, but also the optimal hyperparameters of the model. For the research discussed in this paper, CV also plays an important role in the generation of the training set of the second-level model, which is explained in detail in Section 2. Therefore, we compare two types of CV to select the best process: Random or Blocked CV.

Random CV is the most common Cross Validation process and consists of dividing the training data set into folds whose records are chosen randomly. However, it is not the most appropriate option for practical problems where time dependence between instances is found, as in the case of dams. If Random CV is used in such problems, the error of the Random CV will be too optimistic (over-estimated error), giving very low errors during Cross Validation compared to the validation set. As Roberts et al. note, "The tendency for values of nearby observations to be more similar than distant observations is widespread, if not pervasive" [14], which implies an overly optimistic CV error.

To solve this problem, researchers, such as Bergmeir and Benítez, developed and used different types of CV. Among these methods lies the method based on the last block, used in some papers that will be mentioned below, and also the following methods: Cross Validation with omission of dependent data, where the dependent data are identified and excluded from the training set, and Cross Validation with blocked subsets, which is the CV proposed in this paper, where each fold corresponds to a year of the training set [15]. Although CV based on the last block is also appropriate, more weight should be given to the most recent estimated errors, since following a forward sequence, the fewer data there are, the higher the calculated error [2].

Roberts et al. demonstrate that Blocked CV generates a more robust error estimate than Random CV [14]. On the other hand, Bergmeir and Benítez do not find under- or over-estimated error when applying Random CV, although they recommend using Blocked CV together with an adequate stationarity control [15]. Regarding the research carried out by Herrera et al., the authors prefer a sequential CV as it is more similar to the original problem, where predictions are always made on data in ascending time order [10].

Few of the articles on the behavior of dams contain specific research on the CV employed. Some researchers use only one validation set to estimate the error [11] and emphasize that, if the conditions affecting the dam change, the model will perform poorly in future [4]. This hypothesis always contains some truth, but it is more reliable to give estimates of errors through Blocked CVs because it tests models considering more years. Some authors do not specify the type of CV used [5,7–9], while others divide the data set into training, validation (last two years available) and test (last year available) [3]. Fernando Salazar specified the processes used in two articles, using sequential CVs to estimate errors as averages of weights that decrease geometrically every year [2,6].

Since the movement cycle of the dam is annual, we decided to use Blocked CV, also called Annual CV, in this paper, where each block corresponds to a year. The predictions made, during the CV process of each year, are used to train the meta-learner through Stacking and Blending.

The main interests in the comparison of Stacking and Blending strategies focus on computational cost reduction and error optimization. The development of engineering technology to collect data has led to a very large data set for modeling training, depending on the data collection period. Thus, the computational cost of model training increases. Efficiently decreasing the dimensions of the data set, while being able to maintain model accuracy, is fundamental. However, it is reasonable to expect that a model with more examples would be more accurate.

Regarding this matter, numerous articles have been published in several fields showing the successful results generated from combining experts by linear regression [13], or a multi-response model classifier [16]. These techniques have been applied not only in the scientific domain, but also in business. For instance, Netflix held a Kaggle (a subsidiary of Google LLC, it is an online community of data scientists and machine learning professionals) competition to develop an algorithm to predict user ratings for films, which was won by BellKor's Pragmatic Chaos team thanks to the combination of different experts.

The success of these strategies in other fields [13,17], together with their novelty, explains the interest in their application to the field of dam safety.

Research related to model combination for dam safety encompasses several approaches. Multi-model ensemble strategies using machine learning algorithms have been used to combine the inflow predictions of the Probability Distributed Model, Integrated Flood Analysis System, and Génie Rural à 4 paramètres Horaire models, and improve the accuracy of the predictions [18]. Other authors also use predictions from statistical and time series models as inputs to a second-level model trained by the Extreme Machine Learning algorithm [19], or induced ordered weighted averaging (IOWA) [20].

Other approaches that have been taken include the integration of models that attempt to predict parts of a series caused by external factors with models that attempt to predict the unknown [21,22].

On the other hand, Hong et al. were able to identify a pattern of behavior of two models (Random Forest and Gradient Boosting) to predict dam inflow, where one performed better than the other above a certain cutoff point [23]. However, for most dam problems, the detection of patterns among experts' performance is not straightforward and a more general solution is needed.

All these articles use statistical or time series models to train a second-level predictive model using machine learning algorithms, while we use the predictions of machine learning models as input. Moreover, none of the mentioned articles specifies whether the comparisons have been performed using Stacking or Blending. Our research is innovative because, to our knowledge, it is the first to introduce a combination of experts with these strategies to improve the precision of typical ML models used in the research of the existing literature.

Therefore, this study aims to improve the precision of first-level models by their combination through Stacking and Blending, and to broaden knowledge of both strategies in order to determine the best one. The algorithm chosen to perform such combinations is Generalized Linear Regression (GLM), due to its success in other fields. The selection of the best experts to use as input is made by the Akaike information criterion. We also analyze the differences between Random and Blocked CV.

2. Materials and Methods

2.1. General Approach

This section briefly describes the general approach to the research, with the aim of proposing a strategy to develop prediction models based on a combination of experts (Figure 1).

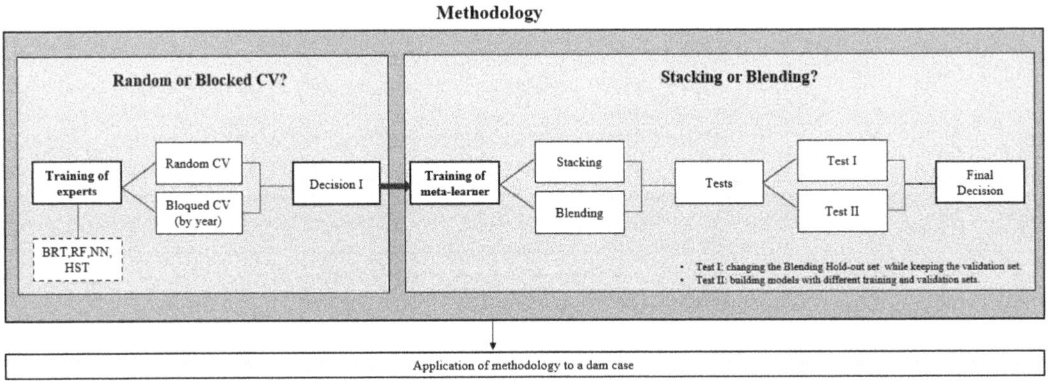

Figure 1. Summary of the methodology.

In the first place, the training of four experts of different algorithmic natures was executed: Boosted Regression Trees (BRT), Random Forest (RF), Neural Networks (NN), Hydrostatic-Seasonal-Time (HST). Each of these experts was trained using two different evaluation and hyperparameter optimization methods: Random and Blocked Cross Validation (CV). In this paper, each block corresponds to a different year, so the term Annual CV will appear throughout this article, referring to a Blocked CV where the blocks are years.

The errors obtained from the two processes of Cross Validation were compared for every expert. The main objective was to observe which one gave a better error estimator and prediction for future data. The strategy that yields a CV error most similar to the validation error is the appropriate estimation strategy.

On the other hand, the optimal hyperparameters of each expert obtained by both strategies were compared and their impact on the error in the validation set was analyzed. Thus, we studied which strategy, Random or Blocked CV, generates a better prediction in the validation set.

The decision made after this analysis is called "Decision I" (Figure 1) and determined which experts and which CV process should be implemented for training the second-level models of combination of experts.

The predictions of the experts chosen were used to train the second-level models through two different strategies: Stacking and Blending. The main objective of this research was to determine the best strategy and improve the accuracy of predictions.

For that purpose, two different sets of tests were performed, Test I and Test II (Figure 1). Test I consisted of changing the Blending Hold-out set while keeping the validation set.

This involved training several models where the training set of each model was a different year. Therefore, it was possible to analyze the differences in the errors in the validation year, depending on the election of the training year (Hold-out set). The variance and mean of the resulting set of errors were used to determine whether Blending is an appropriate strategy or not. Test II was based on the building of models with different training and validation sets. Each training set contained different years and was used to train the experts and the meta-learner. The errors in the validation set of the second-level models, trained through the Stacking and Blending strategies, were compared to determine the best one.

The described methodology was applied to the monitoring data of radial movements of an arch dam.

2.2. Methodology

In this methodology, the experts used to build the meta-learner were chosen based on the quality of their performance and their nature. Sufficiently accurate models of different natures were selected to cover as much information as possible through their prediction vectors. Therefore, we worked with ensembles of decision trees, BRT and RF; Neural Networks, NN; and Hydrostatic-Season-Time, HST. Regarding the integrative method, meta-learners of Stacking and Blending were both built by Generalized Linear Regression. These experts were trained using external and time factors as explanatory variables, including synthetic variables derived from them, such as moving averages, aggregates, and variation rates of different orders.

2.2.1. Random or Blocked Cross Validation?

As mentioned in Section 1, the folds generated during CV are usually composed of randomly chosen instances to achieve an optimal representation of the data set. However, this method is not valid for series where there is time dependence, as is the case for dam behavior. For this reason, the need to partition in a different way arises to better represent the problem. Hence, Blocked CV, or Annual CV, where each block is a year, is introduced to optimize the hyperparameters and estimate the error in future data.

For ease of explanation, it is assumed that there are no hyperparameters to optimize. Therefore, for each split shown in Figure 2, a model is trained on the training folds and predictions are made on the test fold. Thus, we have a vector of predictions for each fold, which together add up to a number of instances, or dates, equal to the total training set. Due to the fact that we have such prediction vectors, we can calculate the error by comparing the observed and predicted values of the test fold using some error measure. The mean error made in all of these folds is the estimated error for future data.

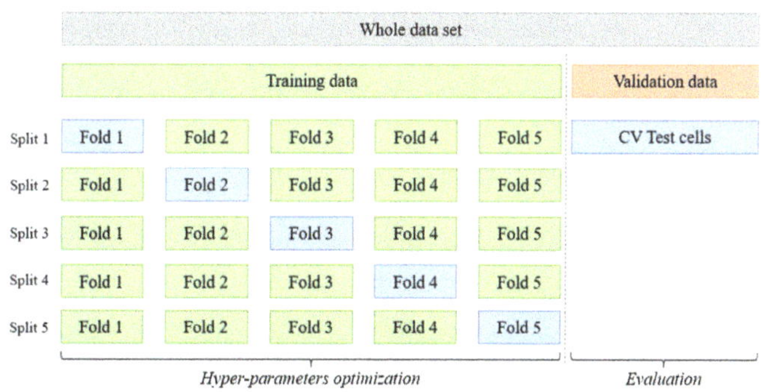

Figure 2. Five-fold Cross Validation, where each fold coincides with an available year. The green color represents the subdivisions for model training, while the blue color represents the test subsets. The validation set is orange in color [24].

It should be noted that, in each of these splits, the data reserved for testing have not been considered during the training phase. This fact is of utmost importance, since it allows one to use the instances corresponding to that fold to make predictions and calculate the error.

Up to this point, the Blocked CV by years has been explained to estimate the model error assuming there are no hyperparameters to optimize.

To introduce the optimization of the hyperparameters, we operate in a similar way. The main difference is that, instead of training only one model for each split, we must train as many models as there are combinations of hyperparameters. These are used to make predictions over the test fold in each split. The combination that gives the lowest mean error among folds is chosen, and its error becomes the estimated value for future data.

The same approach is taken for Random CV, but the instances in each fold are chosen randomly without replacement. In a Random CV, instances from different years can be found, whereas in the Blocked CV, used in our research, all instances necessarily belong to the same year. Decision I in Figure 1 was taken by comparing the errors estimated during Blocked and Random CV to the error in the validation set.

The error measure applied to all estimations was the Root-Mean-Square Error (RMSE), which has the following mathematical form [25]:

$$\text{RMSE} = \sqrt{\sum_{i=1}^{m} \frac{(\hat{y}_i - y_i)^2}{m}}, \tag{1}$$

where $\hat{y}_i$ is the predicted value of instance i, y_i is the observed value of the same instance i, and m is the total number of instances.

2.2.2. Stacking or Blending?

Predictions originating from the test folds of the Blocked CV served as the training data set for the meta-learner built through the Stacking and Blending strategies. If all the available predictions are used to train the meta-learner, we speak of Stacking, while if only 10 or 20% of the data is used, we speak of Blending. The origin of both is detailed below.

We started from an $m \times n$ matrix corresponding to the training set D, where m is the number of rows, or available dates, and n the number of explanatory variables:

$$D = \begin{bmatrix} x_{11} & \cdots & x_{1n} \\ \vdots & \ddots & \vdots \\ x_{m1} & \cdots & x_{mn} \end{bmatrix}. \tag{2}$$

Assume that for each expert a Blocked CV with 5 folds, similar to Figure 2, has been used to find the optimal hyperparameters and estimate the RMSE. Thus, the training set is divided into five blocks, and predictions for each test fold are obtained, so that eventually all instances of the set are used at least once for validation. Then, the following data set shown in Table 1 is obtained:

Table 1. Representation of the data set by train and test folds in tabular form [26].

Split	Fold 1	Fold 2	Fold 3	Fold 4	Fold 5
1	Test_1	Train	Train	Train	Train
2	Train	Test_2	Train	Train	Train
3	Train	Train	Test_3	Train	Train
4	Train	Train	Train	Test_4	Train
5	Train	Train	Train	Train	Test_5

For each fold shown in Table 1, four subsets are used for training (Train) and one subset is used for testing (Test). Train cells are green in Figure 2 and test cells are blue. The prediction vectors of each Test subset are calculated during Cross-Validation for every model. Therefore, for each expert (from 1 to k) we obtain a prediction vector $p(Y)$ that contains a total number of m elements corresponding to the instances of all Test folds together ($\text{Test}_1, \ldots, \text{Test}_5$ in Table 1), which is the same amount as that of matrix D in Equation (2).

By joining these vectors to the target variable $Y = \{y_1, y_2, \ldots, y_m\}^T$, the following training matrix $[m \times (k+1)]$ is obtained:

$$D' = \begin{pmatrix} p_{11} & \cdots & p_{1k} & y_1 \\ \vdots & \ddots & \vdots & \vdots \\ p_{m1} & \cdots & p_{mk} & y_m \end{pmatrix}, \tag{3}$$

where p_{ij} is prediction i of the expert j. The first column in Equation (3) is the prediction vector of expert 1, $p_1(Y) = \{p_{11}, p_{21}, \ldots, p_{m1}\}$, and so on.

The following Figure shows a specific example where 6 years are available in the training set and there are 3 experts built:

The elements on the left of Figure 3 show what has been explained so far in matrix form. Three experts are trained on data from 2000 to 2005, where each year is one fold. The prediction vector for each year is obtained from CV and the matrix D' is formed to train the meta-learner with Stacking strategy. The elements on the right show the number of years used for Stacking and the year used for Blending (2005), called the Hold-out set. Both the experts and the meta-learner are then validated over the validation set (2006). Thus, an error is obtained for every model built.

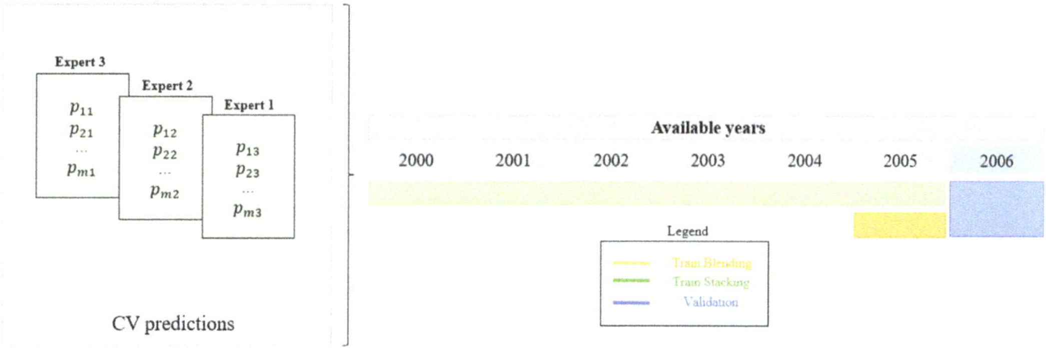

Figure 3. Origin of the data set used to train the Stacking and Blending meta-learners.

More generally, the matrix representing the training set of the experts is D, shown in Equation (2), while that of the Stacking-trained meta-learner is D' (Equation (3)). The matrix representing the instances of the Blending Hold-out set is as follows:

$$B = \begin{pmatrix} p_{m-m_y+1,1} & \cdots & p_{m-m_y+1,k} & y_{m-m_y+1} \\ \vdots & \ddots & \vdots & \vdots \\ p_{m1} & \cdots & p_{mk} & y_m \end{pmatrix} \tag{4}$$

where m is the total number of instances in the training set, k is the total number of experts and m_y is the total number of instances in the last available year of the training set y.

As mentioned in the previous section, two different sets of tests were performed to achieve the main goal of this research. Test I consisted of changing the Hold-out set used to train Blending meta-learners, to analyze the mean and variance of the error made over the validation set.

Figure 4 shows an example of this analysis where 6 years are available in the data set. The usual approach to train the second-level model with the Blending strategy would be using data from 2005. However, to solve the question posed, a meta-learner is trained for each available year in the training set and validated with data from 2006. Thus, it is possible to analyze the variance of the error measure as a function of the year used for its construction and compare its mean to the error made by the Stacking second-level model.

Training						Validation
2000	2001	2002	2003	2004	2005	2006
Year of training			**Holdout set**			$RMSE_{val}$
2000			H6			$RMSE_{H6}$
2001			H5			$RMSE_{H5}$
2002			H4			$RMSE_{H4}$
2003			H3			$RMSE_{H3}$
2004			H2			$RMSE_{H2}$
2005			H1			$RMSE_{H1}$

RMSE over 2006 of Stacking Vs. RMSE MEAN

Figure 4. Test I with 6 years available in the data set. RMSEval is the error over the validation set (2006) and $RMSE_{Hi}$ is the error made by the model trained with the Hold-out set Hi.

On the other hand, Test II consisted of training the experts and the Stacking and Blending meta-learners with different training sets, i.e., with a different number of years. In each case, a validation was performed in the following year of the last training set to compare both strategies.

The flow of the different training processes can be observed in the arrow shown in Figure 5. First, the experts are trained, then, the second-level model with all CV predictions (Stacking), next, the Blending meta-learner is trained with the predictions of the last year of the training set, and, finally, all models are validated. In the table of Figure 5, five different training processes are simulated, where several colored lines are drawn to represent the years used to train and validate the different models. The red color represents the training of experts; the green color represents the Stacking meta-learner; the yellow color is reserved for Blending; and the blue color for the validation year. Thus, in the first round of training, or iteration 1, the training flow is as follows:

1. The experts are first trained with the data corresponding to the instances between 2000 and 2005.
2. The Stacking second-level model is trained with the CV predictions of these experts, which belong to the same years as them.
3. The Blending meta-learner is built with the CV predictions of the experts of the last year available in the training.
4. Models are validated over the following year.

Through this methodology, we expected to achieve the objectives of this research and be able to draw solid conclusions on the best combination and Cross Validation strategy.

2.3. Dam Case: Description and Available Data

With the aim of testing the proposed methodology, data obtained from an arch-gravity dam more than 60 m high were used.

The dam is equipped with different monitoring devices, including direct pendulums. Since the displacement of the dam is a good indicator of its safety condition, the measurement of radial movement at four direct pendulums was chosen. The devices are located in the upper zone and in the horizontal gallery (Figure 6).

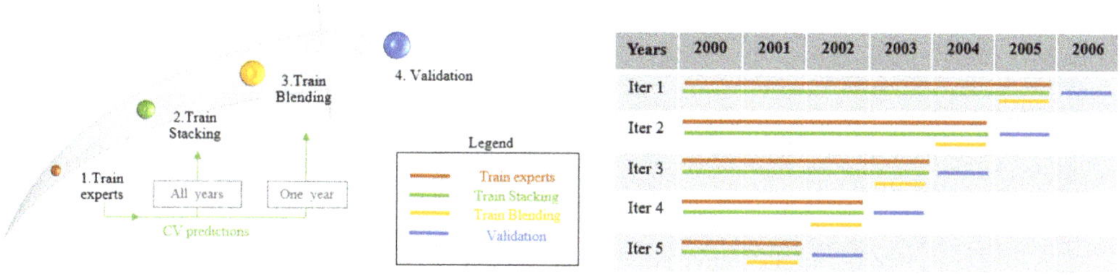

Figure 5. Test II. The arrow on the left indicates the training flow: experts, Stacking, Blending and validation. The table placed on the right points out the years used to train and validate each model.

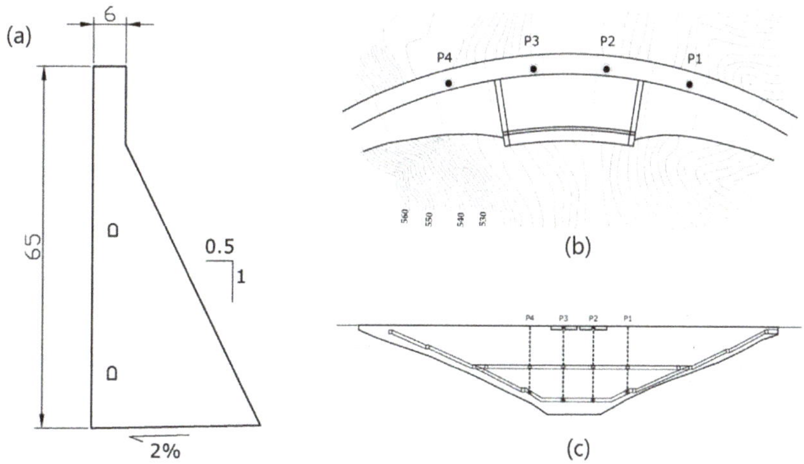

Figure 6. Cross-section (**a**), Plan (**b**) and section (**c**) of the dam. The source of this figure relies on information from the project that has funded this research, called ARTEMISA.

Altogether, there are approximately 3560 instances and 5 explanatory variables related to the external factors affecting the dam: maximum, minimum and mean temperature (Figure 7), reservoir level and rainfall (Figure 8). The data sets include records from 2004 to 2017. The latter was reserved as a validation set and the remaining dates were used for training (Figure 9).

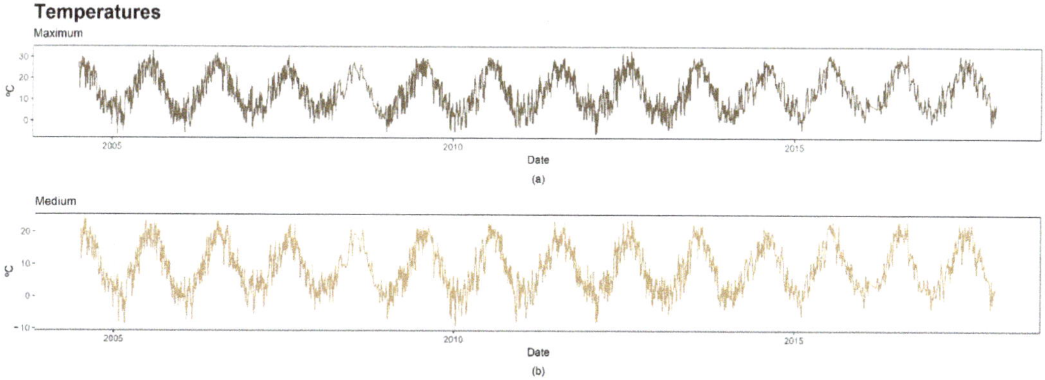

Figure 7. *Cont.*

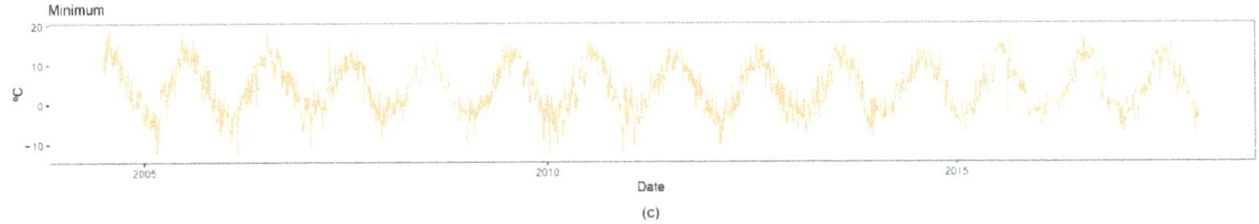

Figure 7. Series of temperatures. (**a**) Series of maximum temperature; (**b**) Series of medium temperature; (**c**) Series of minimum temperature.

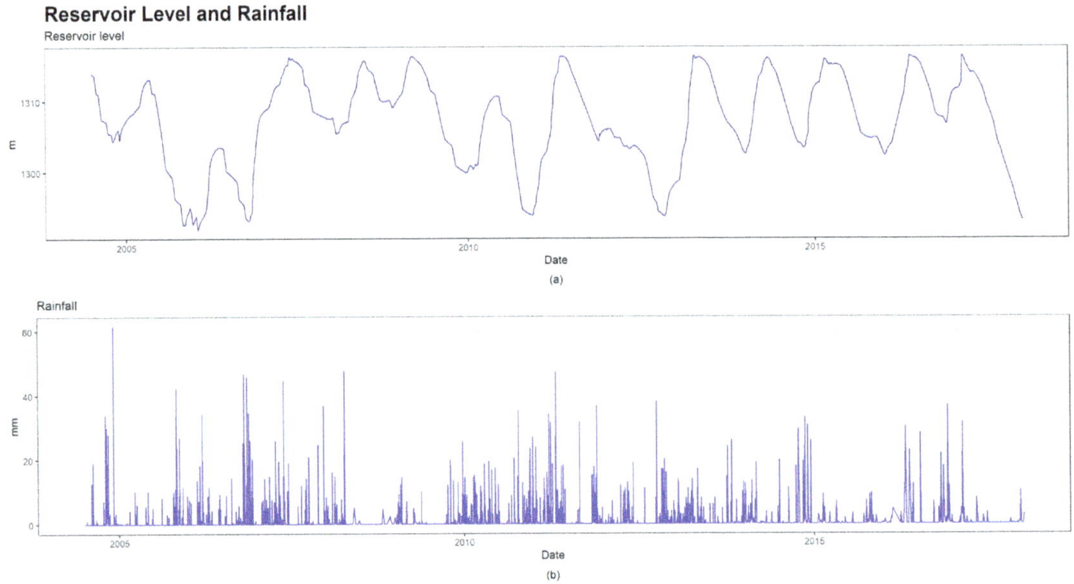

Figure 8. Series of reservoir level and rainfall. (**a**) Series of reservoir level; (**b**) Series of rainfall.

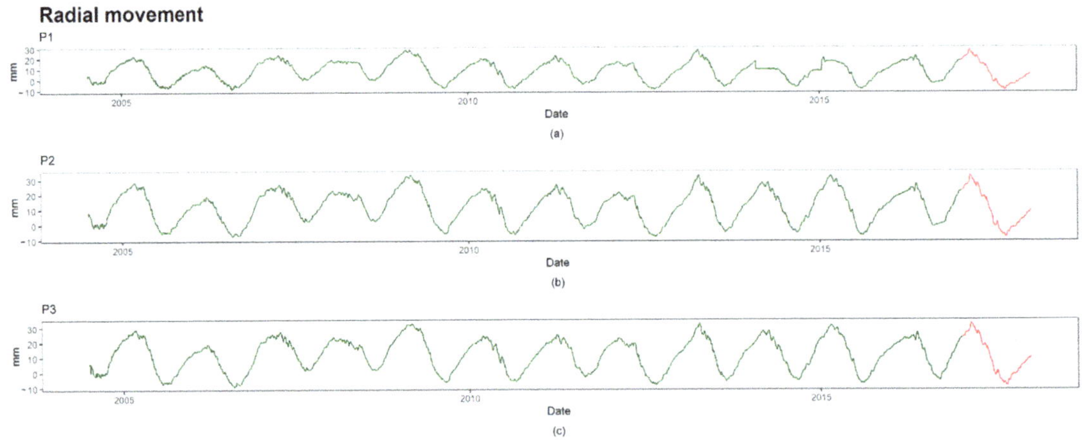

Figure 9. *Cont.*

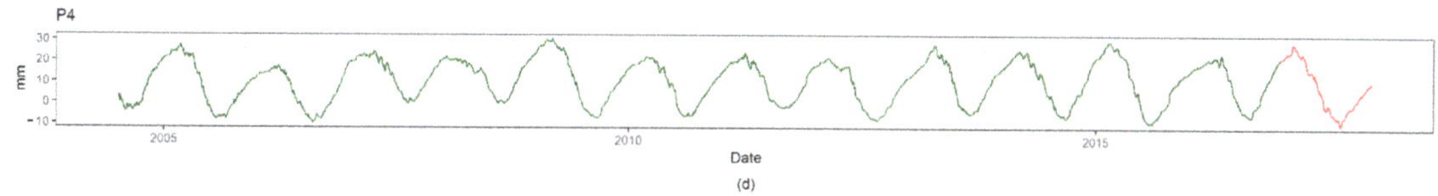

Figure 9. Series of direct pendulums. Training instances are colored in green and validation dates are colored in red. (**a**) Series of pendulum 1; (**b**) Series of pendulum 2; (**c**) Series of pendulum 3; (**d**) Series of pendulum 4.

3. Results and Discussion

This section presents and discusses the results obtained by applying the proposed methodology to the dam described in the previous section. It has been divided into two subsections, following the outline shown in Figure 1.

3.1. Cross Validation

Table 2 shows two different RMSE values for both strategies: $RMSE_{cv}$ and $RMSE_{val}$. The first measures the error (RMSE) during the Cross Validation process, which is the mean of the RMSE values across folds. The second refers to the error over the validation set [24]. The third measure is the relative difference between both errors calculated by the following formula:

$$RD_k = \frac{RMSE_{CV_k} - RMSE_{val_k}}{RMSE_{CV_k}} \times 100; \quad (5)$$

where $k \in \{\text{Random CV}, \text{Blocked CV}\}$.

Table 2. Prediction errors for both strategies: Random and Blocked Cross Validation.

Pendulum	Expert	$RMSE_{cv}$		$RMSE_{val}$		RD	
		Random CV	Blocked CV	Random CV	Blocked CV	Random CV	Blocked CV
1	BRT	0.435	2.2	2.522	2.331	−479.77%	−5.95%
	HST	2.113	2.848	2.113	2.113	0.00%	25.81%
	NN	1.397	2.018	3.254	3.272	−132.93%	−62.14%
	RF	0.506	2.649	3.431	3.494	−578.06%	−31.90%
2	BRT	0.346	1.404	1.487	1.596	−329.20%	−13.65%
	HST	1.853	2.080	2.249	2.249	−21.38%	−8.15%
	NN	0.933	1.446	1.359	1.385	−45.58%	4.22%
	RF	0.437	1.925	2.073	2.117	−374.77%	−9.98%
3	BRT	0.368	1.288	1.478	1.587	−301.63%	−23.21%
	HST	2.224	2.438	2.433	2.433	−9.40%	0.21%
	NN	0.884	1.473	1.316	1.367	−48.87%	7.20%
	RF	0.439	1.832	2.015	1.998	−359.00%	−9.06%
4	BRT	0.361	1.299	1.433	1.314	−296.95%	−1.15%
	HST	2.203	2.436	2.385	2.385	−8.26%	2.09%
	NN	0.882	1.416	1.35	1.371	−53.06%	3.18%
	RF	0.426	1.652	1.832	1.805	−330.05%	−9.26%

As mentioned in previous chapters, it is more robust to use the estimated error obtained during the CV process, because it is calculated using a larger number of instances. However, the RMSE in the validation set should be compared with the $RMSE_{cv}$ value, seeking the smallest difference between them.

The negative values of RD in Table 2 show that the error over the validation set is higher than the one estimated by CV. All these values, except one, are negative and quite high in the case of Random CV. They fall within the range of [−578.06%, −8.26%], excluding the result of HST, which can be considered atypical. Those of Blocked CV are significantly lower, as they fall within the range of [−62.14%, 25.81%]. The mean RD of Random CV is −210.56%, while the mean RD of Blocked CV is −8.23%. Therefore, the error estimated during the Annual CV, or Blocked CV, is considered a preferable estimator for future data.

Looking at the $RMSE_{cv}$ values and the predictions series (Figure 10), BRT is the best expert with respect to the precision of both Random and Blocked CV in all cases, except pendulum 1, where the lowest error is reached by NN trained using the Blocked CV. The predictions of pendulum 1 (a) are less accurate in some of the peaks than those of the other pendulums. Specifically, approximately in 2014, some atypical data are detected, probably generated by a failure in the measuring device. All other experts provide significantly accurate predictions.

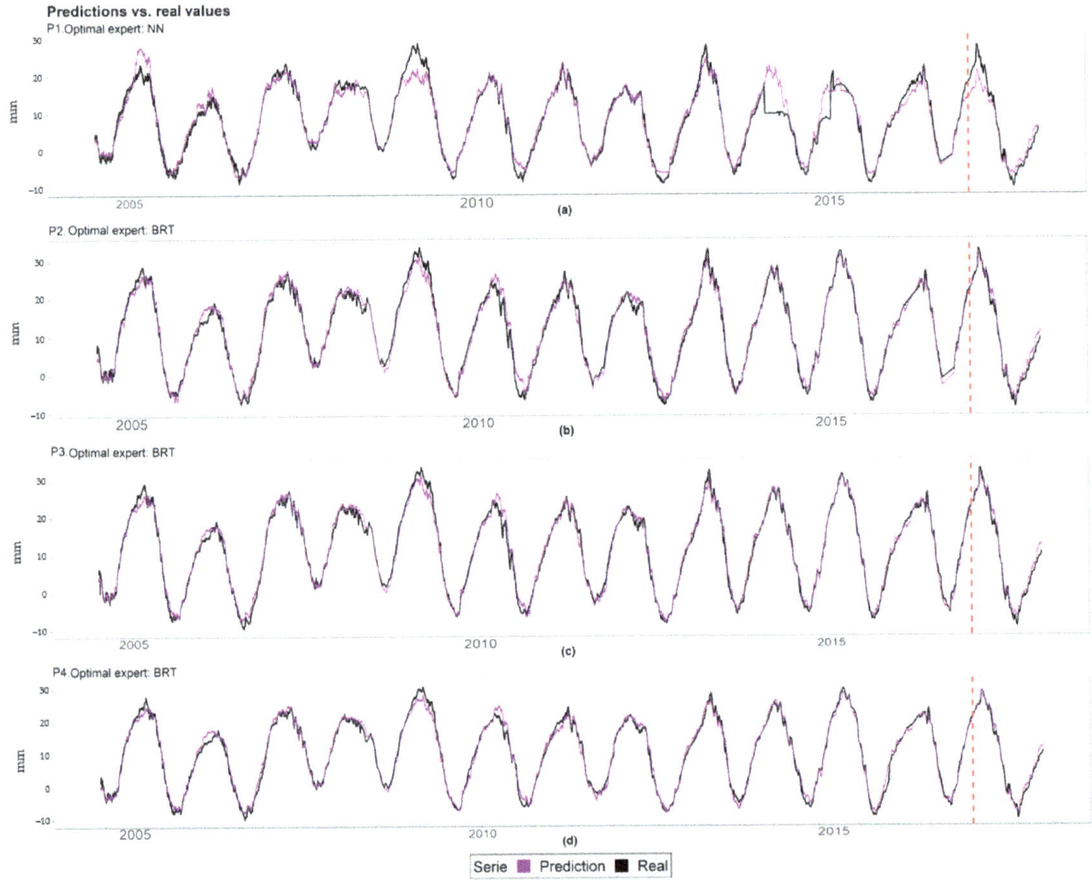

Figure 10. Series of predictions (magenta color) vs. series of measured values (black color) for each target. The red dashed line separates the training (**left**) and validation (**right**) periods. (**a**) Series of predictions of pendulum 1; (**b**) Series of predictions of pendulum 2; (**c**) Series of predictions of pendulum 3; (**d**) Series of predictions of pendulum 4.

In addition to affecting the estimated RMSE, the choice of CV also influences the selection of hyperparameters. We use the example of the expert BRT of pendulum 4 (Table 2) to show the differences between the optimal combination of hyperparameters obtained from both processes.

Two of the hyperparameters differ between the two CV processes (Table 3)This fact, summed up in the difference between the cuts made by the ensemble trees, causes the error in the validation set to be different depending on the CV used for training.

Table 3. Optimal hyperparameters of BRT found by Random and Blocked CV. NTree is the number of trees, nMinObsInNode is the minimum number of observations to consider in one node, shrinkage controls, overfitting and interaction depth control the tree depth.

CV Type	*nTree*	*InteractionDepth*	*Shrinkage*	*nMinObsInNode*
Random	5000	8	0.01	5
Blocked	5000	2	0.01	15

The HST expert generates the same $RMSE_{val}$ regardless of the CV used (Table 2) because there are no hyperparameters to optimize. The results over the validation set of the rest of experts vary among the devices. The results shown in Table 2 indicate that Blocked CV only gives a lower $RMSE_{val}$ than the random approach in 4 out of the 12 cases, excluding HST. In the case of pendulums 1, 2 and 3, the $RMSE_{val}$ generated by Blocked CV is generally higher. However, the models of pendulum 4 are more precise when trained with such CV.

Although the $RMSE_{val}$ of the Random CV is lower 8 out of 12 times, the difference between the two CV types is not significantly high, between 1 and 8% in absolute value. Furthermore, considering that it is better to give an estimation where more years are included, it is convenient to use the estimated error during the CV. Since the $RMSE_{cv}$ of the Random CV does not serve as a good estimator for future data, as the results shown in Table 2 indicate, Annual CV is recommended as the adequate strategy to estimate the error of the model and find the optimal hyperparameters.

3.2. *Experts Committee*

Once the decision of applying Blocked CV to train models was taken, Decision I in Figure 1, the relevant tests were performed to determine the best strategy of combination of experts.

The results of the Test I trainings, which consist of changing the Hold-out set used to train Blending, are shown in Table 4. The column titled Training Year indicates the year used as the Hold-out set to train the Blending model. $RMSE_{val}$ is the error measure in the validation set: 2017. The mean and variance of these values are presented as Mean and Variance for each device, respectively. On the other hand, the error committed by the Stacking meta-learner on the same validation set is represented by $RMSE_{valS}$.

The results (Table 4) reveal that the error depends on the Hold-out set used to train the Blending meta-learner. It reaches the maximum variance in pendulum 1, with a value of 0.868. The error has lower variation in the case of pendulum 2, where the minimum variance is found.

On the other hand, the mean error of Blending meta-learners is higher than the value of $RMSE_{valS}$ in most cases. It is only lower in the case of pendulum 4. However, looking at the errors made by the models trained with each Hold-out set, it is clear that Blending accuracy is higher than Stacking in most cases: 69.2% in pendulum 1, 61.5% in pendulum 3 and 76.9% in pendulum 4. Only in the case of pendulum 2 is there a lower error obtained with Stacking most of the time: 15.4% of cases.

These results were not sufficient to draw a solid conclusion on the issue addressed in this paper, although they do show the dependence of the meta-learner's error on the Hold-out set chosen for training. Hopefully, if the training year is similar to the validation set, the error will be low. The drawback lies in the uncertainty of this fact.

Therefore, further experiments were performed with different training sets and different validation sets (Test II) to make the final decision.

Table 4. Results of Test I, which consists of changing the training Hold-out set of Blending models.

Training Year	$RMSE_{val}$			
	Pendulum 1	Pendulum 2	Pendulum 3	Pendulum 4
2004	4.184	2.524	2.866	2.175
2005	1.342	1.257	1.090	1.203
2006	1.843	1.28	1.215	0.581
2007	2.232	1.622	1.611	1.534
2008	1.775	1.527	1.167	1.181
2009	1.793	1.583	1.313	1.168
2010	1.416	1.611	1.098	1.720
2011	1.485	1.357	1.021	0.576
2012	2.505	1.447	1.351	0.750
2013	3.336	1.086	0.937	0.810
2014	1.495	1.126	0.960	0.757
2015	1.112	1.213	0.902	1.198
2016	1.967	1.309	1.412	0.974
Mean	2.037	1.457	1.303	1.125
Variance	0.868	0.367	0.513	0.469
$RMSE_{valS}$	2.035	1.1813	1.257	1.257

The summary of the results obtained during Test II are shown in Table 5. The column named Available years indicates the number of years contained in the simulated training set. T years Stacking and experts represents the years used for the training of experts and the Stacking meta-learner. T year Blending shows the year used for Blending training, while V year is the validation set. $RMSE_{valS}$ and $RMSE_{valB}$ are the errors made by the Stacking and Blending meta-learners, respectively. The last column refers to the percentage improvement, reduction of the error (ER) of Stacking over Blending, calculated as follows:

$$ER = \frac{RMSE_{valB} - RMSE_{valS}}{RMSE_{valB}} \times 100 \quad (6)$$

The error of the meta-learner constructed through Stacking was lower than Blending in 60% of the cases, for different validation years. For pendulums 1 and 3, the error of Stacking was lower in 3 out of 5 validations (60%). In the case of pendulum 2, the maximum percentage is found, where Stacking is better in 80% of the cases. In contrast, this was met in only 40% of the validation sets in the case of pendulum 4. The results in Table 4 show that for this pendulum, Blending is better than Stacking 76.9% of the time over the validation of 2017.

The maximum advantage of Stacking over Blending is found in 2014 in pendulum 3, where the Stacking meta-learner is 36.1% better than Blending. Regarding the advantage of Blending over Stacking, its maximum is reached in the validation of 2006 in pendulum 1, where the error reduction is 47.6%.

The evident dependence of the Blending meta-learner result on the Hold-out set and the results obtained in Test II where, overall, Stacking gives a lower error, makes Stacking a more robust strategy. It might be due to the fact that Stacking strategy uses more data to train the model.

The final decision was taken, and the second-level models were finally trained following the Stacking strategy and using Blocked CV, where each block was one year.

The results in Table 6 show that for all pendulums, besides pendulum 1, the meta-learner reduces the error of the optimal expert. The maximum relative difference is reached at 14.48% in pendulum 2. Furthermore, regarding individual experts, BRT achieves the greatest accuracy for all pendulums, except pendulum 1, where NN is the most precise. Figure 11 shows parts of the series where the meta-learner performs better than the best expert.

Table 5. Results of Test II, which consists of training the experts, Stacking and Blending meta-learners, with different training sets. Positive values of ER imply a lower error on the Stacking meta-learner than Blending.

Pendulum	Available Years	T years Stacking and Experts	T Year Blending	V Year	$RMSE_{valS}$	$RMSE_{valB}$	ER
1	3 years: [2004–2006]	2 years: [2004–2005]	2005	2006	2.443	1.655	−47.6%
	5 years: [2004–2008]	4 years: [2004–2007]	2007	2008	0.960	0.988	2.8%
	6 years: [2004–2009]	5 years: [2004–2008]	2008	2009	1.454	1.317	−10.4%
	11 years: [2004–2014]	10 years: [2004–2013]	2013	2014	6.864	7.444	7.8%
	13 years: [2004–2016]	12 years: [2004–2015]	2015	2016	3.171	3.330	4.8%
2	3 years: [2004–2006]	2 years: [2004–2005]	2005	2006	23.830	28.095	15.2%
	5 years: [2004–2008]	4 years: [2004–2007]	2007	2008	1.389	1.152	−20.6%
	6 years: [2004–2009]	5 years: [2004–2008]	2008	2009	1.948	2.714	28.2%
	11 years: [2004–2014]	10 years: [2004–2013]	2013	2014	1.232	1.435	14.1%
	13 years: [2004–2016]	12 years: [2004–2015]	2015	2016	1.055	1.157	8.8%
3	3 years: [2004–2006]	2 years: [2004–2005]	2005	2006	3.736	3.488	−7.1%
	5 years: [2004–2008]	4 years: [2004–2007]	2007	2008	0.908	0.974	6.8%
	6 years: [2004–2009]	5 years: [2004–2008]	2008	2009	1.393	1.139	−22.4%
	11 years: [2004–2014]	10 years: [2004–2013]	2013	2014	1.048	1.639	36.1%
	13 years: [2004–2016]	12 years: [2004–2015]	2015	2016	1.238	1.347	8.1%
4	3 years: [2004–2006]	2 years: [2004–2005]	2005	2006	3.024	2.412	−25.4%
	5 years: [2004–2008]	4 years: [2004–2007]	2007	2008	0.892	0.797	−11.9%
	6 years: [2004–2009]	5 years: [2004–2008]	2008	2009	1.433	2.096	31.6%
	11 years: [2004–2014]	10 years: [2004–2013]	2013	2014	1.094	0.954	−14.7%
	13 years: [2004–2016]	12 years: [2004–2015]	2015	2016	1.088	1.241	12.3%

Table 6. Results of experts and meta-learners trained through Stacking strategy and Blocked CV. $D_{i,S}$ is the relative difference between the expert (i) and the second-level model (S).

Pendulum			**1**		
Model	**BRT**	**NN**	**RF**	**HST**	**Meta-Learner**
$RMSE_{cv}$	2.200	2.018	2.649	2.848	2.019
$D_{i,S}$	−9.02%	0.00%	−31.27%	−41.13%	−0.05%
			2		

Table 6. *Cont.*

Pendulum			1		
Model	BRT	NN	RF	HST	Meta-Learner
$RMSE_{cv}$	1.404	1.446	1.925	2.080	1.201
$D_{i,S}$	0.00%	−2.97%	−37.07%	−48.08%	14.48%
			3		
Model	BRT	NN	RF	HST	Meta-Learner
$RMSE_{cv}$	1.288	1.473	1.832	2.438	1.174
$D_{i,S}$	0.00%	−14.36%	−42.24%	−89.29%	8.85%
			4		
Model	BRT	NN	RF	HST	Meta-Learner
$RMSE_{cv}$	1.299	1.416	1.652	2.436	1.223
$D_{i,S}$	0.00%	−9.01%	−27.17%	−87.53%	5.85%

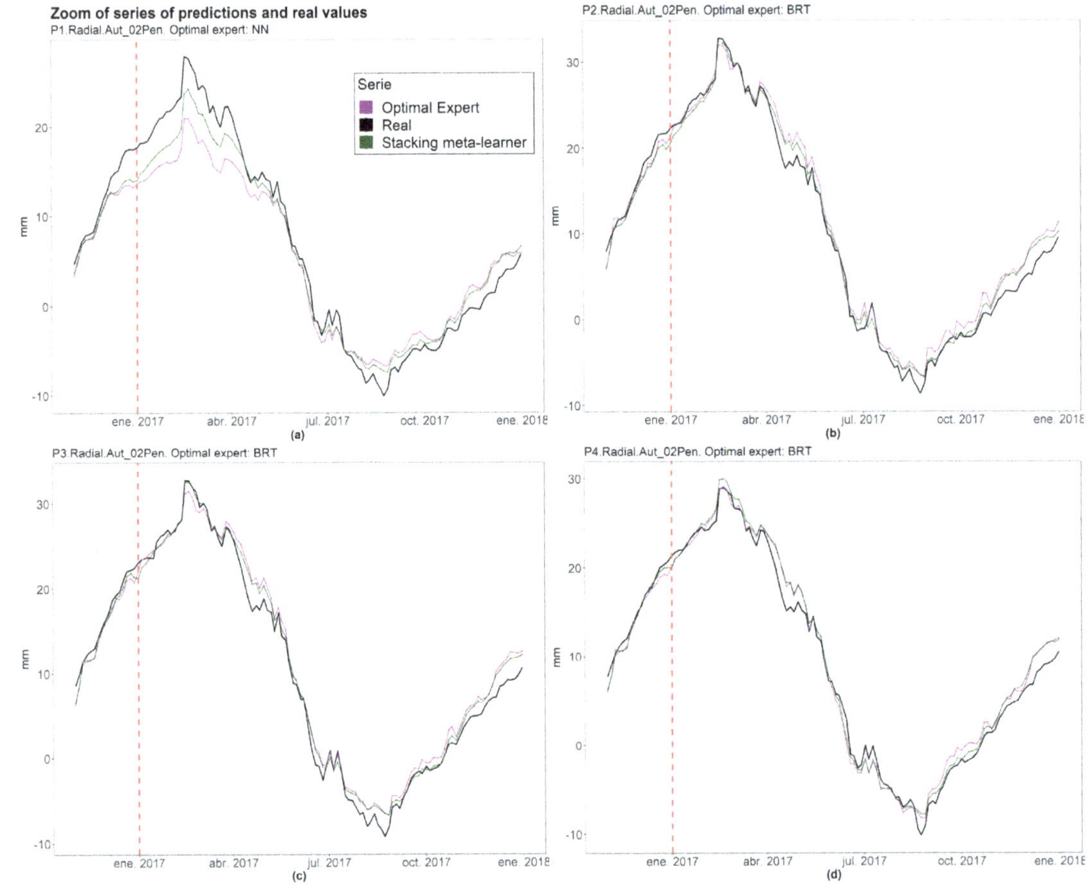

Figure 11. Zoom series of the optimal expert, the real value and the Stacking meta-learner. The red dashed line indicates the first date of validation. (**a**) Zoom series of pendulum 1; (**b**) Zoom series of pendulum 2; (**c**) Zoom series of pendulum 3; (**d**) Zoom series of pendulum 4.

Overall, these results suggest that a combination of experts can improve the optimal expert's precision through Stacking combination, and that Blocked CV gives the best estimate of the model error on future data compared to the random approach.

4. Conclusions

We presented a methodology that successfully combined experts to improve the accuracy and robustness of a machine learning model of the movement of a concrete dam. This paper provides new insights into the optimal strategy for performing combinations in the field of dam behavior and security. Furthermore, we highlighted the importance of the appropriate choice of the type of Cross Validation process.

Blocked CV was preferable to Random CV to estimate the model error on future data. It was observed that Random CV generates an error estimator significantly different from that obtained in the validation set, with an average difference (Equation (5)) of −210.56%, presumably due to time dependence, which makes it an unreliable strategy. On the contrary, the differences (Equation (5)) regarding Blocked CV have a mean value of −8.23%, which is significantly lower than Random CV. The RMSE values in the validation set for both types of CV are similar. Regarding the training of experts through Blocked CV, we achieved models with good prediction accuracy for all target variables, with a $RMSE_{cv}$ of the optimal experts lying within the range of 1.288 mm and 2.018 mm.

Stacking was considered a better strategy than Blending, since clear dependence of the Blending model on the Hold-out set used in the training was observed, with a variance value of up to 0.868 in the case of pendulum 1. Since a model trained using the Blending strategy involves using 10% or 20% of the data, the model is subject to the peculiarities of the year used in its training. The results in Table 4 emphasized that, by changing the Hold-out set in blending, the RMSE committed in the validation set significantly varies, and is higher, on average, than when adopting the Stacking strategy.

Regarding the results obtained by training experts and meta-learners on different sets and validated over different years, it was noted that the Stacking meta-learner was more accurate in most cases (60% on average). Consequently, Stacking was considered a more robust strategy for training second-level models, presumably due to superiority in the number of instances used for training.

Finally, comparing the series of predictions of the meta-learner built by generalized linear regression and the optimal expert, the second-level model improves the accuracy of the best expert in all the pendulums, with improvement percentages of up to 14.8%. Only one exception is found where accuracy is almost identical. Future research should aim to train the meta-learner through different algorithms using the Stacking strategy to determine the best meta-learner algorithm.

As a global conclusion, a methodology is proposed in which experts of different natures are trained using Blocked CV combined with a Stacking strategy.

Author Contributions: Conceptualization, P.A., M.Á.F.-C. and M.Á.T.; methodology, P.A.; software, P.A.; formal analysis, P.A., M.Á.F.-C. and M.Á.T.; investigation, P.A.; data curation, P.A.; writing—original draft preparation, P.A.; writing—review and editing, M.Á.F.-C. and M.Á.T.; visualization, P.A.; supervision, M.Á.T. All authors have read and agreed to the published version of the manuscript.

Funding: This research was funded by the project ARTEMISA (grant number RIS3-2018/L3-218), funded by the Regional Authority of Madrid (CAM) and partially funded by the European Regional Development Fund (ERDF) of the European Commission.

Institutional Review Board Statement: Not applicable.

Informed Consent Statement: Not applicable.

Data Availability Statement: The data are not available due to confidential reasons.

Acknowledgments: The authors thank Canal de Isabel II (CYII) for providing the data. In particular, we acknowledge the collaboration of David Galán from CYII.

Conflicts of Interest: The authors declare no conflict of interest.

Nomenclature

BRT	Boosted Regression Trees
CV	Cross Validation
D	matrix representation of the data set
D'	matrix form of prediction vectors of the experts
$D_{i,S}$	relative difference between the expert (i) and the second-level model (S)
ER	reduction of the error of Stacking over Blending
GLM	Generalized Linear Regression
HST	Hydrostatic-Seasonal-Time
interactionDepth	hyperparameter of BRT that controls the depth of the tree
k	number of experts
m	number of instances of the data set
n	number of explanatory variables
nMinObsInNode	hyperparameter of BRT representing the minimum number of observations to consider in one node
NN	Neural Networks
nTree	hyperparameter of BRT representing the number of trees
RD	relative difference between the Rooted Squared Error over CV and the Validation set
RF	Random Forest
RMSE	Root-Mean-Square Error
$RMSE_{cv}$	Root-Mean-Square Error over Cross Validation
$RMSE_{val}$	Root-Mean-Square Error over validation
$RMSE_{valB}$	RMSE of Blending approach in Test II
$RMSE_{valS}$	the RMSE committed by the Stacking meta-learner in the validation set
shrinkage	hyperparameter of BRT that controls overfitting
SVM	Support Vector Machine

References

1. Hariri-Ardebili, M.A.; Pourkamali-Anaraki, F. Support vector machine based reliability analysis of concrete dams. *Soil Dyn. Earthq. Eng.* **2018**, *104*, 276–295. [CrossRef]
2. Salazar, F.; Toledo, M.; Gonzalez, J.M.; Oñate, E. Early detection of anomalies in dam performance: A methodology based on boosted regression trees. *Struct. Control Health Monit.* **2017**, *24*, e2012. [CrossRef]
3. Tsihrintzis, G.A.; Virvou, M.; Sakkopoulos, E.; Jain, L.C. Machine Learning Paradigms Applications of Learning and Analytics in Intelligent Systems. Available online: http://www.springer.com/series/16172 (accessed on 13 April 2021).
4. Salazar, F.; Morán, R.; Toledo, M.; Oñate, E. Data-Based Models for the Prediction of Dam Behaviour: A Review and Some Methodological Considerations. *Arch. Comput. Methods Eng.* **2015**, *24*, 1–21. [CrossRef]
5. Salazar, F.; Toledo, M.; Oñate, E.; Suárez, B. Interpretation of dam deformation and leakage with boosted regression trees. *Eng. Struct.* **2016**, *119*, 230–251. [CrossRef]
6. Salazar, F.; González, J.M.; Toledo, M.Á.; Oñate, E. A Methodology for Dam Safety Evaluation and Anomaly Detection Based on Boosted Regression Trees. 2016. Available online: https://www.researchgate.net/publication/310608491 (accessed on 5 March 2020).
7. Salazar, F.; Toledo, M.; Oñate, E.; Morán, R. An empirical comparison of machine learning techniques for dam behaviour modelling. *Struct. Saf.* **2015**, *56*, 9–17. [CrossRef]
8. Rankovic, V.; Grujović, N.; Divac, D.; Milivojević, N. Development of support vector regression identification model for prediction of dam structural behaviour. *Struct. Saf.* **2014**, *48*, 33–39. [CrossRef]
9. Mata, J. Interpretation of concrete dam behaviour with artificial neural network and multiple linear regression models. *Eng. Struct.* **2011**, *33*, 903–910. [CrossRef]
10. Herrera, M.; Torgo, L.; Izquierdo, J.; Pérez-García, R. Predictive models for forecasting hourly urban water demand. *J. Hydrol.* **2010**, *387*, 141–150. [CrossRef]
11. Kang, F.; Li, J.; Zhao, S.; Wang, Y. Structural health monitoring of concrete dams using long-term air temperature for thermal effect simulation. *Eng. Struct.* **2018**, *180*, 642–653. [CrossRef]
12. Wolpert, D.H. Stacked generalization. *Neural Netw.* **1992**, *5*, 241–259. [CrossRef]
13. Wu, T.; Zhang, W.; Jiao, X.; Guo, W.; Hamoud, Y.A. Evaluation of stacking and blending ensemble learning methods for estimating daily reference evapotranspiration. *Comput. Electron. Agric.* **2021**, *184*, 106039. [CrossRef]

14. Roberts, D.R.; Bahn, V.; Ciuti, S.; Boyce, M.; Elith, J.; Guillera-Arroita, G.; Hauenstein, S.; Lahoz-Monfort, J.J.; Schroder, B.; Thuiller, W.; et al. Cross-validation strategies for data with temporal, spatial, hierarchical, or phylogenetic structure. *Ecography* **2017**, *40*, 913–929. [CrossRef]
15. Bergmeir, C.; Benítez, J.M. On the use of cross-validation for time series predictor evaluation. *Inf. Sci.* **2012**, *191*, 192–213. [CrossRef]
16. Džeroski, S.; Ženko, B. Is Combining Classifiers with Stacking Better than Selecting the Best One? *Mach. Learn.* **2004**, *3*, 255–273. [CrossRef]
17. Dou, J.; Yunus, A.P.; Bui, D.T.; Merghadi, A.; Sahana, M.; Zhu, Z.; Chen, C.-W.; Han, Z.; Pham, B.T. Improved landslide assessment using support vector machine with bagging, boosting, and stacking ensemble machine learning framework in a mountainous watershed, Japan. *Landslides* **2019**, *17*, 641–658. [CrossRef]
18. Mohd, L.; Gasim, S.; Ahmed, H.; Mohd, S.; Boosroh, H. Water Resources Development and Management ICDSME 2019 Proceedings of the 1st International Conference on Dam Safety Management and Engineering. Available online: http://www.springer.com/series/7009 (accessed on 13 April 2021).
19. Cheng, J.; Xiong, Y. Application of Extreme Learning Machine Combination Model for Dam Displacement Prediction. *Procedia Comput. Sci.* **2017**, *107*, 373–378. [CrossRef]
20. Bin, Y.; Hai-Bo, Y.; Zhen-Wei, G. A Combination Forecasting Model Based on IOWA Operator for Dam Safety Monitoring. In Proceedings of the 2013 5th Conference on Measuring Technology and Mechatronics Automation, ICMTMA 2013, Hong Kong, China, 16–17 January 2013; pp. 5–8. [CrossRef]
21. Wei, B.; Yuan, D.; Li, H.; Xu, Z. Combination forecast model for concrete dam displacement considering residual correction. *Struct. Health Monit.* **2017**, *18*, 232–244. [CrossRef]
22. Wei, B.; Yuan, D.; Xu, Z.; Li, L. Modified hybrid forecast model considering chaotic residual errors for dam deformation. *Struct. Control Health Monit.* **2018**, *25*, e2188. [CrossRef]
23. Hong, J.; Lee, S.; Bae, J.H.; Lee, J.; Park, W.J.; Lee, D.; Kim, J.; Lim, K.J. Development and Evaluation of the Combined Machine Learning Models for the Prediction of Dam Inflow. *Water* **2020**, *12*, 2927. [CrossRef]
24. Pedregosa, F.; Varoquaux, G.; Gramfort, A.; Michel, V.; Thirion, B.; Grisel, O.; Blondel, M.; Prettenhofer, P.; Weiss, R.; Dubourg, V.; et al. Scikit-learn: Machine learning in Python. *J. Mach. Learn. Res.* **2011**, *12*, 2825–2830.
25. Chai, T.; Draxler, R.R. Root mean square error (RMSE) or mean absolute error (MAE)?—Arguments against avoiding RMSE in the literature. *Geosci. Model Dev.* **2014**, *7*, 1247–1250. [CrossRef]
26. Berrar, D. Cross-Validation. In *Encyclopedia of Bioinformatics and Computational Biology: ABC of Bioinformatics*; Elsevier: Amsterdam, The Netherlands, 2018; Volume 1–3, pp. 542–545. [CrossRef]

water

MDPI

Article

Influences on the Seismic Response of a Gravity Dam with Different Foundation and Reservoir Modeling Assumptions

Chen Wang [1], Hanyun Zhang [2,*], Yunjuan Zhang [1], Lina Guo [3], Yingjie Wang [1] and Thiri Thon Thira Htun [1]

1 College of Water Conservancy and Hydropower Engineering, Hohai University, Nanjing 210098, China; iwangchen@hhu.edu.cn (C.W.); zhangyunjuan@hhu.edu.cn (Y.Z.); 201302020065@hhu.edu.cn (Y.W.); thirithonthonhtun@gmail.com (T.T.T.H.)
2 State Key Laboratory of Simulation and Regulation of Water Cycle in River Basin, Beijing 100048, China
3 Development Research Center, The Ministry of Water Resource of P.R. China, Beijing 100038, China; guolina@waterinfo.com.cn
* Correspondence: zhanghanyun@hhu.edu.cn; Tel.: +86-137-7095-7562

Abstract: The seismic design and dynamic analysis of high concrete gravity dams is a challenge due to the dams' high levels of designed seismic intensity, dam height, and water pressure. In this study, the rigid, massless, and viscoelastic artificial boundary foundation models were established to consider the effect of dam–foundation dynamic interaction on the dynamic responses of the dam. Three reservoir water simulation methods, namely, the Westergaard added mass method, and incompressible and compressible potential fluid methods, were used to account for the effect of hydrodynamic pressure on the dynamic characteristics and seismic responses of the dam. The ranges of the truncation boundary of the foundation and reservoir in numerical analysis were further investigated. The research results showed that the viscoelastic artificial boundary foundation was more efficient than the massless foundation in the simulation of the radiation damping effect of the far-field foundation. It was found that a foundation size of 3 times the dam height was the most reasonable range of the truncation boundary of the foundation. The dynamic interaction of the reservoir foundation had a significant influence on the dam stress.

Keywords: hydraulic structure; high gravity dams; dam-foundation-reservoir dynamic interaction; earthquake input mechanisms; hydrodynamic pressure; foundation size; reservoir length

Citation: Wang, C.; Zhang, H.; Zhang, Y.; Guo, L.; Wang, Y.; Thira Htun, T.T. Influences on the Seismic Response of a Gravity Dam with Different Foundation and Reservoir Modeling Assumptions. *Water* **2021**, *13*, 3072. https://doi.org/10.3390/w13213072

Academic Editors: Miguel Á. Toledo, Rafael Morán and Paolo Mignosa

Received: 8 August 2021
Accepted: 29 October 2021
Published: 2 November 2021

1. Introduction

Concrete gravity dams have received increasing attention in recent years because of their reliable structures, simple design and construction techniques, and high adaptability to topographic and geological conditions. Several concrete gravity dams of over 200 m in height are planned to be constructed in high seismic regions of Western China. However, it is still challenging to deal with problems such as high levels of designed seismic intensity, dam height, and high water pressure in the seismic design and seismic response analysis of concrete gravity dams.

The dam–foundation dynamic interaction needs to be carefully considered in the seismic response analysis of concrete gravity dams. Ghaedi et al. [1] compared the acceleration, displacement, stress, and dynamic damage of the 81.8 m high Kinta roller compacted concrete (RRC) gravity dam in models of the dam, dam–reservoir, and dam–foundation–reservoir, and the results showed that foundation flexibility significantly affected the seismic response of the RCC dam–reservoir–foundation system. Bayraktar et al. [2] investigated the effect of base-rock characteristics on the dynamic response of dam–foundation interaction systems subjected to three different earthquake input mechanisms, and the simulation results with a 90 m high concrete gravity dam showed that the rigid-base input model was inadequate to describe the dynamic interaction of dam–foundation systems, whereas the massless foundation input model could be used for practical analysis.

Ghaemian et al. [3] compared the relative crest displacement and principal stress of the 103 m high Koyna concrete gravity dam between rigid, massless, and massed foundation models, and concluded that the massless foundation input model overestimated the dam dynamic response. Salamon et al. [4] compared the dam horizontal acceleration responses between the massless foundation model and the massed foundation model. The results revealed that the horizontal acceleration response at the dam crest obtained from a model with the massless foundation was about 1.5 times greater than the response from the model with the mass foundation. Chopra [5] revealed that seismic demands are considerably overestimated by assuming the foundation rock to be massless. Hariri-Ardebili et al. [6] investigated the seismic responses of the dam under near-fault and far-field ground motion using three different types of foundation models, and revealed that considering the radiation damping (massed foundation) decreases the response values compared to the standard massless model. Hariri-Ardebili et al. [7] also compared the seismic responses of coupled arch dam–reservoir–foundation systems with three types of foundation model. The results showed that the smaller seismic responses obtained from the massed foundation model compared with the massless foundation model, and the stress responses obtained from either viscous boundary model or infinite elements model, were quite similar. Burman et al. [8] presented a simplified direct method incorporating the effect of soil–structure interaction (SSI), and carried out a time domain transient analysis of a concrete gravity dam and its foundation in a coupled status. The 3D full dam models with different foundation densities were used to analyze the seismic responses of a concrete gravity dam [9], and the results indicated that the dynamic interaction between the dam and the foundation significantly reduced the responses of the monoliths on the river bed but increased the responses of the monoliths on the steep slopes of both banks.

Both foundation and dam are included in finite element models to analyze the seismic response of the concrete gravity dams considering dam–foundation dynamic interaction, but significant differences exist regarding the boundary simulated methods and the foundation range in the previous research. The radiation damping of an infinite foundation is an important factor affecting the structure–foundation interaction, which can be simulated by setting artificial boundary conditions at the foundation truncation. Various artificial boundaries have been proposed to date, such as viscous boundary [10,11], viscous-spring boundary [12,13], scaled boundary finite element method [14,15], infinite elements [7,16], and perfectly matched layers [17]. Hariri-Ardebili and Saouma [18] investigated the effect of different foundation numerical models and corresponding boundary conditions on the seismic responses of arch dam–foundation systems under near-fault and far-field ground motions. The results indicated that the massed foundation model with infinite elements at far-end boundaries would be a more appropriate method than the massless model, and the rigid foundation model would not be suitable for simulating the seismic behavior of arch dams. Pan et al. [19] proposed that installing a series of viscous dampers at the dam–foundation interface in the massless foundation model could accurately simulate the seismic response of the gravity dam. Salamon et al. [4] compared the seismic responses of the Pine Flat dam under free-field boundary and non-reflection boundary conditions, and the results showed that the free-field boundary condition was essential to obtain realistic ground motions. Chen et al. [20] investigated the influences of two boundary conditions (the viscous-spring boundary and the viscous boundary) in their earthquake input models on the seismic analysis of the Pine Flat and Jin'anqiao gravity dam–foundation–reservoir systems, and the results revealed that the agreement between the two boundary conditions was good. Wang et al. [21] investigated the seismic damage development and potential failure pattern of the 142 m high Guandi concrete gravity dam using incremental dynamic analysis (IDA), in which the massless foundation model was used to simulate the dam–foundation dynamic interaction, and the truncation boundary of the foundation was set to 1.5 times the dam height in the upstream and downstream directions, and 2 times the dam height in the depth direction. Wang et al. [22] studied the seismic duration effect of the gravity dam–foundation–reservoir system under horizontal and vertical ground motions using the Koyna gravity dam as a numerical example. It was noted that the truncation boundary of the foundation was set to 2 times the dam height in the upstream direction, and 1 times the dam height in the downstream and depth directions. Gorai and Maity [23] investigated the seismic response of the concrete gravity dam–reservoir–foundation system to near- and far-field ground motions, also using the Koyna gravity dam as a numerical example. Here, the truncation boundary of the foundation was set to 1 times the width at the dam base in the upstream and downstream directions, and 3 times the width at the dam base in the depth direction. Chen and Yang et al. [24] studied the damage process and potential damage modes of the 112 m high Jin'anqiao concrete gravity dam under seismic loads with different peak accelerations. The viscoelastic artificial boundary conditions and corresponding free-field input mechanisms were introduced to account for the dam–foundation dynamic interaction and, in this case, the truncation boundary of the foundation extended 3 times the dam height in each direction. Salamon et al. [4] revealed that the variation of foundation length had a very limited influence on seismic responses when the free-field boundary condition was used. To locate the free field boundaries, Asghari et al. [25] modeled and analyzed several models with various foundation sizes from 2 to 10 h (where h is the height of the dam) in all directions, and the results showed that 5 h can be interpreted as the relatively appropriate distance for truncating the boundaries. The foundation was assumed to be massless [26], and the far-end boundary of the foundation was at a distance from the dam of about 2 times the dam height in all directions.

The hydrodynamic pressure is another key factor that should also be considered in the seismic response analysis of concrete gravity dams. Westergaard [27] assumed that the dam and foundation were rigid and then derived formulas of added masses for hydrodynamic pressures on the vertical upstream face of the dam. Chopra [5] revealed that using added mass to simulate hydrodynamic effects ignores the water compressibility, which would lead to unreliable decisions in the seismic analysis of concrete dams. Khiavi et al. [28] investigated the hydrodynamic response of a concrete gravity dam and reservoir under vertical vibration using an analytical method. Amina et al. [29] conducted a series of modal analyses of the Brezina concrete arch dam based on the Lagrangian and added mass approaches. The results indicated that the higher coupled frequencies would be obtained from the added mass approach as compared to the actual ones, whereas the more approximate coupled frequencies would be obtained from the Lagrangian approach. Altunisik and Sesli [30] used three different reservoir water modelling methods—Westergaard, Lagrange and Euler—to calculate the dynamic hydrodynamic pressures on the 90 m high Sariyar gravity dam. The reservoir length was 3 times the dam height in both the Lagrange and Euler methods. It was concluded that more general results could be obtained by the Westergaard method, whereas the results obtained by the Lagrange and Euler methods were closer to the actual behaviors of the dam. The Eulerian approach for hydrodynamic pressures was used to obtain the seismic performance of concrete gravity dam structures in their research [31]. Bayraktar et al. [32] investigated the effect of reservoir length (1–4 times the dam height) on the seismic response of the 82.45 m high Folsom gravity dam to near- and far-fault ground motions using the Lagrange method. Given the similar maximum principal tensile stress and performance curves for 3 and 4 times the dam height, a reservoir length of 3 times the dam height is sufficient to evaluate the seismic performance of concrete gravity dams. Kartal et al. [33] arrived at a similar conclusion for the cases of linear and non-linear analysis of a 2D roller-compacted concrete dam. They showed that the reservoir with the length of 3 times the dam height was adequate to assess the seismic response of RCC dams. Moreover, Hariri-Ardebili et al. [6] claimed that the reservoir with a length of 3 times the dam height may be the computationally optimal model. According to the hydrodynamic pressure distribution of the upstream dam surface under seismic load in different reservoir length models, Pelecanos et al. [34] revealed that, for concrete gravity dams, the upstream reservoir length should be 5 times the height of the reservoir.

Studies have also been conducted on the seismic response of concrete gravity dams in which dam–foundation–reservoir dynamic interactions were simultaneously considered. Mandal et al. [35] proposed a two-dimensional direct coupling method for the linear dynamic response analysis of the dam–foundation–reservoir system considering both soil-structure and fluid–structure interactions. They also concluded that the dynamic responses of these respective subsystems would be affected by the dam–foundation–reservoir interaction. Løkke and Chopra [36] presented a direct finite element method for nonlinear earthquake analysis of concrete dams interacting with the fluid and foundation, where the semi-unbounded fluid and foundation domains were truncated by absorbing boundaries with viscous dampers. This direct finite element method for earthquake analysis of dam–reservoir–foundation systems was simplified for easy implementation in commercial finite element software [37]. Chopra [38] revealed that the semi-unbounded size of the reservoir and foundation–rock domains, dam–foundation interaction, dam–reservoir interaction, water compressibility, hydrodynamic wave absorption at the reservoir boundary, and spatial variations in ground motion at the dam–rock interface should be included in the earthquake analysis of arch dams. A comprehensive procedure was proposed to analyze the nonlinear earthquake response of arch dams [39], and the following factors were considered: dynamic dam–reservoir and dam–foundation interactions, the semi-unbounded size of the foundation, compressible water, the opening of contraction joints, the cracking of the dam body, and the spatial variation of ground motions. Wang et al. [40] developed a nonlinear analysis procedure for earthquake response analysis of arch dam–reservoir–foundation systems, and the effects of the earthquake input mechanism, joint opening, water compressibility, and radiation damping on the earthquake response of the Ertan arch dam were analyzed using the proposed procedure. The results showed that such factors should be considered in the earthquake safety evaluation of high arch dams. Amini et al. [41] revealed that the consideration of dam–reservoir–foundation interaction in nonlinear analysis of concrete dams is of great importance.

Despite numerous studies on the seismic response of concrete gravity dams, it should also be noted that a wide variety of models have been developed to simulate dam–foundation, dam–reservoir, and dam–foundation–reservoir dynamic interactions, and no consensus has been reached on the foundation and reservoir water simulation methods and ranges of the truncation boundary of the foundation and reservoir in numerical analysis. In this study, both dam–foundation and dam–reservoir dynamic interactions were considered in the seismic response of concrete gravity dams. A 203 m high concrete gravity dam in Southwest China was taken as the numerical example, and rigid, massless, and viscoelastic artificial boundary foundation models were established to account for the effect of dam–foundation dynamic interactions on the dynamic characteristics and seismic response of the dam. Three reservoir water simulation methods, namely, the Westergaard added mass method, the incompressible potential fluid method, and the compressible potential fluid method, were used in the massless foundation model and the viscoelastic artificial boundary model to account for the effect of dam–foundation–reservoir dynamic interactions on the dynamic characteristics and seismic responses of the dam. The ranges of the truncation boundary of the foundation and reservoir in finite element models were further investigated.

2. Hydrodynamic Pressure Modelling Approaches

2.1. Westergaard Added Mass Method

Westergaard [27] derived a theoretical solution to simulate the hydrodynamic pressure of reservoir water using added mass, which was later improved by Clough [42] in 1982. The generalized Westergaard formula can be expressed as Equation (1). It is applicable to arbitrarily shaped surfaces subjected to hydrodynamic pressure, and the seismic acceleration in three directions can be considered in this formula.

$$M_{\alpha i} = \frac{7}{8}\rho_w A_i \sqrt{H_i(H_i - Z_i)}\lambda_i^T \lambda_i \tag{1}$$

where i is a node on the structural surface subjected to hydrodynamic pressure, ρ_w is the mass density of water, A_i is the effective area of i, H_i is the total water depth of the vertical surface at which i is located, Z_i is the height from i to the bottom of the structural surface subjected to hydrodynamic pressure, and λ_i is the normal vector of i, $\lambda_i = \{\lambda_{ix}, \lambda_{iy}, \lambda_{iz}\}$.

2.2. Potential-Based Fluid Formulation

The following assumptions and constraints are made for the potential-based fluid elements in ADINA (Automatic Dynamic Incremental Nonlinear Analysis, a finite element analysis program) [43]: inviscid, irrotational flow with no heat transfer; slightly compressible or almost incompressible flow; relatively small displacement of the fluid boundary; actual fluid flow with velocities below the sound speed or no actual fluid flow. The structure–fluid interaction is described as follows.

The finite element equation of motion for low velocity fluid is expressed as:

$$\begin{bmatrix} 0 & 0 \\ 0 & -M_{FF} \end{bmatrix}\begin{bmatrix} \Delta\ddot{u} \\ \Delta\ddot{\phi} \end{bmatrix} + \begin{bmatrix} C_{UU} & C_{UF} \\ C_{FU} & -(C_{FF} + (C_{FF})_S) \end{bmatrix}\begin{bmatrix} \Delta\dot{u} \\ \Delta\dot{\phi} \end{bmatrix} + \begin{bmatrix} K_{UU} & K_{UF} \\ K_{FU} & -(K_{FF} + (K_{FF})_S) \end{bmatrix}\begin{bmatrix} \Delta u \\ \Delta\phi \end{bmatrix} = \begin{bmatrix} 0 \\ 0 \end{bmatrix} - \begin{bmatrix} F_U \\ F_F + (F_F)_S \end{bmatrix} \tag{2}$$

where Δu is the unknown displacement vector increment; $\Delta\phi$ is the increment of the unknown potential vector; M_{FF} is the fluid element mass matrix; $C_{UU}, C_{FU}, C_{UF}, C_{FF}$ are the damping matrices of the structure, the fluid caused by the structure, the structure caused by the fluid, and the fluid on the fluid–solid coupling interface, respectively; K_{UU} , K_{FU}, K_{UF}, K_{FF} are the stiffness matrices of the structure, the fluid caused by the structure, the structure caused by the fluid, and the fluid on the fluid–solid coupling interface, respectively; $F_U, F_F, (F_F)_S$ are the fluid pressure on the structure boundary, the volume integral term, and the area integral term corresponding to the fluid continuity equation, respectively.

Equation (2) does not include any structural system matrices, and it only gives the contribution of the potential-based fluid elements to the system matrices. The contribution of the structural term is added to Equation (2) to obtain the finite element equation of motion for fluid–structure interaction, as follows:

$$\begin{bmatrix} M_{SS} & 0 \\ 0 & -M_{FF} \end{bmatrix}\begin{bmatrix} \Delta\ddot{u} \\ \Delta\ddot{\phi} \end{bmatrix} + \begin{bmatrix} C_{UU} + C_{SS} & C_{UF} \\ C_{FU} & -(C_{FF} + (C_{FF})_S) \end{bmatrix}\begin{bmatrix} \Delta\dot{u} \\ \Delta\dot{\phi} \end{bmatrix} + \begin{bmatrix} K_{UU} + K_{SS} & K_{UF} \\ K_{FU} & -(K_{FF} + (K_{FF})_S) \end{bmatrix}\begin{bmatrix} \Delta u \\ \Delta\phi \end{bmatrix} = \begin{bmatrix} 0 \\ 0 \end{bmatrix} - \begin{bmatrix} F_U \\ F_F + (F_F)_S \end{bmatrix} \tag{3}$$

In Equation (3), the structural element matrix of mass, damping, and stiffness, and the load vector, can be defined as:

$$M_{SS} = \int_{V_S} \rho_S \overline{N}^T \overline{N} dV, C_{SS} = \alpha M_{SS} + \beta K_{SS}, K_{SS} = \int_{V_S} B^T D B dV, F_S = \int_{V_S} \overline{N}^T P dV + \int_S \overline{N}^T \overline{T} dS \tag{4}$$

where V_S is the solid region of the calculation; ρ_s is the density of the solid region; $\overline{N}$ is the nodal shape function of the solid region; α and β are the structural mass and stiffness matrix coefficients, respectively; B and D are the displacement-strain matrix and the elastic stiffness matrix of the solid region, respectively; P, $\overline{T}$, and S are the physical force, surface force and boundary surface of the solid region, respectively.

3. Viscoelastic Artificial Boundary and Earthquake Input Mechanisms

The vibration energy of a dam subjected to earthquake will propagate through the infinite foundation to the far field, causing a radiation damping effect on the dynamic characteristics of the dam. In this study, the radiation damping effect is simulated by imposing the viscoelastic artificial boundary condition at the foundation truncation, and then converting the displacement and velocity time history of seismic wave motion into equivalent nodal loads applied to the viscoelastic artificial boundary to complete the input of ground motion.

3.1. Viscoelastic Artificial Boundary Condition

In this paper, the viscoelastic artificial boundary condition is implemented using linear spring-damping elements in ADINA. Figure 1 shows a schematic diagram of a two-dimensional spring-damping element. The stiffness coefficient of the spring element and the damping coefficient of the damper element are:

$$\begin{cases} K_N = \alpha_N \frac{G}{r} \sum A_i, & C_N = \rho c_p \sum A_i \\ K_T = \alpha_T \frac{G}{r} \sum A_i, & C_T = \rho c_s \sum A_i \end{cases} \quad (5)$$

where K_N and K_T are the normal and tangential stiffness of the spring, respectively; C_N and C_T are the normal and tangential damping coefficients of the damper, respectively; α_N and α_T are the normal and tangential correction coefficients of the viscoelastic artificial boundary, respectively, which are set to $\alpha_N = 1.0$ and $\alpha_T = 0.5$; c_p and c_s are the wave velocities of the P-wave and S-wave, respectively; G and ρ are the shear modulus and mass density of the medium, respectively; r is the distance between the wave source and the node on the viscoelastic artificial boundary; $\sum A_i$ is the effective area of the node on the viscoelastic artificial boundary, which usually is the effective length for a two-dimensional finite element model.

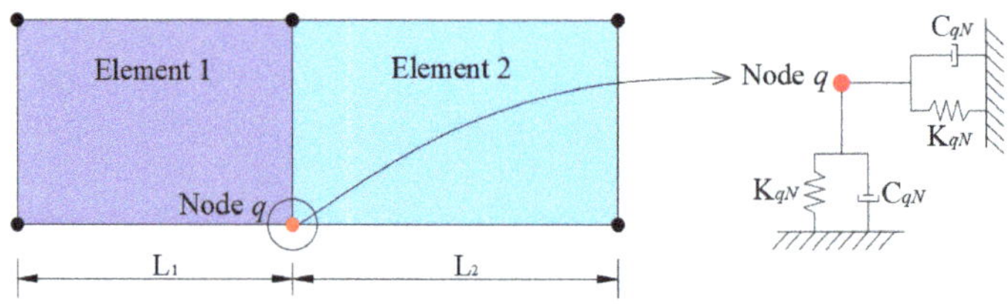

Figure 1. Schematic diagram of a two-dimensional spring-damping element.

3.2. Earthquake Input Mechanisms

According to the characteristics of wave fields on different viscoelastic artificial boundaries, the total wave field on the bottom boundary is decomposed into an incident field and a scattered field, whereas that on the side boundary is decomposed into a free field and a scattered field. The energy of the scattered field is absorbed by the viscoelastic artificial boundaries, whereas that of incident and free fields can be transformed into equivalent nodal loads and then applied to the boundaries. Figure 2 shows a schematic diagram of wave input for a viscoelastic artificial boundary.

The displacement, velocity, and acceleration time history of wave fields are expressed as $u^i(t)$, $\dot{u}^i(t)$, and $\ddot{u}^i(t)$, respectively, in which $i = m$ denotes the total wave field, $i = r$ denotes the incident wave field, $i = f$ denotes the free wave field, and $i = s$ denotes the scattered field. According to the displacement continuity condition and the mechanical equilibrium condition, the motion equation for node q on the bottom boundary can be expressed as:

$$M_q \ddot{u}_q^m(t) + C_q \dot{u}_q^m(t) + K_q u_q^m(t) = F_q^r(t) + F_q^s(t) \quad (6)$$

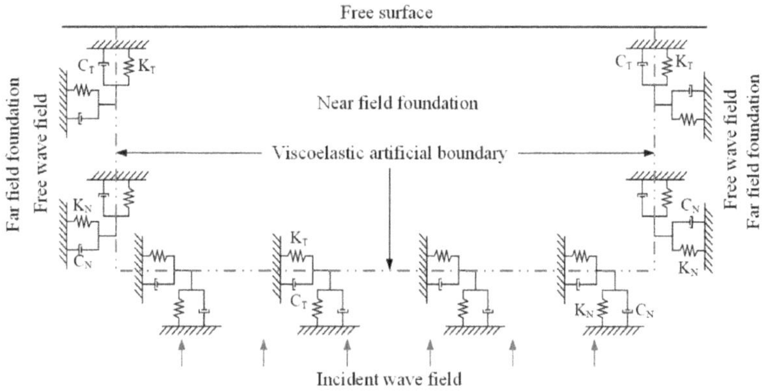

Figure 2. Schematic diagram of wave input for a viscoelastic artificial boundary.

The motion equation for node q on the side boundary can be expressed as:

$$M_q\ddot{u}_q^m(t) + C_q\dot{u}_q^m(t) + K_q u_q^m(t) = F_q^f(t) + F_q^s(t) \tag{7}$$

where K_q and C_q are the artificial boundary parameters of q; $F_q^r(t)$, $F_q^f(t)$ and $F_q^s(t)$ are the equivalent nodal loads to be applied at q to simulate the incident, free and scattered wave field, respectively. For seismic wave motion input, only equivalent nodal loads $F_q^r(t)$ and $F_q^f(t)$ need to be applied to the bottom and side boundary, which are solved using Equations (6) and (7) based on the seismic wave motion propagation pattern and the stress state of wave fields, respectively. The equivalent nodal loads that should be applied at q are calculated as follows:

When the primary wave is incident:

$$\begin{cases} F_{qz}^{r-z}(t) = A_q\left[K_{qN}u_{qp}^r(t) + C_{qN}\dot{u}_{qp}^r(t) + \rho c_p\dot{u}_{qp}^r(t)\right] \\ F_{qy}^{f-y}(t) = A_q\frac{\lambda}{c_p}\left[\dot{u}_{qp}^r(t-\Delta t_1) - \dot{u}_{qp}^r(t-\Delta t_2)\right] \\ F_{qz}^{f-y}(t) = A_q\left[K_{qT}u_{qp}^r(t-\Delta t_1) + C_{qT}\dot{u}_{qp}^r(t-\Delta t_1) + K_{qT}u_{qp}^r(t-\Delta t_2) + C_{qT}\dot{u}_{qp}^r(t-\Delta t_2)\right] \\ F_{qy}^{f-y}(t) = -F_{qy}^{f+y}(t) \quad , \quad F_{qz}^{f-y}(t) = F_{qz}^{f+y}(t) \end{cases} \tag{8}$$

When the shear wave vibrating along the Y-axis is incident:

$$\begin{cases} F_{qy}^{r-z}(t) = A_q\left[K_{qT}u_{qs}^r(t) + C_{qT}\dot{u}_{qs}^r(t) + \rho c_s\dot{u}_{qs}^r(t)\right] \\ F_{qy}^{f-y}(t) = A_q\left[K_{qN}u_{qs}^r(t-\Delta t_3) + C_{qN}\dot{u}_{qs}^r(t-\Delta t_3) + K_{qN}u_{qs}^r(t-\Delta t_4) + C_{qN}\dot{u}_{qs}^r(t-\Delta t_4)\right] \\ F_{qz}^{f-y}(t) = A_q\rho c_s\left[\dot{u}_{qs}^r(t-\Delta t_3) - \dot{u}_{qs}^r(t-\Delta t_4)\right] \\ F_{qy}^{f+y}(t) = F_{qy}^{f-y}(t) \quad , F_{qz}^{f+y}(t) = -F_{qz}^{f-y}(t) \end{cases} \tag{9}$$

in which

$$\Delta t_1 = h/c_p, \qquad \Delta t_2 = 2H_s - h/c_p, \qquad \Delta t_3 = h/c_s, \qquad \Delta t_4 = 2H_s - h/c_s \tag{10}$$

where ρ, c_p, c_s, λ are the foundation density, P-wave velocity, S-wave velocity, and Lame constant, respectively; H_s is the vertical distance from the wave source to the bottom boundary; h is the vertical distance from q to the bottom boundary; Δt_1, Δt_2 , Δt_3, and Δt_4 are the time delay of the incident P-wave at q, the reflected P-wave at the foundation surface, the incident S-wave at q, and the reflected S-wave at the foundation surface, respectively. The subscripts of the equivalent nodal loads represent the node number and component direction, and the superscripts represent the wave field for calculating the equivalent nodal loads and the outer normal direction of the boundary surface at which q is located, which is positive if the direction is the same as the coordinate axis and negative if the direction is opposite to the coordinate axis.

3.3. Verification Test

The viscoelastic artificial boundary is verified by a two-dimension test [44]. As shown in Figure 3, the model size is 905.5 m × 370 m, and the finite element mesh size is 5 m × 10 m. The modulus of elasticity of the medium is 1.05×10^{10} N/m^2, the mass density is 2777 kg/m^3, the Poisson's ratio is 0.23, the S-wave velocity is 1239.8 m/s, and the P-wave velocity is 2093.6 m/s. The dynamic time-history analysis is performed with a total calculation time of 1 s and a time step of 0.01 s. The input displacement, velocity, and acceleration time history are determined by Equations (11)–(13) respectively, and their time-history curves are shown in Figure 4.

$$u(t) = \begin{cases} \frac{t}{2} - \frac{\sin(2\pi f t)}{4\pi f t} & 0 \le t \le 0.25 \\ 0.125 & t > 0.25 \end{cases} \tag{11}$$

$$v(t) = \begin{cases} \frac{1}{2} - \frac{\cos(2\pi f t)}{2} & 0 \le t \le 0.25 \\ 0 & t > 0.25 \end{cases} \tag{12}$$

$$a(t) = \begin{cases} \pi f \sin(2\pi f t) & 0 \le t \le 0.25 \\ 0 & t > 0.25 \end{cases} \tag{13}$$

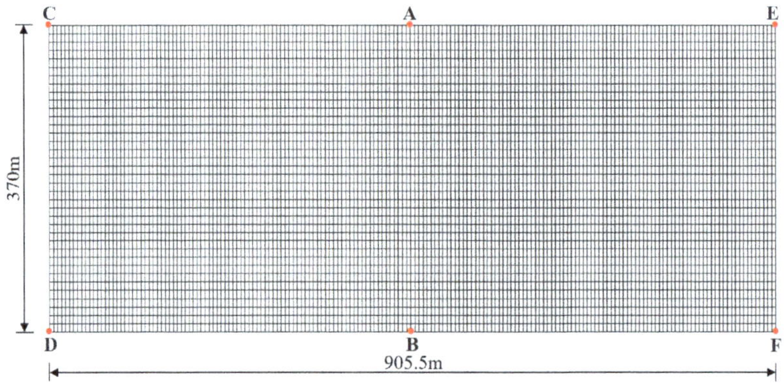

Figure 3. The finite element model of the verification test.

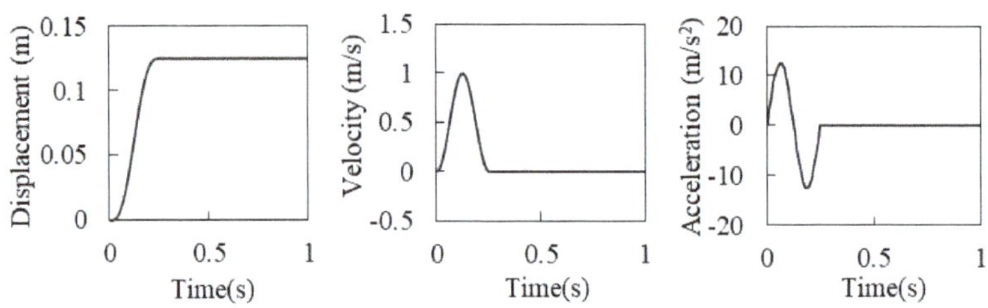

Figure 4. The input time–history curves of displacement, velocity, and acceleration.

The time histories of displacement, velocity, and acceleration at each observation point (A and B, C and D, E and F in Figure 3) are calculated. Figure 5 shows the time history curves of horizontal displacement, velocity, and acceleration at position A on the free surface and position B on the bottom boundary directly below position A. It is seen that the seismic wave is input from the bottom truncated boundary, and the amplitude of the seismic wave is doubled when the incident wave reaches the free surface. The seismic wave reflected from the free surface is absorbed by the viscoelastic boundary after reaching the bottom surface, without reflection on the truncated boundary. Similar dynamic responses are observed at other observation positions (C and D, E and F). These results indicate that the viscoelastic artificial boundary condition and the corresponding earthquake input mechanism are feasible.

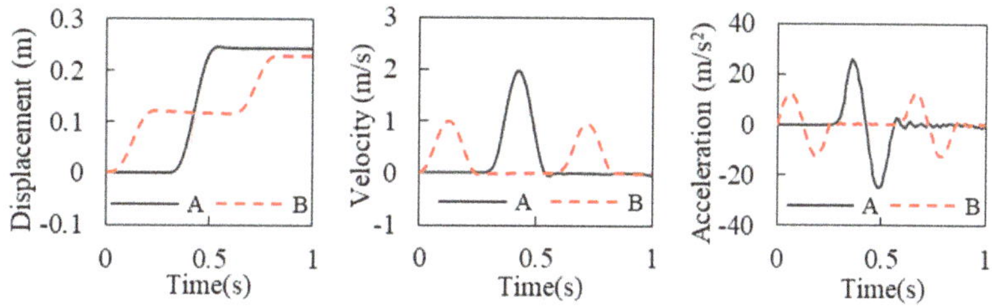

Figure 5. The time–history curves of horizontal displacement, velocity, and acceleration.

4. General Description of the Numerical Example

4.1. General Information

A concrete gravity dam in Southwest China was selected as the study case. The concrete gravity dam block has a crest elevation of 1625 m, a heel elevation of 1422 m, a dam height of 203 m, a crest width of 16 m, and a normal water level of 1619 m. The calculation cross-section of the dam block is shown in Figure 6. In this study, the following material parameters were considered. Dam: static modulus of elasticity = 2.5×10^{10} N/m^2, mass density = 2400 kg/m^3, Poisson's ratio = 0.167 according to Design Code for Hydraulic Concrete Structures (SL 191-2008) [45], structural damping = 5%. Foundation: modulus of elasticity = 1.5×10^{10} N/m^2, mass density = 2700 kg/m^3, Poisson's ratio = 0.24. Reservoir: mass density of water = 1000 kg/m^3, acoustic wave speed = 1440 m/s.

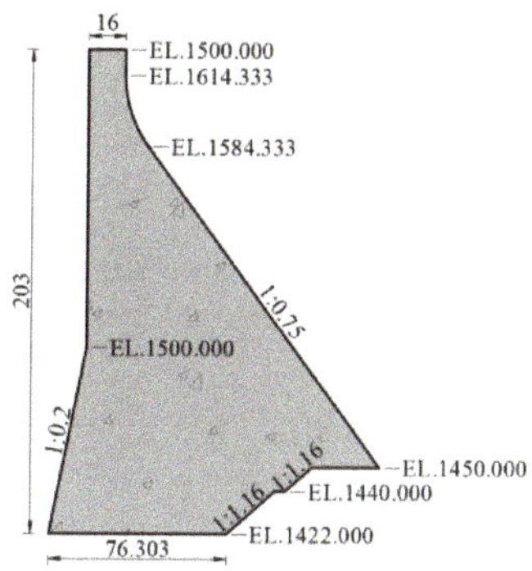

Figure 6. Geometry of the dam (m).

The geometry of the dam–foundation–reservoir system is shown in Figure 7. The mesh size of the two-dimension finite element model was controlled to be about 2 m × 2 m. The dam was simulated with 4640 plane stress elements, and the foundation was simulated with 207,900 plane strain elements. Both static and dynamic effects were taken into account by the static–dynamic superposition method. The loads include the self-weight, hydrostatic pressure, hydrodynamic pressure, and seismic load, where the hydrodynamic pressure is simulated by added mass or potential-based fluid elements. The acceleration record of Koyna earthquake with the United States Geological Survey (USGS) site classification of A and the magnitude of 6.3 M was selected as the seismic loading, the site of which is similar to the example dam. Figure 8 shows the Koyna earthquake acceleration time history, and only downstream (Y) and vertical (Z) ground motions are considered in the calculations, where the PGA in the Y and Z direction is 0.474 g and 0.304 g, respectively.

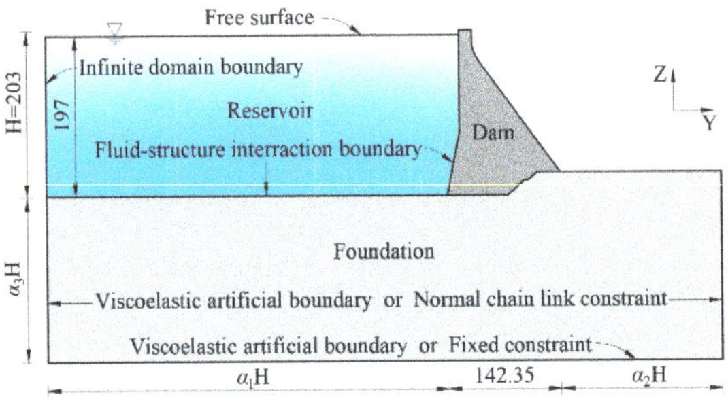

Figure 7. Overview of the dam–foundation–reservoir system (m).

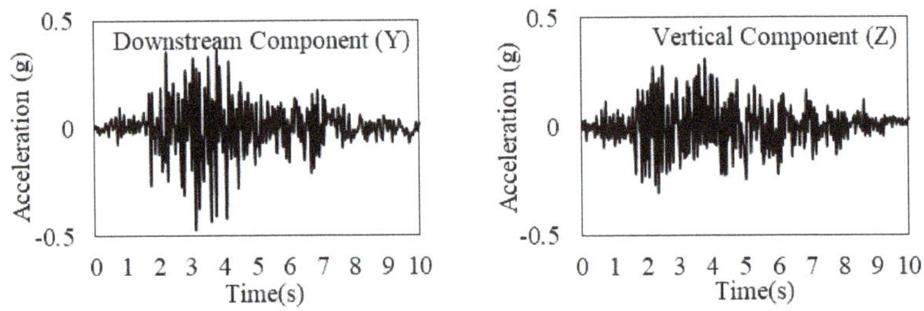

Figure 8. The Koyna earthquake acceleration time–history (g).

4.2. Cases of the Numerical Analysis

A series of cases were designed to identify factors affecting the dynamic response of concrete gravity dams, such as the simulation method of the foundation, foundation size, radiation damping effect, compressibility of reservoir water, and reservoir water length. Table 1 summarizes the dam–foundation–reservoir finite element models used for all cases, where RF = rigid foundation, MLF = massless foundation, VABF = foundation with viscoelastic artificial boundary, WAMR = reservoir water simulated by the Westergaard added mass method, IPFR = reservoir water simulated by incompressible potential-based fluid elements, and CPFR = reservoir water simulated by compressible potential-based fluid elements.

Table 1. Summary of models used for all cases.

Cases	Dam	Foundation				Reservoir	
		Simulation Methods	Foundation Sizes (H = Dam Height)			Simulation Methods	Reservoir Lengths
			Upstream	Downstream	Depth		
A-1	Linear	RF	/	/	/	WAMR	/
B-1	Linear	MLF	1H	1H	1H	WAMR	/
B-2	Linear	MLF	1.5H	1.5H	1.5H	WAMR	/
B-3	Linear	MLF	2H	2H	2H	WAMR	/
B-4	Linear	MLF	3H	3H	3H	WAMR	/
C-1	Linear	VABF	1H	1H	1H	WAMR	/
C-2	Linear	VABF	1.5H	1.5H	1.5H	WAMR	/
C-3	Linear	VABF	2H	2H	2H	WAMR	/
C-4	Linear	VABF	3H	3H	3H	WAMR	/
D-1	Linear	VABF	3H	3H	3H	IPFR	3H
D-2	Linear	VABF	3H	3H	3H	CPFR	3H
D-3	Linear	MLF	3H	1H	1H	WAMR	3H
D-4	Linear	MLF	3H	1H	1H	IPFR	3H
D-5	Linear	MLF	3H	1H	1H	CPFR	3H
D-6	Linear	MLF	4H	1H	1H	CPFR	4H
D-7	Linear	MLF	5H	1H	1H	CPFR	5H

5. Dynamic Characteristics

The foundation is assumed to be massless in all cases, and thus case B is the same as case C in modal analyses. A fixed constraint is applied to the bottom boundary of the foundation and a normal chain link constraint is applied to the side boundary. A zero potential boundary condition is set at the reservoir free surface [46], the infinite domain boundary condition is applied at the far end of reservoir, and the fluid–solid coupling boundary conditions are set at the interfaces of reservoir–dam and reservoir–foundation. The first 10 natural frequencies of the dam are listed in Table 2, and their distributions are shown in Figure 9.

Table 2. The first ten natural frequencies of the dam (Hz).

Case	1st	2nd	3rd	4th	5th	6th	7th	8th	9th	10th
A-1	1.926	4.261	6.951	7.251	10.147	12.751	14.612	15.029	16.732	17.064
B-1/C-1	1.286	2.943	3.306	5.082	7.949	10.771	11.046	12.752	13.752	14.344
B-2/C-2	1.247	2.838	3.032	4.994	7.906	10.681	11.023	12.726	13.730	14.310
B-3/C-3	1.223	2.747	2.892	4.941	7.881	10.633	11.011	12.713	13.719	14.295
B-4/C-4	1.194	2.582	2.775	4.877	7.853	10.583	10.997	12.701	13.709	14.280
D-1	1.294	1.968	2.405	3.183	3.574	5.263	5.710	7.394	9.151	10.025
D-2	1.243	1.474	1.921	2.607	3.022	3.227	3.912	4.510	4.843	4.997
D-3	1.282	2.919	3.303	5.047	7.927	10.770	11.032	12.750	13.747	14.344
D-4	1.387	2.976	3.492	3.499	4.304	5.580	5.993	7.548	9.222	10.116
D-5	1.322	1.751	2.152	2.810	3.392	3.484	3.986	4.903	4.929	5.150
D-6	1.322	1.708	1.970	2.415	2.962	3.419	3.484	3.907	4.595	4.903
D-7	1.322	1.681	1.871	2.185	2.604	3.060	3.440	3.484	3.856	4.407

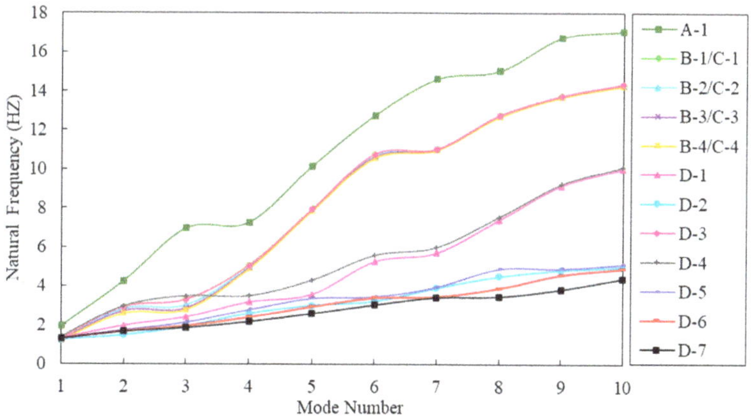

Figure 9. The distributions of the first ten natural frequencies of the dam.

In cases A and B, the hydrodynamic pressure of reservoir water is simulated by the added mass method considering only the unidirectional effect of reservoir water on the dam. It is found that in case A in which the RF model is used without considering the dam–foundation interaction, the natural frequency is the highest in all cases and the 10th natural frequency is about 9 times that of the first one. In cases B in which the dam–foundation interaction is considered, the first 10 natural frequencies are greatly reduced. Compared with the RF model, the first three natural frequencies of the MLF model are reduced by about 35%, 35%, and 57%, respectively. Comparison of cases B-1, B-2, B-3, and B-4 shows that the natural frequency decreases with the increase in foundation size, which appears to be more pronounced for low-order natural frequencies. The first natural frequency on the foundation with a size of 3H is reduced by 7.12% compared that with the size of 1 H, whereas the 10th natural frequency is reduced by only 0.44%.

In cases D, except D-3, reservoir water is simulated using potential-based fluid elements considering the dam–foundation–reservoir dynamic interaction. Comparing C-4, D-1, and D-2, it is found that the WAMR model gives the highest natural frequencies of the dam, followed by the IPFR model and then the CPFR model, and the difference increases as the mode number increases. As a result, the 10th natural frequency of the CPFR model is about 65% lower than that of the WAMR model. The simulation methods of reservoir water may have an effect on the increase in the natural frequency with the mode number. The highest increase is achieved by the WAMR model, and it is decreased by 30% for the IPFR model and by 70% for the CPFR model. It is seen in cases D-5, D-6, and D-7 that increasing the reservoir water length has little effect on the first three natural frequencies, but a greater effect on the higher order natural frequencies. The first natural frequencies are basically the same for models of 3H and 5H, but the 10th natural frequency for models of 5H are reduced by 14.43% compared with that of 3H.

Thus, it is concluded that the natural frequency of the dam decreases greatly when the dam–foundation interaction is considered, and decreases slightly with the increase in the foundation size. The simulation methods of reservoir water have significant effects on the natural frequency of the dam, whereas the reservoir water lengths have no obvious effect.

6. Dam-Foundation Interaction

6.1. Simulation Methods of Foundation

In this study, the RF, MLF, and VABF models were used to analyze the effects of different foundation simulation methods on the dynamic response of concrete gravity dams. Table 3 lists the extreme values of dynamic responses for the three simulation methods, in which the acceleration magnification factor is the multiple of the maximum acceleration of the dam crest over the maximum input acceleration, and the relative displacement is obtained by subtracting the displacement of the dam heel from that of the dam crest.

Table 3. The extreme values of dam dynamic response for different foundation simulation methods.

Case		A-1	B-1	C-1
Simulation Methods of Foundation		**RF**	**MLF**	**VABF**
Downstream direction	Acceleration magnification factor	5.894	4.680	2.112
	Relative displacement (m)	0.081	0.148	0.100
Vertical direction	Acceleration magnification	5.483	3.447	1.781
	Relative displacement (m)	−0.031	−0.037	−0.022
Dam stress (MPa)	Vertical normal tensile stress (σ_{zz})	7.369	9.227	3.042
	Vertical normal compressive stress (σ_{zz})	−8.255	−20.05	−15.08
	Principal tensile stress (σ_1)	7.969	16.64	10.56
	Principal compressive stress (σ_3)	−10.03	−28.00	−24.89

As can be seen from Table 3, the RF model gives the largest acceleration response, and the acceleration is reduced by about 30% for the MLF model and 65% for the VABF model. The MLF model gives the maximum relative displacement of the dam crest, as it takes into account the elasticity of the foundation, and the deformation of the foundation during the earthquake may tilt the gravity dam as a whole. Compared with the VABF model, the radiation damping effect of the infinite foundation is not adequately considered in the MLF model. For the RF model, the maximum values of vertical normal tensile and compressive stress appear at the slope of the upstream dam face with little difference, whereas the maximum values of vertical normal tensile and compressive stress appear at the dam heel for both MLF and VABF models, and the vertical normal compressive stress is significantly larger than the vertical normal tensile stress. The RF model gives the lowest principal tensile and compressive stress of the dam, and the maximum principal tensile stress appears at the dam heel and the maximum principal compressive stress appears at the neck of the downstream dam face. For both MLF and VABF models, the maximum principal tensile stress appears at the dam heel, and the maximum principal compressive stress appears at the dam toe. However, it is noted that the maximum principal tensile stress under the VABF model is smaller than that under the MLF model.

6.2. Sensitivity Analysis of Foundation Size

Cases B-1, B-2, B-3, and B-4 with different foundation sizes (1H, 1.5H, 2H, and 3H in the upstream, downstream and depth directions) were analyzed to elucidate their effects on the dynamic response of gravity dams. Each of the four cases were modeled with the MLF model and calculated by applying fixed constraints to the bottom boundary of the foundation and normal chain link constraints to the side boundary. The hydrodynamic pressure of reservoir water was simulated by the added mass method. Table 4 lists the extreme values of dam dynamic responses under for different foundation sizes.

Table 4. The extreme values of dam dynamic response for different foundation sizes (MLF).

	Case	B-1	B-2	B-3	B-4
	Foundation Size	**1H**	**1.5H**	**2H**	**3H**
Downstream direction	Acceleration magnification factor	4.680	5.057	5.789	3.477
	Relative displacement (m)	0.148	0.150	0.154	0.141
Vertical direction	Acceleration magnification	3.447	3.056	3.856	4.220
	Relative displacement (m)	−0.037	−0.038	−0.048	−0.042
Dam stress (MPa)	Vertical normal tensile stress (σ_{zz})	9.227	9.903	8.770	6.754
	Vertical normal compressive stress (σ_{zz})	−20.05	−20.79	−21.15	−18.72
	Principal tensile stress (σ_1)	16.64	16.37	13.70	10.66
	Principal compressive stress (σ_3)	−28.00	−29.52	−27.21	−28.01

Table 4 clearly shows that foundation size has a significant effect on the acceleration response. Specifically, the downstream acceleration at the dam crest reaches a maximum at a foundation size of 2H, which is 5.789 times that of the input; and a minimum at a foundation size of 3H, which is 3.477 times that of the input. The vertical acceleration at the dam crest reaches a maximum at a foundation size of 3H, which is 4.220 times that of the input. However, foundation size has little effect on displacement, and both downstream and vertical relative displacement at the dam crest reach a maximum at a foundation size of 2H. The tensile stress can also be significantly affected by foundation size, and the maximum tensile stress decreases with the increase in foundation sizes. As a result, the maximum vertical normal tensile stress and the maximum principal tensile stress at a foundation size of 3H are decreased by about 26.8% and 35.9% compared with that at a foundation size of 1H, respectively.

Figure 10 shows the principal stress envelopes of the dam in the four cases. It is found that foundation size has little effect on the distributions of principal tension and compression stress, and in all cases, the maximum principal tension stress appears at the dam heel and the maximum principal compressive stress appears at the dam toe.

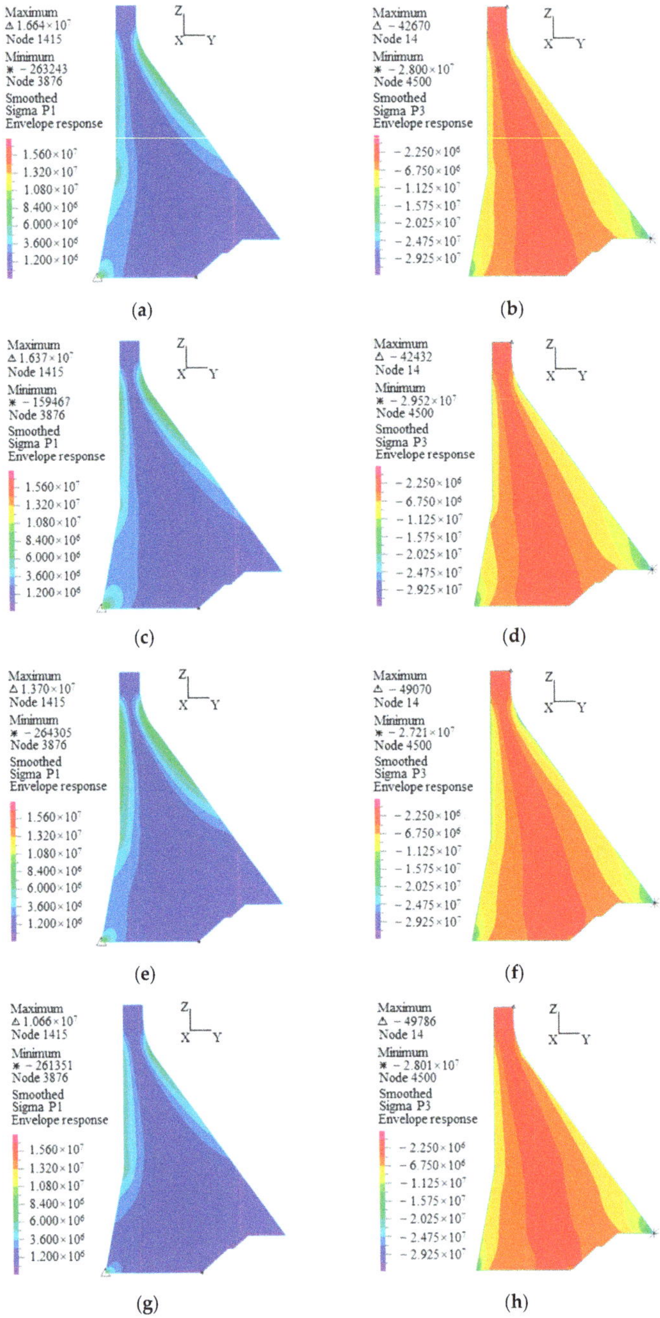

Figure 10. The principal stress envelopes of the dam for different foundation sizes obtained using the MLF model (Pa). (**a**) The distribution of σ_1 for B-1 (1H), (**b**) the distribution of σ_3 for B-1 (1H), (**c**) the distribution of σ_1 for B-2 (1.5H), (**d**) the distribution of σ_3 for B-2 (1.5H), (**e**) the distribution of σ_1 for B-3 (2H), (**f**) the distribution of σ_3 for B-3 (2H), (**g**) the distribution of σ_1 for B-4 (3H), (**h**) the distribution of σ_3 for B-4 (3H).

6.3. The Radiation Damping Effect of Infinite Foundation

6.3.1. Verification of the Foundation Model

The finite element model of the foundation (Figure 11) was analyzed using the method described in Section 3. The foundation sizes were also assumed to be 1H, 1.5H, 2H, and 3H, and the material properties are described in Section 4. The sinusoidal waves given by Equations (11)–(13) are input from the bottom boundary of the foundation with a peak of 12.542 m/s^2 in both downstream and vertical directions. Three observation points are set on the foundation surface: point M which is closest to the upstream side, point N which is the intermediate point of the upstream side, and point O where there is the wave source. The PGA values at these three points are listed in Table 5. It is seen that the amplitude of the wave is almost doubled when it reaches the free surface of the foundation, which is consistent with the theory and further verifies the applicability of the viscoelastic artificial boundary and corresponding earthquake input mechanism.

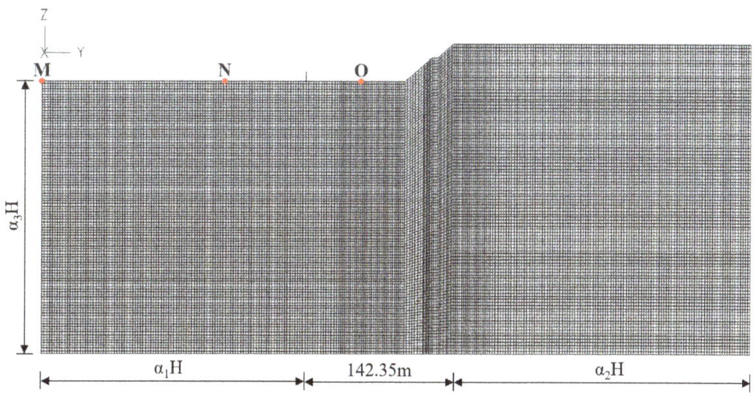

Figure 11. The finite element model of the foundation.

Table 5. The PGA values at the observation points under sinusoidal excitation (m/s^2).

Foundation Size	Downstream			Vertical		
	M	N	O	M	N	O
1H	25.185 (2.008)	26.065 (2.078)	24.103 (1.922)	25.712 (2.050)	−27.683 (2.207)	23.948 (1.909)
1.5H	−25.773 (2.055)	26.833 (2.139)	23.757 (1.894)	−25.197 (2.009)	−26.204 (2.089)	−24.580 (1.960)
2H	25.141 (2.005)	−27.651 (2.205)	23.269 (1.855)	25.277 (2.015)	26.125 (2.083)	25.152 (2.005)
3H	25.451 (2.029)	−27.316 (2.178)	23.394 (1.865)	25.234 (2.012)	26.589 (2.120)	25.749 (2.053)

Note: The values in ( ) are the multiples of input amplitude at the bottom boundary of the foundation.

The Koyna earthquake record was used to further verify the foundation model. It was found that the input from the bottom boundary of the foundation is folded in half, with a PGA of 2.324 m/s^2 in the downstream direction and 1.528 m/s^2 in the vertical direction. The displacement and velocity time histories are given in Figure 12, and the PGA values at each observation point are shown in Table 6.

Theoretically, the PGA at the observation points should be twice that of the input as the seismic excitation is transmitted to the foundation surface. However, Table 6 shows that there are some discrepancies between the theoretical and observed PGAs, and in general the error is the smallest for the foundation with a size of 3H.

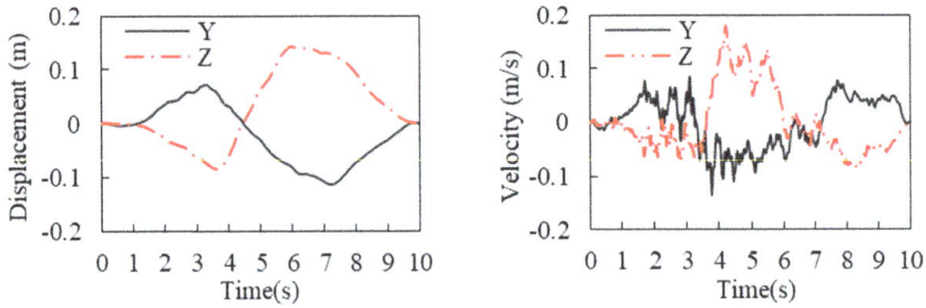

Figure 12. The Koyna earthquake displacement and velocity time history.

Table 6. The PGA values at the observation points under the Koyna earthquake excitation (m/s^2).

Foundation Size	Downstream			Vertical		
	M	N	O	M	N	O
1H	3.506 (−24.6%)	4.133 (−11.1%)	4.121 (−11.3%)	2.807 (−8.1%)	3.410 (+11.6%)	3.125 (+2.3%)
1.5H	−4.259 (−8.4%)	−4.572 (−1.6%)	−3.819 (−17.8%)	3.541 (+15.9%)	2.898 (−5.2%)	3.178 (+4.0%)
2H	−4.200 (−9.6%)	−4.716 (+1.5%)	−5.092 (+9.6%)	3.027 (−0.9%)	3.031 (−0.8%)	3.888 (+27.2%)
3H	4.493 (−3.3%)	−4.472 (−3.8%)	4.509 (−3.0%)	2.792 (−8.6%)	−3.267 (+6.9%)	3.405 (+11.4%)

Note: The values in ( ) are the percentage changes compared with the theoretical value, − indicates decrease, + indicates increase.

6.3.2. The Radiation Damping Effect

In contrast with the results of the MLF model in Section 6.2, four cases C-1, C-2, C-3, and C-4, which have the same foundation sizes as cases B-1, B-2, B-3, and B-4, respectively, were designed and analyzed. The viscoelastic artificial boundary condition is applied at the truncation boundary of the foundation, and the hydrodynamic pressure of reservoir water is simulated by added mass.

Comparison between Tables 4 and 7 indicates that the dynamic responses of the dam obtained using the VABF model are significantly smaller than those obtained using the MLF model. Specifically, the VABF model leads to 45–67% reductions in acceleration, 38–50% reductions in displacement, 28–56% reductions in principal tensile stress, and 11–24% reductions in principal compressive stress. Figure 13 shows the principal stress envelopes of the dam at different foundation sizes obtained using the VABF model. As compared with Figure 10, the high stress zone is significantly reduced, indicating that more conservative dynamic responses are obtained by the MLF model, whereas the dam dynamic responses can be reduced effectively when the VABF model is used.

Table 7 shows that the downstream acceleration at the dam crest decreases as the foundation size increases, and the minimum acceleration is 1.910 times that of the input at a foundation size of 3H. The vertical acceleration is 1.654 times that of the input at a foundation size of 2H and 3H. However, the relative displacement at the dam crest is less affected by foundation size. The downstream relative displacement decreases slightly with the increase in foundation size, and the displacement at a foundation size of 3H is 25.4% lower than that at a foundation size of 1H. It should be noted that the tensile and compressive stresses of the dam decrease most significantly with the increase in foundation size. As a result, the foundation size of 3H leads to a 22.1% decrease in the maximum vertical normal tensile stress, a 20.2% decrease in the maximum vertical normal compressive stress, a 55.5% decrease in the maximum principal tensile stress, and a 14.9% decrease in the maximum principal compressive stress compared to the foundation size of 1H.

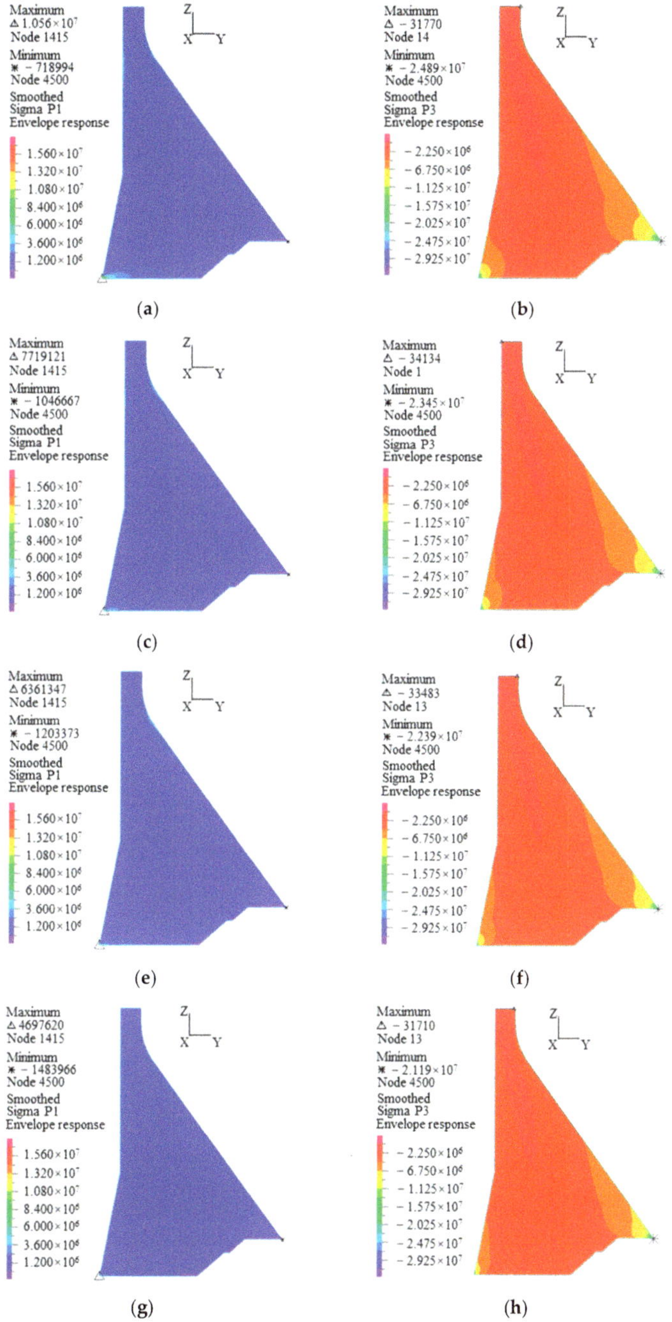

Figure 13. The principal stress envelopes of the dam for different foundation sizes obtained using the VABF model (Pa). (**a**) The distribution of σ_1 for C-1 (1H), (**b**) the distribution of σ_3 for C-1 (1H), (**c**) the distribution of σ_1 for C-2 (1.5H), (**d**) the distribution of σ_3 for C-2 (1.5H), (**e**) the distribution of σ_1 for C-3 (2H), (**f**) the distribution of σ_3 for C-3 (2H), (**g**) The distribution of σ1 for C-4 (3H), (**h**) the distribution of σ_3 for C-4 (3H).

Table 7. The extreme values of dam dynamic responses for different foundation sizes (VABF).

	Case	C-1	C-2	C-3	C-4
	Foundation Size	**1H**	**1.5H**	**2H**	**3H**
Downstream direction	Acceleration magnification factor	2.112	2.052	1.992	1.910
	Relative displacement (m)	0.100	0.088	0.081	0.074
Vertical direction	Acceleration magnification	1.781	1.529	1.654	1.654
	Relative displacement (m)	−0.022	−0.023	−0.024	−0.022
Dam stress (MPa)	Vertical normal tensile stress (σ_{zz})	3.042	2.524	2.368	2.403
	Vertical normal compressive stress (σ_{zz})	−15.08	−13.78	−12.97	−12.03
	Principal tensile stress (σ_1)	10.56	7.719	6.361	4.698
	Principal compressive stress (σ_3)	−24.89	−23.45	−22.39	−21.19

7. Dam–Reservoir Interaction

7.1. Simulation Methods of Reservoir

In the above cases, the hydrodynamic pressure of reservoir water was simulated using the Westergaard added mass method without considering its compressibility. In order to deal with this problem, the potential-based fluid elements in ADINA were used to simulate the incompressible and compressible reservoir water, as presented in this section. Accordingly, the free surface boundary condition is applied on the reservoir surface, the infinite domain boundary condition is applied at the far end of the reservoir, and the fluid–solid coupling boundary condition is applied at the junctions of reservoir–dam and reservoir–foundation. Table 8 shows the extreme values of dam dynamic responses in the six cases, and Figure 14 shows the distributions of maximum hydrodynamic pressure along the dam–reservoir interface obtained by different reservoir simulation methods.

Table 8. The extreme values of dam dynamic responses for different reservoir simulation methods.

	Case	C-4	D-1	D-2	D-3	D-4	D-5
	Simulation Method of Foundation		**VABF**			**MLF**	
	Simulation Method of Reservoir Water	**WAMR**	**IPFR**	**CPFR**	**WAMR**	**IPFR**	**CPFR**
Downstream direction	Acceleration magnification factor	1.910	2.429	2.219	4.575	4.833	5.249
	Relative displacement (m)	0.074	0.069	0.068	0.149	0.112	0.172
Vertical direction	Acceleration magnification	1.654	1.822	1.715	3.433	4.085	2.660
	Relative displacement (m)	−0.022	−0.017	−0.018	−0.037	−0.032	−0.027
Dam stress (MPa)	Vertical normal tensile stress (σ_{zz})	2.403	5.477	6.374	8.122	9.161	11.224
	Vertical normal compressive stress (σ_{zz})	−12.03	−8.797	−9.121	−20.27	−11.84	−13.65
	Principal tensile stress (σ_1)	4.698	12.11	13.98	14.48	15.07	17.91
	Principal compressive stress (σ_3)	−21.19	−19.09	−19.50	−27.99	−22.55	−24.76
	Hydrodynamic pressure of dam heel (KN/m^2)	716.8	429.8	371.4	1188	1127	994.3

According to Table 8, the reservoir water simulation methods have no significant effects on the acceleration and displacement of the dam, but have significant effects on the stress. The Westergaard added mass method leads to a significant reduction in the vertical normal tensile stress at the dam heel, because it only considers the unidirectional effect of reservoir water on the dam and thus neglects the movement of reservoir water and its interaction with the foundation. In contrast, the potential-based fluid simulation method takes into account both reservoir–dam and reservoir–foundation interactions. Under the influence of gravity, the hydrodynamic pressure is applied to the upstream foundation surface as the reservoir water flows, which may cause downward deformation of the foundation and consequently affect the normal tensile stress at the dam heel. The maximum principal tensile stress of the dam appears at the dam heel. It is noted that when the VABF model is used, the maximum principal tensile stress of the CPFR model is 15%

higher than that of the IPFR model, which is approximately 3 times that of the WAMR model, whereas when the MLF model is used, the maximum principal tensile stress of the CPFR model is 19% higher than that of the IPFR model and 24% higher than that of the WAMR model. The maximum principal compressive stress of the dam appears at the dam toe, and is less affected by the reservoir water simulation methods. When the VABF model is used, the maximum principal compressive stress of the CPFR model is 8% lower than that of the WAMR model but 2% higher than that of the IPFR model, whereas when the MLF model is used, the maximum principal compressive stress of the CPFR model is 12% lower than that of the WAMR model but 10% higher than that of the IPFR model. As shown in Figure 14, the hydrodynamic pressure along the dam–reservoir interface of the WAMR model is significantly higher compared to the other two models. Below 100 m, the hydrodynamic pressure of the IPFR model is significantly higher than that of the CPFR model.

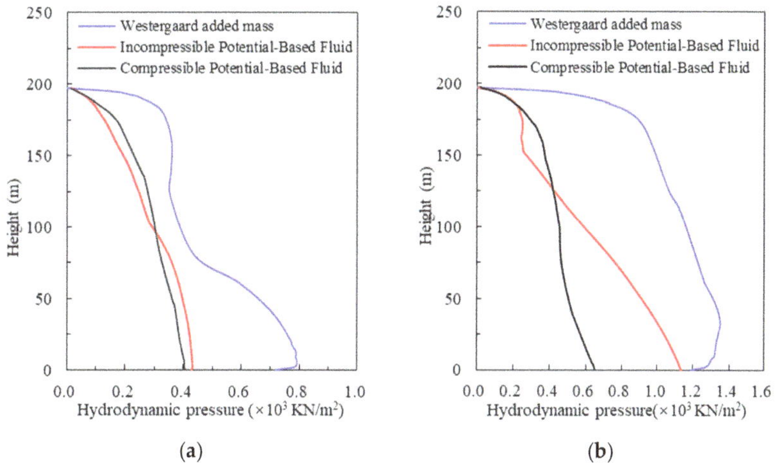

Figure 14. Hydrodynamic pressure distributions along the dam–reservoir interface. (**a**) The VABF model, (**b**) the MLF model.

7.2. Reservoir Water Length

As presented in this section, reservoir water with a length of 3H, 4H, and 5H [47] was simulated using compressible potential-based fluid elements to investigate its effect on the dynamic response of the dam, and the extreme values of dynamic responses are given in Table 9.

Table 9. The extreme values of the dam dynamic response at different reservoir water lengths.

Case		D-5	D-6	D-7
Reservoir Water Length		**3H**	**4H**	**5H**
Downstream direction	Acceleration magnification factor	5.249	4.275	4.303
	Relative displacement (m)	0.172	0.129	0.140
Vertical direction	Acceleration magnification	2.660	2.235	2.172
	Relative displacement (m)	−0.027	−0.027	−0.024
Dam stress (MPa)	Vertical normal tensile stress (σ_{zz})	11.224	9.52	10.15
	Vertical normal compressive stress (σ_{zz})	−13.65	−12.25	−11.96
	Principal tensile stress (σ_1)	17.91	15.23	15.62
	Principal compressive stress (σ_3)	−24.76	−24.11	−23.35
Hydrodynamic pressure of dam heel (KN/m^2)		994.3	704.0	615.9

The results show that the reservoir water length of 3H leads to a 7–23% increase in acceleration, a 9–33% increase in displacement, a 15–18% increase in the maximum principal tensile stress, a 3–6% increase in the maximum principal compressive stress, and a 41–61% increase in the maximum hydrodynamic pressure at the dam heel compared to the reservoir water length of 4H and 5H.

8. Conclusions

In this study, some key factors, such as the dynamic interactions of dam–foundation and dam–reservoir, were considered in the dynamic numerical analysis of a 200 m high gravity dam. The rigid, massless, and viscoelastic artificial boundary foundation models were established to account for the effect of dam–foundation dynamic interaction on the dynamic responses of the dam. Three reservoir water simulation methods, namely, the Westergaard added mass method, and the incompressible and compressible potential fluid methods, were used to account for the effect of hydrodynamic pressure on the dynamic characteristics of the dam. The ranges of the truncation boundary of the foundation and reservoir in numerical analysis were further investigated. The following conclusions can be drawn:

(1) The natural frequency of the dam decreases greatly in numerical analysis when the dam–foundation interaction is considered, and decreases slightly with the increase in the foundation size. The simulation methods of reservoir water have significant effects on the natural frequency of the dam, whereas the reservoir water lengths have no significant effect.
(2) The dynamic interaction of the dam and the foundation cannot be ignored. The radiation damping effect should be considered in the dynamic numerical analysis. The viscoelastic artificial boundary foundation is more efficient than the massless foundation in simulating the radiation damping effect of the far-field foundation. It was found that a foundation range of 3 times the dam height in all directions, such as upstream, downstream, and depth, is the most reasonable range of the truncation boundary of the foundation.
(3) The methods used for reservoir water simulation have no significant effects on the acceleration and displacement of the dam, but have a significant effect on the stress. Compared with the Westergaard added mass method, the potential-based fluid simulation method simultaneously takes into account the reservoir–dam and reservoir–foundation interactions. The static and dynamic water pressure was applied to the upstream foundation surface as the reservoir water, which may cause downward deformation of the foundation and consequently increase the normal tensile stress at the dam heel. It was found that a reservoir length of 3 times the dam height is feasible for the truncation boundary of the reservoirs.

Author Contributions: Conceptualization, H.Z.; methodology, H.Z.; validation, C.W., Y.Z., Y.W. and T.T.T.H.; resources, H.Z. and L.G.; data curation, C.W.; writing—original draft preparation, C.W.; writing—review and editing, H.Z. and C.W.; supervision, H.Z., L.G., Y.W. and T.T.T.H. All authors have read and agreed to the published version of the manuscript.

Funding: This work was supported by the Open Research Fund of State Key Laboratory of Simulation and Regulation of Water Cycle in River Basin (China Institute of Water Resources and Hydropower Research) (grant number IWHR-SKL-KF201816), the National Key Research and Development Program of China (grant number 2017YFC0404903), and the Natural Science Foundation of Jiangsu Province (grant number BK20170884).

Institutional Review Board Statement: Not applicable.

Informed Consent Statement: Not applicable.

Data Availability Statement: The data presented in this study are available on request from the corresponding author.

Conflicts of Interest: The authors declare that they have no known competing financial interests or personal relationships that could have appeared to influence the work reported in this paper.

References

1. Ghaedi, K.; Hejazi, F.; Ibrahim, Z.; Khanzaei, P. Flexible foundation effect on seismic analysis of roller compacted concrete (RCC) dams using finite element method. *KSCE J. Civ. Eng.* **2017**, *22*, 1275–1287. [CrossRef]
2. Bayraktar, A.; Hançer, E.; Akköse, M. Influence of base-rock characteristics on the stochastic dynamic response of dam–reservoir–foundation systems. *Eng. Struct.* **2005**, *27*, 1498–1508. [CrossRef]
3. Ghaemian, M.; Noorzad, A.; Mohammadnezhad, H. Assessment of foundation mass and earthquake input mechanism effect on dam–reservoir–foundation system response. *Int. J. Civ. Eng.* **2019**, *17*, 473–480. [CrossRef]
4. Salamon, J.W.; Wood, C.; Hariri-Ardebili, M.A.; Malm, R.; Faggiani, G. *Seismic Analysis of Pine Flat Concrete Dam: Formulation and Synthesis of Results*; Springer International Publishing: Cham, Switzerland, 2020; pp. 3–97.
5. Chopra, A. Earthquake analysis of concrete dams: Factors to be considered. In Proceedings of the Proceedings of the 10th National Conference in Earthquake Engineering, Anchorage, AK, USA, 21–25 July 2014.
6. Hariri-Ardebili, M.A.; Kolbadi, S.M.S.; Kianoush, M. FEM-based parametric analysis of a typical gravity dam considering input excitation mechanism. *Soil Dyn. Earthq. Eng.* **2016**, *84*, 22–43. [CrossRef]
7. Hariri-Ardebili, M.A.; Mirzabozorg, H. A comparative study of seismic stability of coupled arch dam-foundation-reservoir systems using infinite elements and viscous boundary models. *Int. J. Struct. Stab. Dyn.* **2013**, *13*, 1350032. [CrossRef]
8. Burman, A.; Nayak, P.; Agrawal, P.; Maity, D. Coupled gravity dam–foundation analysis using a simplified direct method of soil–structure interaction. *Soil Dyn. Earthq. Eng.* **2012**, *34*, 62–68. [CrossRef]
9. Wang, H.; Feng, M.; Yang, H. Seismic nonlinear analyses of a concrete gravity dam with 3D full dam model. *Bull. Earthq. Eng.* **2012**, *10*, 1959–1977. [CrossRef]
10. Lysmer, J.; Kuhlemeyer, R.L. Finite dynamic model for infinite media. *J. Eng. Mech. Div.* **1969**, *95*, 859–878. [CrossRef]
11. Hariri-Ardebili, M.A. Impact of foundation nonlinearity on the crack propagation of high concrete dams. *Soil Mech. Found. Eng.* **2014**, *51*, 72–82. [CrossRef]
12. Liu, J.; Du, Y.; Du, X.; Wang, Z.; Wu, J. 3D viscous-spring artificial boundary in time domain. *Earthq. Eng. Eng. Vib.* **2006**, *5*, 93–102. [CrossRef]
13. Liang, H.; Tu, J.; Guo, S.; Liao, J.; Li, D.; Peng, S. Seismic fragility analysis of a High Arch Dam-Foundation System based on seismic instability failure mode. *Soil Dyn. Earthq. Eng.* **2020**, *130*, 105981. [CrossRef]
14. Song, C.; Wolf, J.P. The scaled boundary finite-element method—alias consistent infinitesimal finite-element cell method—for elastodynamics. *Comput. Methods Appl. Mech. Eng.* **1997**, *147*, 329–355. [CrossRef]
15. Qu, Y.; Chen, D.; Liu, L.; Ooi, E.T.; Eisenträger, S.; Song, C. A direct time-domain procedure for the seismic analysis of dam–foundation–reservoir systems using the scaled boundary finite element method. *Comput. Geotech.* **2021**, *138*, 104364. [CrossRef]
16. Hariri-Ardebili, M.A.; Kianoush, M. Integrative seismic safety evaluation of a high concrete arch dam. *Soil Dyn. Earthq. Eng.* **2014**, *67*, 85–101. [CrossRef]
17. Basu, U.; Chopra, A.K. Erratum to'Perfectly matched layers for transient elastodynamics of unbounded domains'(Int. J. Numer. Meth. Engng 2004; 59:1039–1074). *Int. J. Numer. Methods Eng.* **2004**, *61*, 156–157. [CrossRef]
18. Harirri-Ardebili, M.A.; Saouma, V.E. Impact of near-fault vs. far-field ground motions on the seismic response of an arch dam with respect to foundation type. *Dam Eng.* **2013**, *24*, 19–52.
19. Pan, J.W.; Song, C.M.; Zhang, C.H. Simple model for gravity dam response considering radiation damping effects. *J. Hydraul. Eng.* **2010**, *41*, 493–498. [CrossRef]
20. Chen, D.; Hou, C.; Wang, F. Influences on the seismic response of the gravity dam-foundation-reservoir system with different boundary and input models. *Shock. Vib.* **2021**, *2021*, 1–15. [CrossRef]
21. Wang, G.; Wang, Y.; Lu, W.; Zhou, C.; Chen, M.; Yan, P. XFEM based seismic potential failure mode analysis of concrete gravity dam–water–foundation systems through incremental dynamic analysis. *Eng. Struct.* **2015**, *98*, 81–94. [CrossRef]
22. Wang, G.; Wang, Y.; Lu, W.; Yan, P.; Zhou, W.; Chen, M. A general definition of integrated strong motion duration and its effect on seismic demands of concrete gravity dams. *Eng. Struct.* **2016**, *125*, 481–493. [CrossRef]
23. Gorai, S.; Maity, D. Seismic response of concrete gravity dams under near field and far field ground motions. *Eng. Struct.* **2019**, *196*, 109292. [CrossRef]
24. Chen, D.-H.; Yang, Z.-H.; Wang, M.; Xie, J.-H. Seismic performance and failure modes of the Jin'anqiao concrete gravity dam based on incremental dynamic analysis. *Eng. Fail. Anal.* **2019**, *100*, 227–244. [CrossRef]
25. Asghari, E.; Taghipour, R.; Bozorgnasab, M.; Moosavi, M. Seismic analysis of concrete gravity dams considering foundation mass effect. *KSCE J. Civ. Eng.* **2018**, *22*, 4988–4996. [CrossRef]
26. Mirzabozorg, H.; Akbari, M.; Hariri-Ardebili, M.A. Wave passage and incoherency effects on seismic response of high arch dams. *Earthq. Eng. Eng. Vib.* **2012**, *11*, 567–578. [CrossRef]
27. Westergaard, H.M. Water pressures on dams during earthquakes. *Trans. Am. Soc. Civ. Eng.* **1933**, *98*, 418–433. [CrossRef]
28. Khiavi, M.P.; Sari, A. Evaluation of hydrodynamic pressure distribution in reservoir of concrete gravity dam under vertical vibration using an analytical solution. *Math. Probl. Eng.* **2021**, *2021*, 1–9. [CrossRef]

29. Amina, T.B.; Mohamed, B.; André, L.; Abdelmalek, B. Fluid–structure interaction of Brezina arch dam: 3D modal analysis. *Eng. Struct.* **2015**, *84*, 19–28. [CrossRef]
30. Altunişik, A.C.; Sesli, H. Dynamic response of concrete gravity dams using different water modelling approaches: Westergaard, lagrange and euler. *Comput. Concr.* **2015**, *16*, 429–448. [CrossRef]
31. Altunisik, A.C.; Sesli, H.; Hüsem, M.; Akköse, M. Performance Evaluation of gravity dams subjected to near- and far-fault ground motion using Euler approaches. *Iran. J. Sci. Technol. Trans. Civ. Eng.* **2018**, *43*, 297–325. [CrossRef]
32. Bayraktar, A.; Türker, T.; Akköse, M.; Ateş, Ş. The effect of reservoir length on seismic performance of gravity dams to near- and far-fault ground motions. *Nat. Hazards* **2010**, *52*, 257–275. [CrossRef]
33. Kartal, M.E.; Cavusli, M.; Sunbul, A.B. Assessing seismic response of a 2D roller-compacted concrete dam under variable reservoir lengths. *Arab. J. Geosci.* **2017**, *10*, 488. [CrossRef]
34. Pelecanos, L.; Kontoe, S.; Zdravković, L. Numerical modelling of hydrodynamic pressures on dams. *Comput. Geotech.* **2013**, *53*, 68–82. [CrossRef]
35. Mandal, K.; Maity, D. Transient response of concrete gravity dam considering dam-reservoir-foundation interaction. *J. Earthq. Eng.* **2016**, *22*, 211–233. [CrossRef]
36. Løkke, A.; Chopra, A.K. Direct finite element method for nonlinear earthquake analysis of 3-dimensional semi-unbounded dam-water-foundation rock systems. *Earthq. Eng. Struct. Dyn.* **2018**, *47*, 1309–1328. [CrossRef]
37. Løkke, A.; Chopra, A.K. Direct finite element method for nonlinear earthquake analysis of concrete dams: Simplification, modeling, and practical application. *Earthq. Eng. Struct. Dyn.* **2019**, *48*, 818–842. [CrossRef]
38. Chopra, A.K. Earthquake analysis of arch dams: Factors to be considered. *J. Struct. Eng.* **2012**, *138*, 205–214. [CrossRef]
39. Wang, J.-T.; Lv, D.-D.; Jin, F.; Zhang, C.-H. Earthquake damage analysis of arch dams considering dam–water–foundation interaction. *Soil Dyn. Earthq. Eng.* **2013**, *49*, 64–74. [CrossRef]
40. Wang, J.-T.; Zhang, C.-H.; Jin, F. Nonlinear earthquake analysis of high arch dam-water-foundation rock systems. *Earthq. Eng. Struct. Dyn.* **2012**, *41*, 1157–1176. [CrossRef]
41. Amini, A.; Motamedi, M.H.; Ghaemian, M. The Impact of dam-reservoir-foundation interaction on nonlinear response of concrete gravity dams. In *AIP Conference Proceedings*; American Institute of Physics: College Park, MD, USA, 2008; pp. 585–594.
42. Clough, R.W. Reservoir interaction effects on the dynamic response of arch dams. In Proceedings of the US-PRC Bilateral workshop on Earthquake Engineering, Harbin, China, 27–31 August 1982; pp. 58–84.
43. Sussman, T.; Sundqvist, J. Fluid–structure interaction analysis with a subsonic potential-based fluid formulation. *Comput. Struct.* **2003**, *81*, 949–962. [CrossRef]
44. Yan, C.L. *Study on the Mechanism of Strong Earthquake Failure of Concrete Gravity Dams Considering Material Tensile and Com-Pression Damage*; Chain Institute of Water Resources and Hydropower Research: Beijing, China, 2020. [CrossRef]
45. Ministry of Water Resources of the People's Republic of China. *Design Code for Hydraulic Concrete Structures (SL 191-2008)*; China Water Power Press: Beijing, China, 2008; p. 23.
46. Zeinizadeh, A.; Mirzabozorg, H.; Noorzad, A.; Amirpour, A. Hydrodynamic pressures in contraction joints including waterstops on seismic response of high arch dams. In *Structures*; Elsevier: Amsterdam, The Netherlands, 2018; pp. 1–14. [CrossRef]
47. Mahmoodi, K.; Noorzad, A.; Mahboubi, A.; Alembagheri, M. Seismic performance assessment of a cemented material dam using incremental dynamic analysis. In *Structures*; Elsevier: Amsterdam, The Netherlands, 2021; pp. 1187–1198. [CrossRef]

water

MDPI

Article

Discharge Flow Rate for the Initiation of Jet Flow in Sky-Jump Spillways

Raffaella Pellegrino [1,*], Miguel Ángel Toledo [1] and Víctor Aragoncillo [2]

1 Departamento de Ingeniería Civil: Hidráulica, Energía y Medio Ambiente, Universidad Politécnica de Madrid, 28040 Madrid, Spain; miguelangel.toledo@upm.es
2 Ingeniero de Caminos, Canales y Puertos; varagonc@gmail.com
* Correspondence: r.pellegrino@upm.es

Received: 25 May 2020; Accepted: 22 June 2020; Published: 24 June 2020

Abstract: The sky-jump spillway is an economical and effective solution to return water to a river, eventually complemented by a pre-excavated basin. However, an inappropriate design could endanger spillways and even the dam itself. For the design of a sky-jump it is necessary to evaluate the position and dimensions of the potential pre-excavated basin based on the characteristics of the water flow to be evacuated and the geometric configuration of the sky-jump. The jump of the water jet occurs when a certain flow rate is reached. This flow rate for the initiation of the jet flow determines the position of the impact area closest to the spillway. We propose a new formula for the determination of the flow rate for the initiation of the jet flow, which incorporates as a novelty the influence of the curvature of the flip bucket. A methodology for the direct determination of the flow rate for the initiation of the jet flow is also presented. The new formula and methodology, based on experimental laboratory work and numerical modeling, will support the designer to choose the energy dissipation way, in the riverbed or inside the flip bucket, for low and frequent discharge flows.

Keywords: hydraulic structure; sky-jump; spillway; flip bucket; chute; basin; erosion; flow rate; jet flow

1. Introduction and Background

The sky-jump spillway was built for the first time in France in 1930 on the Dordogne's hydraulic scheme [1]. Rhone and Peterka [2] studied and improved the design of this type of spillway, which today is commonly used in dams when the speed of the water, in the final stretch of the chute if adequately high [3]. With this type of hydraulic structure, it is possible both to dissipate the energy and direct the jet to a certain area of the riverbed [4]. In fact, the water jet can be directed to the most convenient area and also can be mixed as much as possible with the air along its path to the riverbed. In such a way, the jet loses some of its energy and, if the sky-jump spillway was properly designed, the local scour does not compromise the safety of the dam.

Determining the position of the scour basin during the design stage allows to control the impact zone. The shape and size of the scour basin depend on both the geometrical configuration of the flip bucket and the operating flow rates. Furthermore, the plunge pool water level upstream of impact location is important regarding riverbed scour. The water level range can be from 3 m [5] to 8 m [6].

Three options are available in order to minimize the damage: (a) To completely avoid scour; (b) to design the spillway so that the scour occurs far away from dam foundation and abutments; and (c) to limit the scour extent [7].

Different bucket types are used to achieve an optimal combination with the chute end layout to reach an appropriate impact location of the jet into the plunge pool. When the flow discharge is small, the flip bucket works as a stilling basin and a hydraulic jump appears on it. In this condition the flow passing the sill falls close to bucket foundation might jeopardize the bucket stability. When the spillway is unregulated, the type of bucket with a drainage channel is used to avoid choking at the small discharge which could endanger dam stability. In this type of bucket, the hydraulic jump does not occur for relatively small discharge, so that a free jet emerges [8].

However, it is important to notice that the hydraulic jump might be used to dissipate the energy for small discharges before the jump occurs, and remaining energy would be easy to dissipate as the falling height from the lip is usually low. In this way, the hydraulic jump inside the flip bucket might be useful to dissipate most of the flow energy in the range of low discharges.

An inadequate flip bucket could jeopardize the downstream area of the dam, as well as the dam itself. Hence there is a need to evaluate the geometric characteristics of the scour basin based on the features of both the water flow to be evacuated and the flip bucket. The first thing that is needed to be able to predict the position of the scour basin is to determine the flow rate for the initiation of the jet flow, which allows to establish with sufficient approximation the position of the scour basin at its extreme closest to the spillway.

Rouve [9] concluded that for a low flow rate range, the jet does not form, and the flip bucket works as a stilling basin ended with an overflow weir. The flow of water arrives through the chute to the flip bucket in supercritical regime and produces a hydraulic jump when contacting the water mass on the flip bucket, which is under subcritical regime. As the flow increases, the hydraulic jump moves towards the lip of the bucket, which is working as an overflow weir, until the flow rate for the initiation of the jet flow Qi (with increasing flow) is reached. The hydraulic jump is then dragged out of the flip bucket, the jet flow occurs and the water if thrown out. The jet flow is maintained, with the sky-jump operating in supercritical regime, until reaching the maximum flow rate. Later on, as the flow rate descends, until the flow rate is reduced to the flow rate for the finishing of the jet flow Q_f (with decreasing flow). A hydraulic jump is then re-established on the flip bucket, which works again as a stilling basin. Therefore, a hysteresis phenomenon occurs, since the flow rates for the initiation and finishing of the jet flow, Q_i and Q_f, are different, being Q_i slightly higher than Q_f.

The hysteresis phenomenon has been studied, both experimentally [10,11] and theoretically [12,13], by several authors [14,15]. The formulas obtained by these authors were developed for flat approach channels, setting obstacles or deflectors at their end, but without considering the effect of the curvature of the flip bucket.

Abecasis and Quintela [16,17] obtained formulas for determining Q_i and Q_f. They applied the momentum Equation to the control volume between two sections, the first located upstream of the obstacle, with a flow depth y_0, and the second on the obstacle, where a section with the critical flow depth y_c was assumed. The formula obtained for Q_f, neglecting the horizontal component of the friction resistance, considering uniform velocity, parallel flow, and assuming a linear distribution of the hydrostatic pressure in the initial and final sections of the control volume, was:

$$\frac{z}{y_0} = \frac{1 + 2F_0^2 - 3F_0^{4/3}}{1 + F_0^{2/3}} \tag{1}$$

where, y_0 is the flow depth and F_0 the Froude number both in the initial section, upstream of the obstacle, and z is the vertical height of the obstacle, equivalent to the depth of the flip bucket.

Applying the momentum Equation again, but assuming a pressure on the upstream face of the obstacle, exerted by a critical flow depth y_c, they obtained the formula for the initiation of the jet flow:

$$\frac{z}{y_0} = \frac{1 + 2F_0^2 - 3F_0^{4/3}}{2F_0^{2/3}} \tag{2}$$

Muskatirovic and Batinic [18] applied the momentum theorem, assuming a hydrostatic distribution on the upstream face of the obstacle, with flow depth y_0, and a hydrostatic distribution over the obstacle corresponding to the critical flow depth y_c. The formula obtained for the end of the jet flow was:

$$\frac{z}{y_0} = \sqrt{1+2F_0^2-2F_0^{4/3}} - F_0^{2/3} \tag{3}$$

For the initiation of the jet flow, they combined the Equation of the hydraulic jump with the Bernoulli's Equation, obtaining the following relationship:

$$\frac{z}{y_0} = \frac{1}{2}\left(\sqrt{1+8F_0^2}-1\right) + \frac{1+4F_0^2+\sqrt{1+F_0^2}}{16F_0^2} - 1.5F_0^{2/3} \tag{4}$$

Heller, Hager, and Minor [19] applied the momentum theorem by considering a baffle with height z at the end of a channel. For the end of the jet flow they proposed:

$$\frac{z}{y_0} = \frac{1}{2}\left(\sqrt{1+8F_0^2}-1\right) - F_0^{2/3} \tag{5}$$

and for the initiation:

$$\frac{z}{y_0} = \sqrt{1+2F_0^2-2F_0^{4/3}} - F_0^{2/3}. \tag{6}$$

Our research was focused on the development of a new formula that includes the effect of the curvature of the flip bucket on the initiation of the jet flow, and also the proposal of a new conceptual framework that leads to a novel procedure for the direct determination and analysis of the flow rate for the initiation of the jet flow.

2. General Approach

To take into account the effect of the flip bucket's radius of curvature a new formula was deduced, that incorporates this parameter by applying the momentum Equation. This formula establishes a relationship between the flow depth y_0 and the number of Froude F_0 at the lowest point of the flip bucket that must be met for jet flow to initiate, and it is the relationship that allows to determine the flow rate for the initiation of the jet flow.

The spillway is considered a system consisting of two elements: The chute and the flip bucket. For each of these elements, it is possible to define a *characteristic curve* that link the flow depth at the lowest point of the flip bucket y_0 and the flow rate Q. The flow rate for the initiation of the jet flow Qi, and the corresponding flow depth y_{0i} are given by the point of intersection of both *characteristic curves*. The approach is analogous to that of determining a pump operating point, when the system consists of the pump, or group of pumps, and the pipe. The characteristic curves link the pump head or head loss and the pumped flow.

The flow rate for the initiation of the jet flow must simultaneously satisfy the *flip bucket characteristic curve*, ratio between y_0 and Q (which can be determined from the relationship between z/y_0 and F_0 obtained by the momentum Equation), and the *chute characteristic curve*, which expresses the ratio between y_0 and the flow rate Q compatible with the energy loss that occurs along the channel, for being physically feasible.

The *chute characteristic curve* $y_0 = f(Q)$ may be constructed point to point by applying the Bernoulli theorem to successive sections of the chute, or by using a Computational Fluid Dynamic numerical model. We followed the second option, using the commercial software Flow3D. Three different tools were used along the workflow: Experimental work at the hydraulic laboratory, numerical Computational Fluid Dynamic modeling, and analytical deduction based on the momentum theorem (Figure 1). An empirical flow rate for the initiation of the jet flow was obtained from the physical models at the laboratory. The results of these laboratory tests were used to calibrate and validate the numerical models performed with Flow3D, regarding the flow characteristics along the chute. Once validated, Flow3D models served to elaborate the *chute characteristic curves* and to determine the *flow rate for the initiation of the jet flow* in the absence of air within the flow.

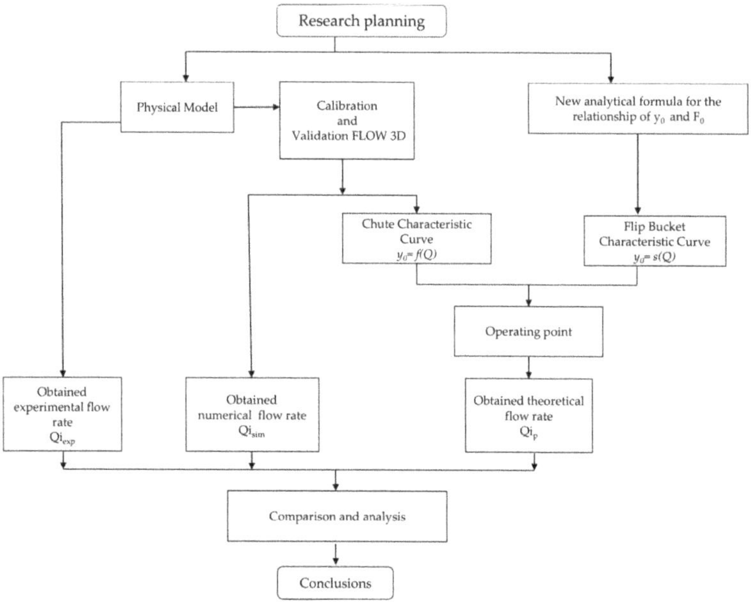

Figure 1. Workflow of the research.

Later on, the proposed theoretical formula, obtained by means of the momentum theorem, allowed to elaborate the *flip bucket characteristic curve*, taking into consideration the curvature of the bucket. The flow rate for the initiation of the jet flow was determined at the operating point, intersection of both characteristic curves of the chute and the flip bucket.

The work ends up comparing and analyzing the three flow rates for the initiation of the determined jet flow: Experimental or empirical, numerical, and analytical or theoretical, in order to discuss the validity of the proposed methodology and formula.

3. New Formula for The Initiation of the Jet Flow

The flip bucket operates like a weir and presents a hydraulic jump for low flow rates, so it is not possible to define a clear flow depth y_0 at the lowest point of the flip bucket (Figure 2). As the supercritical flow from the chute increases, the hydraulic jump moves towards the lip of the flip bucket until the energy of the flow is enough to completely sweep the hydraulic jump out of the flip bucket and the jet flow occurs. When the stream of water jumps, the flow regime is supercritical all along the chute–bucket system. Therefore, the flow depth y_0 at the lowest point of the flip bucket is determined by the upstream conditions, with control section at the ogee crest of the spillway.

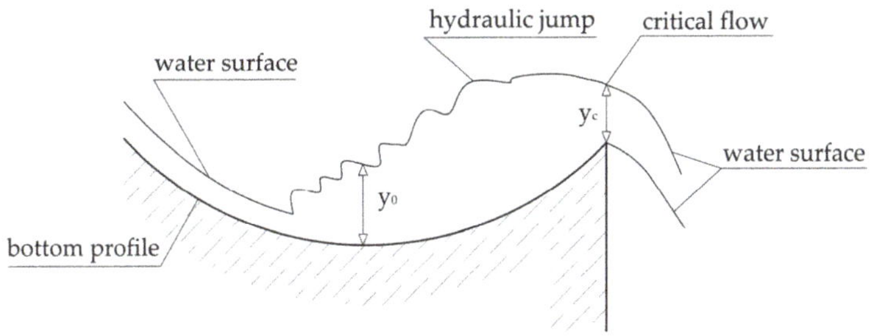

Figure 2. Flip bucket working as a weir for low flow rates.

We apply the momentum theorem in the horizontal direction [20]. The considered control volume is between the two vertical sections: (A) Passing through the lowest point of the flip bucket and (B) containing the lip (Figure 3). We accept, as Heller et al. [19], critical flow on the lip of the flip bucket before the jet flow occurs, since it is working as a weir, with a regimen change from subcritical to supercritical.

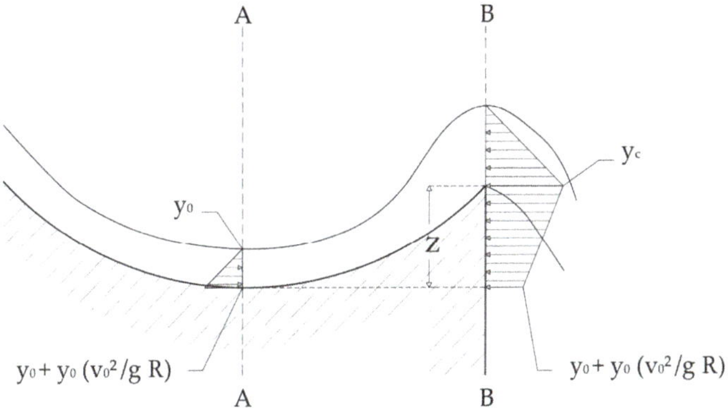

Figure 3. Head of pressure laws on the initial and final sections at the moment of the jet flow initiation.

In section (A), y_0 is the flow depth at the bottom of the flip bucket, where the corresponding speed is v_0. In section (B) the flow is critical (depth y_c) and the corresponding speed v_c. The radius of curvature R and depth of the flip bucket z are also relevant parameters.

In the initial section (A), the hydrostatic pressure was considered plus the increase of the dynamic pressure due to the curvature of the flow in the flip bucket [21]. For simplicity, linear distribution was assumed. The resulting force on section (A) is:

$$S_A = \frac{1}{2} y_0 \left(y_0 + y_0 \frac{v_0^2}{gR} \right). \tag{7}$$

In the final section (B), the hydrostatic pressure corresponding to the critical flow depth on the flip bucket's lip was assumed. The horizontal component of the resulting pressures on the flip bucket was determined considering, for simplicity, a linear distribution of pressure between the lowest point and the lip of the flip bucket. These pressures are applied to the vertical projection of the flip bucket, resulting the force:

$$S_B = \frac{z}{2} \left(y_c + y_0 + \frac{y_0 v_0^2}{gR} \right). \tag{8}$$

Neglecting the horizontal component of friction resistance, considering uniform velocity and parallel flow, and applying the momentum Equation, we obtained Equation (9):

$$\frac{1}{2}y_0\left(y_0+y_0\frac{v_0^2}{gR}\right)+\frac{q^2}{y_0g} = \frac{1}{2}y_c^2+\frac{z}{2}\left(y_c+y_0+\frac{y_0v_0^2}{gR}\right)+\frac{q^2}{y_cg} \tag{9}$$

being q the unit flow rate and g the gravity acceleration. This Equation (9) can be expressed in a different and more compact way using the Froude number at the initial section (A) F_0, and the parameter z/y_0. For that, we divide Equation (9) by y_0 and substitute for:

$$y_0v_0 = y_cv_c \tag{10}$$

$$F_0\frac{v_0}{\sqrt{gv_0}} \tag{11}$$

$$y_c = \sqrt[3]{\frac{y_0^2v_0^2}{g}} \tag{12}$$

and the resulting formula is:

$$\frac{z}{y_0} = \frac{1+2F_0^2-3F_0^{\frac{4}{3}}}{1+F_0^{\frac{2}{3}}} - \frac{F_0^2}{1+F_0^{\frac{2}{3}}}\left[\frac{(z-y_0)}{R}\right]. \tag{13}$$

Equation (13) includes the effect of the curvature of the flip bucket by means of the parameter R: Flip bucket radius of curvature. It is a novelty related to the previous existing formulas. It can be observed that the obtained formula differs from that of Abecasis and Quintela (Equation (1)) just in the negative term of the second member of the equality, that includes the radius as a parameter. This term approaches zero when increasing the radius and, at the limit, it becomes zero, as befits a straight bucket, without curvature, which is the hypothesis adopted by Abecasis and Quintela.

4. The Method of the Characteristic Curves

The sky-jump spillway can be considered as a system made up of two elements: The chute and the flip bucket. The lowest point of the bucket deflector (Section A; Figure 4) was defined as the limit between them. It is analogous to a pumping system, consisting of a pump, or group of pumps, and the pump piping. In this analogy, the chute and the pump piping might be considered analogous elements, since the frictional force, along with the gravity force, dominates the behavior in both elements. The flip bucket and the pump group might also be considered analogous. In both cases, energy is required for water elevation. In the case of the group of pumps it is provided by the engines; in the case of the flip bucket the energy is provided by the water stream coming from the chute.

Like in the case of a pumping system, there is an operating point for the chute–bucket system. The operating point is the intersection of the characteristic curves of the elements of the system: Chute and flip bucket. Both characteristic curves express a relationship between the flow depth at the lowest point of the flip bucket y_0 and the flow rate Q. The operating point defines the flow rate for the initiation of the jet flow and the flow depth at the lowest point of the flip bucket at that instant.

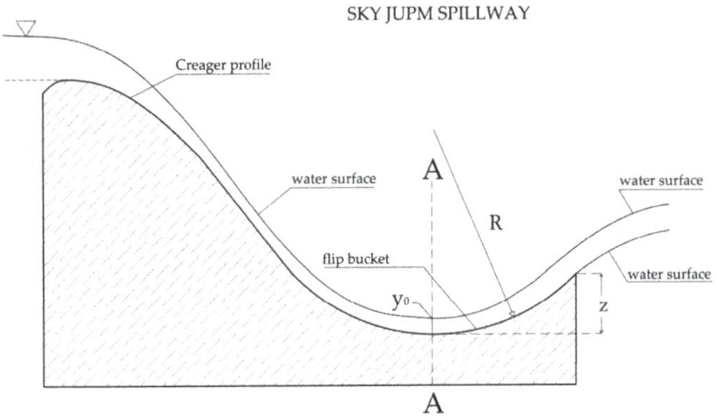

Figure 4. Section of a spillway with a flip bucket with a radius of curvature R and with a depth z, and flow depth y_0 at the lowest point of the flip bucket.

The *characteristic curve of the flip bucket* can be obtained by applying the momentum Equation at the instant of initiation of the jet flow, using the new proposed formula (Equation (13)). It establishes a relationship between the parameter z/y_0, where z is the depth of the flip bucket, and the Froude number F_0 at the lowest point of the bucket. This Equation can also be expressed as a function of the unit flow rate q and the flow depth at the lowest point of the flip bucket y_0, taking into consideration Equations (10)–(12):

$$y_0^3 - y_0^2 z + y_0\left(\frac{q^2}{gR} - 3\frac{q^{\frac{4}{3}}}{g^{\frac{2}{3}}} - z\frac{q^{\frac{2}{3}}}{g^{\frac{1}{3}}}\right) + 2\frac{q^2}{g} - \frac{zq^2}{gR} = 0. \quad (14)$$

The function $y_0 = s(q)$ can be built point by point using Equation (14), under the hypothesis of cylindrical flip bucket and neglecting the effect of the flow aeration. This is the curve we call *flip bucket characteristic curve.*

The *chute characteristic curve* $y_0 = f(q)$ can be built using a CFD numerical model with greater accuracy than using Bernoulli theorem. We determined the value of y_0 for different unit flow rates q using the commercial CFD code Flow3D.

Jet flow initiates when the operation conditions of the chute and the flip bucket are physically compatible, at the intersection of both characteristic curves: The operating point. It provides the flow rate for the initiation of the jet flow and the flow depth at the lowest point of the bucket deflector at that instant.

5. Experimental Work with Physical Models

The testing channel (Figures 5 and 6) has a width of 2.46 m, is 1.3 m high, and 13.7 m long. It is divided into three functional areas. The first zone is for water supply and dissipation of its energy. The second is the testing area: 2.46 m wide and 6.37 m long. The testing area is filled with sand up to a height of 0.5 m, and is limited upstream by a wall. The sky-jump spillway is positioned in the middle of that wall. In the third and final zone of the testing channel there is a decant pond to prevent any dragged material to reach the tank, which is under the laboratory floor. On the left side of the channel, in the direction of the flow, there is a glass window 4.6 m long and 1.1 m high, which allows the visual inspection of the test and also to take photographs and video recording.

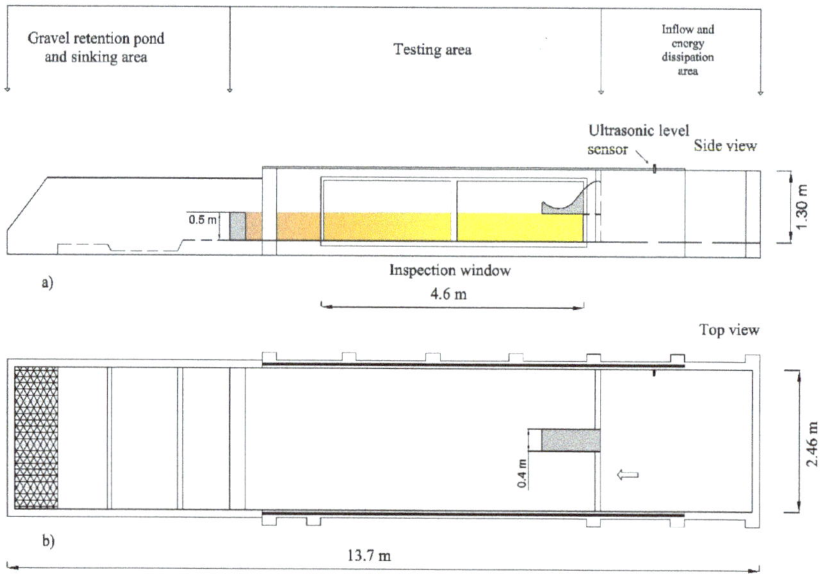

Figure 5. Testing set-up. (**a**) Side view; (**b**) Plan view.

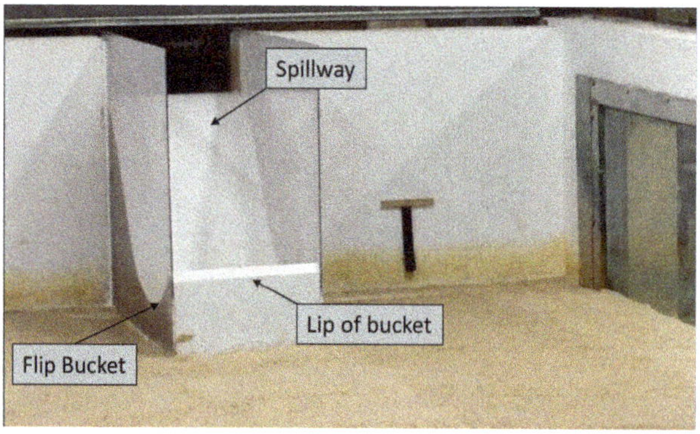

Figure 6. Sky-jump spillway in the testing channel. Case: radius of curvature 0.4 m; flip angle 45°.

The water inlet to the model is regulated by a motorized valve. The flow rate was measured at two locations: Upstream, by means of an ultrasonic flow meter positioned in the supply pipe, and downstream, using a thin-plate rectangular weir located in the return channel that takes water to the tank. Two ultrasonic level probes based on the pulse-eco method, positioned upstream and downstream of the sky-jump spillway, were used for the recording of the water levels. The hardware specifications of the ultrasonic flow meter and the ultrasonic level probe are shown in the (Table 1).

Table 1. Hardware specifications used in the experiments.

Instrumentation	Measuring Range	Accuracy
FLUXUS-ADM7407	0.01–25 m/s	± 1.6 % of reading ±0.1 m/s
Ultrasonic distance measuring system UAS	0.3 m–2 m	>=1 mm

Six cement-made sky-jump spillways were built (Figure 7). The design followed the criteria specified by the United States Bureau of Reclamation [22]. The spillway has a Creager profile. The slopes of the physical model are vertical upstream and 0.8H:1.0V downstream; the height H is 0.6 m; and the distance between the bottom of the flip bucket and elevation spillway is 0.5 m; it is 0.4 m wide. Three different values of radius of curvature R were modeled: 0.2 m, 0.3 m, and 0.4 m; and also three different values of exit angle of flip bucket α: 15°, 30°, and 45°. The radius of curvature and the flip angle allow to determine z, depth of the flip bucket, using the geometric relation $z = R\,(1 - \cos\alpha)$. In order to inspect and measure the water depth, the sides of the physical models were built with methacrylate.

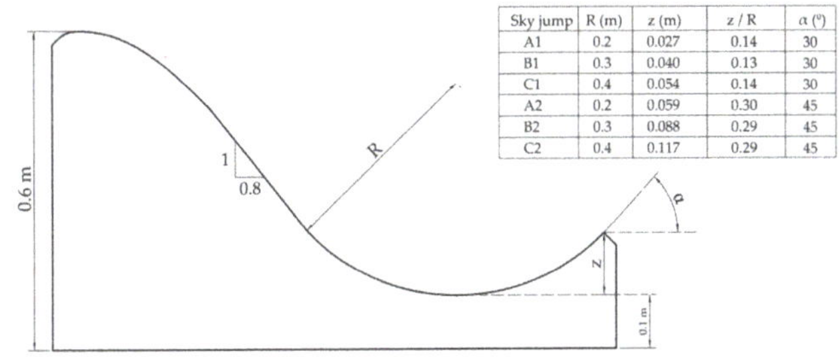

Sky jump	R (m)	z (m)	z / R	α (°)
A1	0.2	0.027	0.14	30
B1	0.3	0.040	0.13	30
C1	0.4	0.054	0.14	30
A2	0.2	0.059	0.30	45
B2	0.3	0.088	0.29	45
C2	0.4	0.117	0.29	45

Figure 7. Geometrical configuration of the physical models: radius of curvature R, depth of the flip-bucket z, flip angle α, and parameter z/R are variable parameters.

A protocol was followed for the experimental determination of the flow rate for the initiation of the jet flow. The level of the reservoir was maintained 0.04 m below the ogee crest threshold to allow the flow to stabilize before allowing the specified flow rate to enter the spillway. Every flow rate was maintained during 10 min flowing along the spillway. After that, if the jet flow did not occur, the procedure was repeated with the next higher flow rate discharge, and so on until the jet flow occurred. The flow rate for the initiation of the jet flow was measured for every geometrical configuration. The flow depth at the lowest point of the flip bucket could only be measured with sufficient accuracy for the sky-jump spillway with radius of curvature of 0.4 m and flip angle of 45° (C2). For the rest of the physical models the depth of water was too small. It was possible to differentiate for the case C2 an intensely aerated area, in the upper part of the flow, from a predominantly liquid area in the lower part (Figure 8).

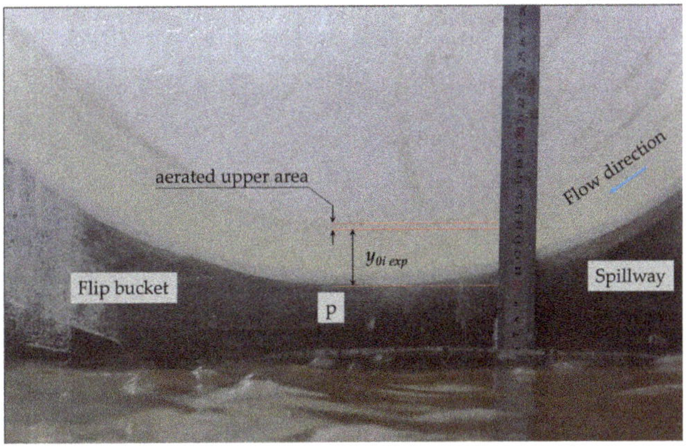

Figure 8. Flow at the lower part of the flip-bucket for case C2, with a discharge of 43.56 L/s. The upper part of the flow is intensely aerated. The lowest point of the flip bucket is "p" and the flow depth without intensely aerated upper area is $y_{0i\ exp}$.

The experimental results (Table 2) are later in this paper compared with the results obtained applying the proposed *method of the characteristic curves*, using the new formula for the characteristic curve of the flip bucket, and with the results obtained from the CFD numerical model (Flow3D), described in the following section. The CFD numerical model was calibrated and validated with the experimental data of the physical model C2.

Table 2. Experimental results: Flow rate for the initiation of the jet flow Qi_{exp} (L/s) for each physical model and total flow depth $y'_{0i\ exp}$ (mm) and flow depth without intensely aerated upper area $y_{0i\ exp}$ (mm).

Physical Model	[1]A1	[2]B1	[3]C1	[4]A2	[5]B2	[6]C2
Qi_{exp} (L/s)	1.65	2.75	3.70	4.80	7.65	11.95
$y'_{0i\ exp}$ (mm)	-	-	-	-	-	*14.50
$y_{0i\ exp}$ (mm)	-	-	-	-	-	**13.00

[1] (R = 0.2 m, α = 30°); [2] (R = 0.3 m and α = 30°); [3] (R = 0.4 m α = 30); [4] (R = 0.2 m α = 45°); [5] (R = 0.3 m α = 45°); [6] (R = 0.4 m α = 45° (C2) The total flow depth $y'_{0i\ exp}$ (mm) with aerated upper area is * 14.50 mm and the flow depth without intensely aerated upper area $y_{0i\ exp}$ (mm) is ** 13.00 mm, corresponding to the measured flow rates for the flip bucket (C2), at the lowest point of the flip bucket.

6. Numerical Models: Chute Characteristic Curve and Flow Rate for The Initiation of the Jet Flow

Every physical model was also modelled numerically. As above mentioned, the *chute characteristic curve* was obtained point by point by means of the CFD numerical modeling code Flow3D. This software solves Reynolds–Navier–Stokes averaged Equations (RANS) in three dimensions along with the FAVOR and VOF algorithms for solid contouring and free surface tracking respectively [23]. Due to a finite difference approach, Flow3D requires the resolution of structured meshes. The software uses the Fractional Area/Volume Obstacle Representation (FAVOR) method, developed by Hirt and Sicilian [24]. This preprocessor is a tool that allows to represent a solid obstacle in a control volume. Control volumes with dead spaces are assumed empty, and value 1 is assigned, and volumes with geometry are assumed solid volumes and a value of 0 is assigned. The Volume of Fluid (VOF) method is described by Nichols and Hirt [25], Nichols et al. [26], and Hirt and Nichols [27]. This method is similar to FAVOR; it defines whether a cell is empty, full, or partially filled with water. Cells without fluid have a value equal to 0. The filled cells have value equal to 1, and between 0 and 1 those partially filled.

The software Flow3D solves in three directions the mass continuity Equation (Equation (15)) and the momentum Equations (Equations (16)–(18)) besides the volume of fluid (VOF) Equation, that ensures that proper boundary conditions are applied at the free surface (Equation (19)). For Cartesian coordinates (x_i, x_j, x_z) and for incompressible fluid (fluid density constant), these Equations are:

$$\frac{\partial}{\partial x_i}(u_i A_i) + \frac{\partial}{\partial x_j}(u_j A_j) + \frac{\partial}{\partial x_k}(u_k A_k) = \frac{R_{SOR}}{\rho} \tag{15}$$

$$\frac{\partial u_i}{\partial t} + \frac{1}{V_F}\left\{u_i A_i \frac{\partial u_i}{\partial x_i} + u_j A_j \frac{\partial u_i}{\partial x_j} + u_k A_k \frac{\partial u_i}{\partial x_k}\right\} = -\frac{1}{\rho}\frac{\partial p}{\partial x_i} + G_{x_i} + f_{x_i} \tag{16}$$

$$\frac{\partial u_j}{\partial t} + \frac{1}{V_F}\left\{u_i A_i \frac{\partial u_j}{\partial x_i} + u_j A_j \frac{\partial u_j}{\partial x_j} + u_k A_k \frac{\partial u_j}{\partial x_k}\right\} = -\frac{1}{\rho}\frac{\partial p}{\partial x_j} + G_{x_j} + f_{x_j} \tag{17}$$

$$\frac{\partial u_k}{\partial t} + \frac{1}{V_F}\left\{u_i A_i \frac{\partial u_k}{\partial x_i} + u_j A_j \frac{\partial u_k}{\partial x_j} + u_k A_k \frac{\partial u_k}{\partial x_k}\right\} = -\frac{1}{\rho}\frac{\partial p}{\partial x_k} + G_{x_k} + f_{x_k} \tag{18}$$

$$V_F \frac{\partial F}{\partial t} + \nabla\,(AUF) = 0 \tag{19}$$

where t is time, ρ is fluid density, p is pressure, (u_i, u_j, u_k) are velocity component in Cartesian coordinates (x_i, x_j, x_k), A_i is fractional area in the i-direction, A_j and A_k are similar area fractions in the j and k direction, respectively, (Gx_i, Gx_j, Gx_k) are body acceleration and (fx_i, fx_j, fx_k) are viscous acceleration. In Equation (19) A is the average flow area, U is the average velocity and F is the volume flow function. When the cell is filled with fluid, the value of F is 1, and when it is empty, F is 0.

The fluid-dynamic variables at a point in space are made up of a series of fluctuations of different scales, for this reason, the analysis of the turbulence is carried out from a statistical point of view, that is, with average velocity and intensity of "Reynolds averaged" fluctuations. This consideration helps solve the closure problem. For the numerical solution of turbulent flows, the Reynolds-Averaged Navier–Stokes Equations are applied, which supplies the turbulence kinetic energy k and the rate of turbulent energy dissipation ε to achieve Reynolds stresses and the turbulent kinematic viscosity.

A quantitative verification, as defined in Jakeman et al. [28] and in Blocken et al. [29], was made for this numerical study, using the available data from Paul Guy Chanel [30].

The turbulence model used in this study is the RNG k-ε [31], which usually provides better performance for swirling flows than standard k-ε model [32,33]. Its formulation is represented in Equations (20) and (21):

$$\frac{\partial}{\partial t}(\rho k) + \frac{\partial}{\partial x_i}(\rho k u_i) = \frac{\partial}{\partial x_j}\left[\left(\mu + \frac{\mu_t}{\sigma_k}\right)\frac{\partial k}{\partial x_j}\right] + P_k - \rho\varepsilon \tag{20}$$

$$\frac{\partial}{\partial t}(\rho \varepsilon) + \frac{\partial}{\partial x_i}(\rho \varepsilon u_i) = \frac{\partial}{\partial x_j}\left[\left(\mu + \frac{\mu_t}{\sigma_\varepsilon}\right)\frac{\partial \varepsilon}{\partial x_j}\right] + C_{1\varepsilon}\frac{\varepsilon}{k}P_k - C_{2\varepsilon}\rho\frac{\varepsilon^2}{k} \tag{21}$$

where μ is dynamic viscosity, μ_t is turbulent dynamic viscosity and P_k is production of turbulence kinetic energy. The remaining terms $C_{1\varepsilon}$, $C_{2\varepsilon}$, σ_k and σ_ε are model parameters whose values can be found in Yakhot et al. [31]. Finally, the turbulence viscosity can be computed using the parameter $C\mu = 0.085$ in the Equation (22):

$$\mu_t = \rho\, C_\mu \frac{k^2}{\varepsilon} \tag{22}$$

The numerical model is implemented by importing the geometry of the entire physical model. A domain of 2.3 m long and 1 m high was defined. Although Flow3D is inherently a 3D software, it was considered a small thickness in the Y direction, transverse to the model section, in such a way that the behavior of the flow could be studied in 2D (Figure 9), since the analyzed phenomenon is essentially two-dimensional.

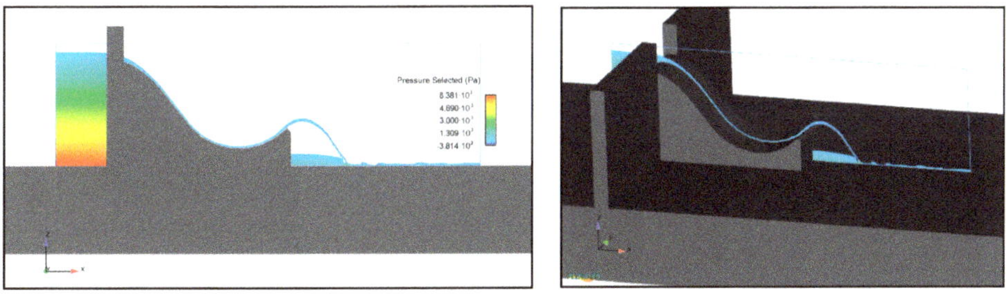

(a) (b)

Figure 9. Geometric model implemented in Flow3D: (**a**) 2D view; (**b**) 3D view.

The flow depths corresponding to the flow rate for the initiation of the jet flow for every modelled case are in the order of millimeters for most of the cases.

The boundary conditions must accurately represent those of the physical phenomenon: Two boundary conditions are fixed for each Cartesian plane: For the plane (XY) the atmospheric pressure and the wall condition are considered, for the plane (YZ) the hydrostatic pressure distribution and the output are defined, for the plane (XZ) the symmetrical smooth walls (free-slip/symmetry) are defined.

The numerical model was calibrated and validated using the experimental data corresponding to the 0.4 m radius of curvature and flip angle of 45° (C2). The test with the smallest flow rate was used to calibrate the roughness of the chute, and the test with the larger flow rate was used to validate the value obtained in the calibration. The flow depth with and without the intensely aerated upper area of the flow were separately considered.

The numerical simulation was started with the Mesh-sensitivity analysis and calibration of the Flow3D model. The measured value of the flow depth without the intensely aerated area ($y_{0\ exp}$) was used, since the proposed analytical formula does not consider the effect of aeration, and the purpose is to compare both results.

The absolute roughness of the chute was used as the calibration parameter. The roughness of the smooth cement is between 0.03 mm and 0.50 mm [21]. We developed numerical models with 0.03 mm, 0.10 mm, 0.25 mm, and 0.50 mm.

The available experimental data refer to the 0.4 m flip bucket's radius and lip angle of 45° (C2), for which we have two flow depth data at the lowest point of the flip bucket and the two related flow rates.

Mesh-sensitivity analysis, categorized as part of the quantitative verification, was performed diminishing the cell size, until changes were not reported in the results (Table 3). However, the results with a 1 mm mesh were computationally unfeasible and a trade-off between computational time and accuracy was made.

Table 3. Calibration results (roughness variation) and mesh-sensitivity analysis (cell side variation).

Roughness (mm)	Q_{sim} (L/s)	$y_{0\ sim\ cell\ 1\ mm}$ (mm)	$y_{0\ sim\ cell\ 1.25\ mm}$ (mm)	$y_{0\ sim\ cell\ 2.5\ mm}$ (mm)
0.03	11.34	12.98	13.01	14.78
0.10	12.21	14.00	14.10	15.35
0.25	12.64	14.25	14.36	15.48
0.50	doesn't jump	doesn't jump	doesn't jump	doesn't jump

The cell with the size of 1.25 mm was adopted because it restores the same water depth value observed experimentally and it also was computationally feasible. Then, different absolute roughness values were considered, as indicated in the (Table 3), checking if the jet flow occurred.

Roughness was set to 0.03 mm and a cell size of 1.25 mm was adopted, because they restore the same water depth value experimentally observed (Table 4).

Table 4. Comparison between experimental and numerical flow rate and flow depth for calibration.

Q_{exp} (L/s)	Q_{sim} (L/s)	*AE_Q (L/s)	**RE_Q (%)	$y'_{0\ exp}$ (mm)	$y_{0\ exp}$ (mm)	$y_{0\ sim\ cell\ 1.25}$ (mm)	***$AEy'_{0\ exp-sim}$ (mm)	****$REy'_{0\ exp-sim}$ (%)
11.95	11.34	0.61	5.07	14.50	13.00	13.01	1.49	10

*AE_Q (Absolute Error) and ** RE_Q (Relative Error) between experimental flow rate and numerical flow rate. *** AEy_0 (Absolute Error) and **** REy'_0 (Relative Error) between experimental flow depth considering the intensely aerated area and numerical flow depth.

This absolute roughness is compatible with the surface of sanded concrete of the physical models. The calibration results were then validated. For that, the flow rates obtained experimentally and numerically that produce the same flow depth at the lowest point of the flip bucket were compared, without considering the intensely aerated area. Validation was performed for an experimental flow rate of 43.56 L/s (Table 5).

Table 5. Comparison between experimental and numerical flow rate and flow depth for validation.

Q_{exp} (L/s)	Q_{sim} (L/s)	*AE_Q (L/s)	**RE_Q (%)	$y'_{0\ exp}$ (mm)	$y_{0\ exp}$ (mm)	$y_{0\ sim}$ (mm)	***$AEy'_{0\ exp\text{-}sim}$ (mm)	****$REy'_{0\ exp\text{-}sim}$ (%)
43.56	42.86	0.70	1.61	42.00	38.00	38.22	3.78	9

*AE_Q (Absolute Error) and **RE_Q (Relative Error) between experimental flow rate and numerical flow rate. ***AEy_0 (Absolute Error) and ****REy'_0 (Relative Error) between experimental flow depth considering the intensely aerated area and numerical flow depth.

The relative error between flow rates observed and obtained from the numerical models is in the order of 5% in calibration and lower in validation. It is in accordance with the expected error in the numerical model [23] and in the laboratory test.

Once the numerical model was calibrated and validated, it was used for two purposes: (a) To build the *chute characteristic curve*; (b) to determine the flow rate for the initiation of the jet flow; both for the different laboratory-tested geometric configurations. For the first purpose, three spillways, corresponding to each considered radius, were numerically modeled (Figure 10). As expected, the differences between the curves of the three cases were minimal, since they only differ slightly on the initial curved part (Table 6). The *chute characteristic curve* was obtained by interpolation from the points defined by numerical modeling, (Figure 11), where the fixed parameters are: total height H of 0.6 m and distance P from the bottom of the flip bucket to the ground of 0.1 m; variable parameters are: radius of curvature R, horizontal distance La from the upstream vertical wall to the point where the straight part of the chute ends, horizontal distance Lc from the upstream vertical wall and measurement point "p", and horizontal distance Lt between the measurement point "p" and the flip bucket's lip. The part of the flip bucket downstream point "p" has no influence on the chute flow, due to the supercritical regime.

The flow rate for the initiation of the jet flow was determined for each of the six laboratory-tested geometric configurations, using the numerical model, by trial and error, increasing the flow discharge in small steps until the jet flow occurred (Figure 12; Table 7).

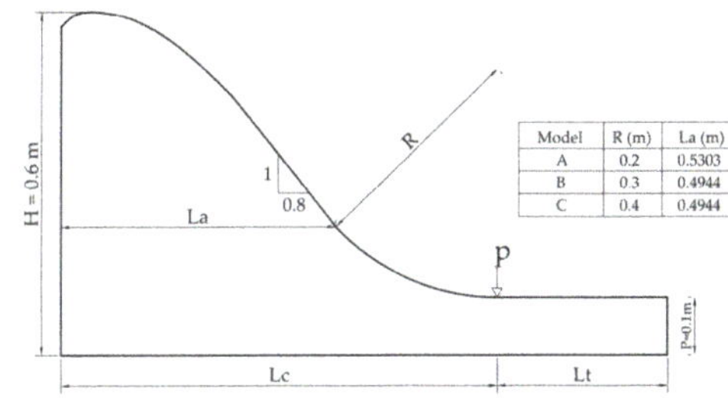

Model	R (m)	La (m)	Lc (m)	Lt (m)
A	0.2	0.5303	0.689	0.306
B	0.3	0.4944	0.737	0.306
C	0.4	0.4944	0.783	0.306

Figure 10. Outline of the geometric configuration used to obtain the *chute characteristic curves* using numerical modeling.

Table 6. Modeled points (Q_{sim}, $y_{0\,sim}$), to draw the *chute characteristic curves* with radius 0.2 m (A), 0.3 m (B), 0.4 m (C).

Q_{sim}_A (L/s)	$y_{0\,sim}$_A (mm)	Q_{sim}_B (L/s)	$y_{0\,sim}$_B (mm)	Q_{sim}_C (L/s)	$y_{0\,sim}$_C (mm)
0.000	0.000	0.000	0.000	0.000	0.000
0.701	2.041	0.701	2.043	0.701	2.044
1.188	2.791	1.188	2.798	1.188	2.873
1.581	3.409	1.581	3.492	1.581	3.474
1.994	3.899	1.994	3.799	1.994	3.892
2.952	4.806	2.952	4.858	2.952	4.861
3.276	5.126	3.276	5.155	3.276	5.211
3.839	5.666	3.839	5.666	3.839	5.712
4.844	6.358	4.844	6.358	4.844	6.557
5.075	6.465	5.075	6.525	5.075	6.707
7.239	8.054	7.239	8.094	7.239	8.094
8.505	8.890	8.505	8.890	8.502	8.977
9.877	9.877	9.877	9.877	9.877	9.917
11.247	10.973	11.247	10.991	11.247	11.027
14.335	13.500	14.335	13.500	14.335	13.433

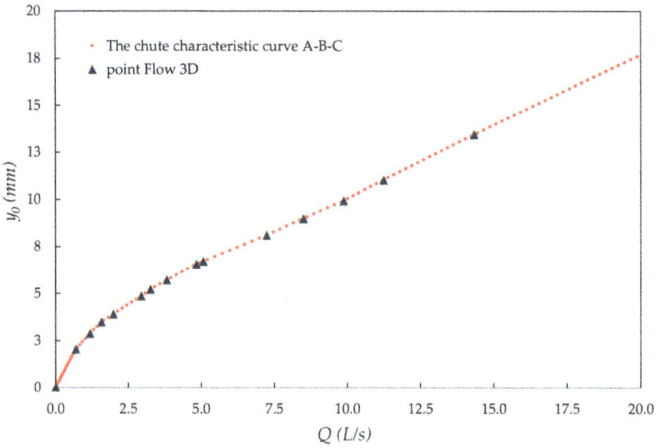

Figure 11. Three *chute characteristic curves*, with radius of 0.2 m (**A**), 0.3 m (**B**) and 0.4 m (**C**), are visually overlapped because they only differ slightly on the initial curved part.

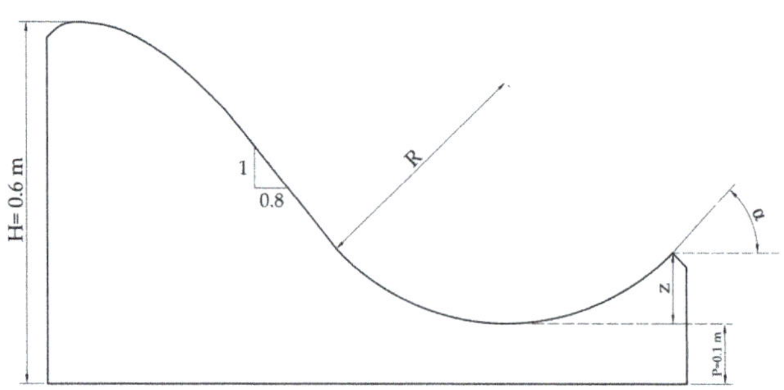

Figure 12. General geometry of the trampoline-type spillway.

Table 7. Results of flow rate for the initiation of the jet flow Qi_{sim} and flow depth $y_{0i\,sim}$ at the lowest point, obtained with numerical simulation for each geometrical configuration.

Tested Geometry	R (m)	α (°)	z (m)	Qi_{sim} (L/s)	$y_{0i\,sim}$ (mm)
A1	0.2	30	0.027	1.12	2.61
B1	0.3	30	0.040	2.03	4.01
C1	0.4	30	0.054	2.86	4.61
A2	0.2	45	0.059	3.79	5.38
B2	0.3	45	0.088	5.76	7.06
C2	0.4	45	0.117	8.49	9.08

7. Results and Discussion

The main purpose of this research is to define and validate a methodology that allows to determine the flow rate for the initiation of the jet flow of a sky-jump spillway, taking into account the effect of the radius of curvature R of the flip bucket.

According to the proposed methodology, the flow rate for the initiation of the jet flow is the abscissa of the *operating point* at the intersection of two curves, *the flip bucket characteristic curve* and *the chute characteristic curve*.

The *flip bucket characteristic curve* was built point by point using the proposed analytical formula (Equation (13)), that includes the effect of the flip bucket curvature. The *chute characteristic curve* was built also point by point using the results of the numerical simulations performed with Flow3D. The operating point was identified for each geometrical configuration considered for the spillway. This way the flow rate Qi_p and flow depth y_{0ip} were determined, and compared to the results of the numerical simulation (Qi_{sim}, $y_{0i\,sim}$) and to the values experimentally observed (Qi_{exp}, $y_{0i\,exp}$). The main results of the different phases of the research are summarized in (Table 8).

Table 8. The flow rate for the initiation of the jet flow and the flow depth at the lowest point of the flip bucket are showed.

Tested Geometry	R (m)	α (°)	z (m)	Qi_p (L/s)	Qi_{sim} (L/s)	Qi_{exp} (L/s)	y_{0ip} (mm)	$y_{0i\,sim}$ (mm)	$y'_{0i\,exp}$ (mm)	$y_{0i\,exp}$ (mm)
A1	0.2	30	0.027	1.20	1.12	1.65	2.81	2.61	-	-
B1	0.3	30	0.040	2.18	2.03	2.75	4.20	4.01	-	-
C1	0.4	30	0.054	2.93	2.86	3.70	4.84	4.61	-	-
A2	0.2	45	0.059	3.86	3.79	4.80	5.67	5.38	-	-
B2	0.3	45	0.088	5.97	5.76	7.65	7.17	7.06	-	-
C2	0.4	45	0.117	8.84	8.49	11.95	9.21	9.08	14.50	13.00

Note: (Qi_p, y_{0ip}) is obtained using the proposed methodology and formula, (Qi_{sim}, $y_{0i\,sim}$) is obtained by numerical simulation and (Qi_{exp}, $y_{0i\,exp}$) experimentally in laboratory. Experimental flow depth was measured with enough accuracy only for the C2 flip bucket, being $y_{0i\,exp}$ the flow depth excluding the intensely aerated upper area.

The flow rates for the initiation of the jet flow obtained by means of the proposed methodology and analytical formula and those obtained by numerical simulation are quite similar, while the values experimentally obtained are somewhat higher. The oscillation of ±0.75 L/s in the flow rates supplied to the physical model does not allow to explain the observed differences. It should be emphasized that both the proposed formula and the Flow3D numerical model do not include the effect of flow aeration, which is present on the physical model tests. This fact might explain the differences observed. If this was the reason, the effect of aeration would be to increase the flow rate for the initiation of the jet flow. In the tests performed the increase was around 26% (Table 9). A greater increase might be expected at prototype size, due to the higher degree of aeration of the flow as compared to the laboratory size.

Table 9. AE: Absolute Error and RE: Relative Error.

Tested Geometry	Mean Value	$AE_{Qi\ (p\text{-}sim)}$ (L/s)	$RE_{Qi\ (p\text{-}sim)}$ (%)	$AEy_{0i\ (p\text{-}sim)}$ (mm)	$REy_{0i\ (p\text{-}sim)}$ (%)	$AE_{Qi\ (exp\text{-}sim)}$ (L/s)	$RE_{Qi\ (exp\text{-}sim)}$ (%)
A1		0.08	6.66	0.20	7.11	0.53	32.12
B1		0.15	6.88	0.19	4.52	0.72	26.18
C1		0.07	2.39	0.24	4.95	0.84	22.70
A2		0.07	1.81	0.29	5.11	1.01	21.04
B2		0.21	3.52	0.12	1.67	1.89	24.70
C2		0.35	3.96	0.13	1.41	3.46	28.95
	*MAE-MRE	0.15	4.20	0.19	4.12	1.41	22.84

*MAE: Mean Absolute Error and MRE: Mean Relative Error between the flow rate obtained by the proposed formulation Qi_p and the Qi_{sim} flow rate determined with Flow3D; between the flow rate obtained experimentally Qi_{exp} and the numerical flow rate Qi_{sim}; and between the flow depth obtained by the proposed formulation y_{0ip} and the flow depth $y_{0i\ sim}$ determined with Flow3D.

Let us now consider the differences between the flow depths at the lower point of the flip bucket obtained using the proposed methodology and formula and those obtained by numerical modeling. It is observed that by increasing the flow rate, the relative error decreases. For the flow rate, the MAE value is 0.15 L/s, the MRE is 4.20%, and the relative error reaches 6.88%. Similarly, for the flow depth it is observed that the MAE is 0.19 mm, and the absolute error does not exceed 0.29 mm; moreover, the MRE is 4.12% and relative error slightly exceeds 7%. For the proposed methodology and formula, it can be said in short, that the mean errors are less than 5%, related to the values obtained by numerical modeling.

Error is generally greater for small values of the flow rate for the initiation of the jet flow. Some uncertainty should be expected due to the markedly sharp angle of the characteristic curves of chute and flip bucket at the intersection point (Figure 13). A small change in any of the two parameters involved may cause a significant variation in the other one, and so in the *chute-flip bucket system*. The high sensitivity of the system to small alterations might explain the difficulty to accurately measure the flow rate for the initiation of the jet flow and the corresponding water depth at the lower point of the flip bucket.

It is observed that the flow rate for the initiation of the jet flow increases with the depth z of the flip bucket. It should be expected z to be the parameter with greater influence in the jet flow triggering. The supercritical flow stream, coming from the chute, must have enough energy to push the water mass out of the flip bucket. The water in the flip bucket has a depth z plus the height of water necessary to discharge over the lip, which acts as a weir before the jet flow occurs. On the other hand, it should be noticed that when the obstacle is not a simple wall, but a flip bucket with a certain radius of curvature, the geometric configuration of the flip bucket would be expected to influence the overspill conditions, and therefore to affect the flow rate for the initiation of the jet flow. The influence of the flip bucket curvature is included into the formula through the parameter R, radius of curvature of the flip bucket.

The effect of the radius of curvature of the flip bucket on the flow rate for the initiation of the jet flow can be easily analyzed by comparing the flow rate for the initiation of the jet flow obtained using the proposed formula with the actual R radius of each flip bucket and that obtained with the same formula considering an infinite radius, equivalent to the absence of curvature. The goodness of the result is assessed by comparison with the results of the numerical simulations performed with Flow3D, since the influence of aeration is present in physical models (Figure 14).

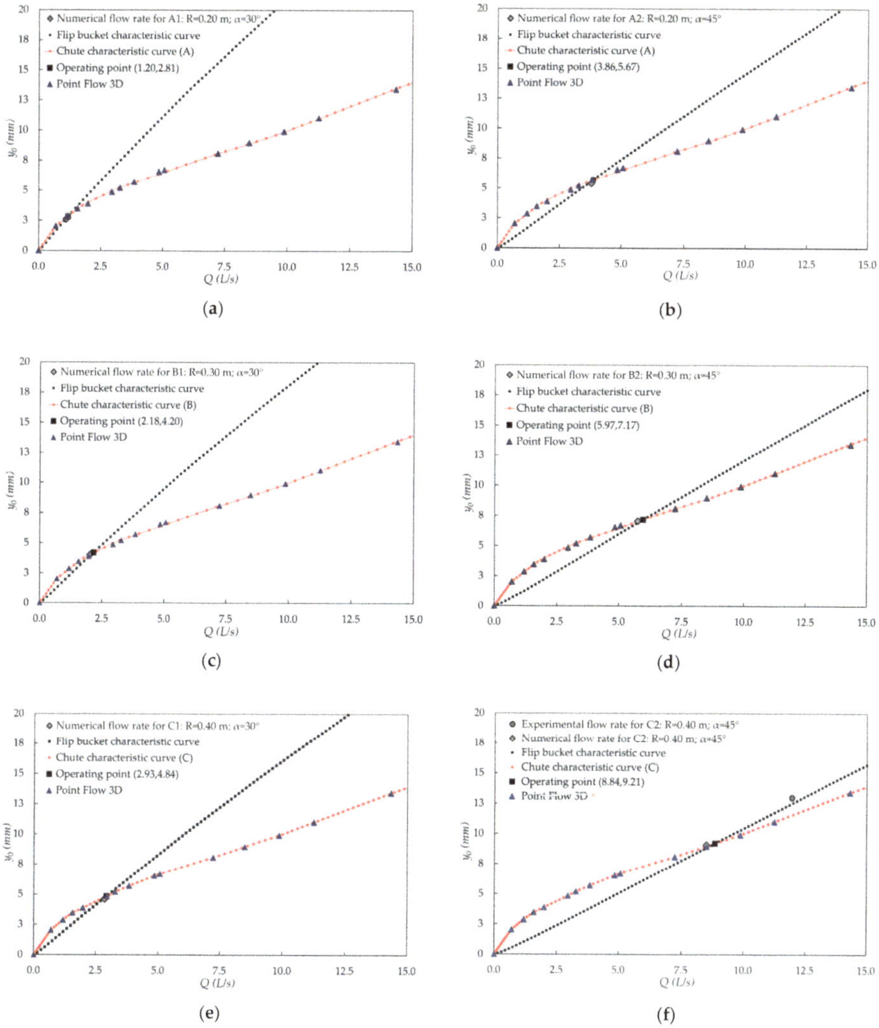

Figure 13. Characteristic curves and operating point of the chute-flip bucket system: (**a**) A1 (R = 0.2 m; α = 30°); (**b**) A2 (R = 0.2 m; α = 45°); (**c**) B1 (R = 0.3 m; α = 30°); (**d**) B2 (R = 0.3 m; α = 45°); (**e**) C1 (R = 0. 4 m; α = 30°); (**f**) C2 (R = 0.4 m; α = 45°).

It was determined the difference between the flow rate for the initiation of the jet flow obtained by numerical simulation, and the values obtained by applying the proposed analytical formula with real radius and infinite radius, at the limit (without curvature). For infinite radius the term of the formula containing the radius becomes zero in Equation (13). Considering the results of the numerical models as the reference, the Mean Absolute Error and the Mean Relative Error were determined (Table 10). It is noted that the Relative Error that is committed considering the complete proposed formula, with the term that includes the radius, does not exceed 8%, while if the proposed formula is considered by removing that term, the Relative Error reaches 24%. Consequently, the inclusion of radius in the formulation is relevant.

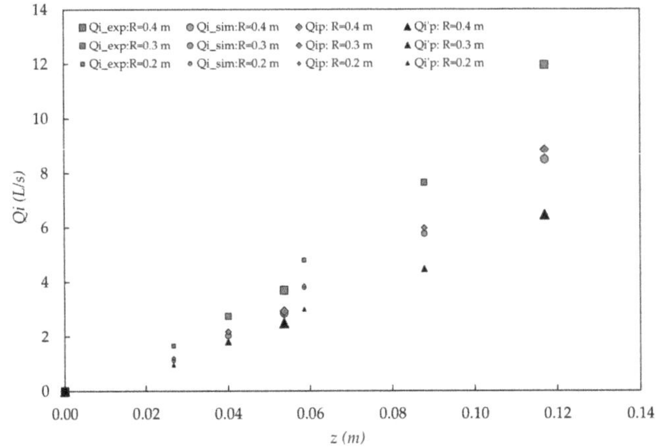

Figure 14. Flow rate for the initiation of the jet flow based on z and R. Qi_p theoretical flow rate considering R; Qi'_p theoretical flow rate not considering R; Qi_{exp} experimental flow rate; Qi_{sim} numerical flow rate.

Table 10. Comparison using the Absolute Error and the Relative Error of the numerical flow rate and theoretical flow rate for the initiation of the jet flow considering or not R.

Tested Geometry	Qi_p (L/s)	Qi'_p (L/s)	Qi_{sim} (L/s)		$AE_{Qi\ (sim-p)}$ (L/s)	$RE_{Qi\ (sim-p)}$ (%)	$AE_{Qi'(sim-p)}$ (L/s)	$RE_{Qi'\ (sim-p)}$ (%)
A1	1.20	0.95	1.12		0.08	7.14	0.17	15.18
B1	2.18	1.80	2.03		0.15	7.39	0.23	11.33
C1	2.93	2.50	2.86		0.07	2.45	0.36	12.59
A2	3.86	3.00	3.79		0.07	1.85	0.79	20.84
B2	5.97	4.47	5.76		0.21	3.65	1.29	22.40
C2	8.84	6.45	8.49		0.35	4.12	2.04	24.30
				*MAE-MRE	0.15	4.43	2.44	17.80

*MAE: Mean Absolute Error and MRE: Mean Relative Error between the flow rate for the initiation of the jet flow, for the different physical models, obtained by the proposed formulation Qi_p and Qi'_p, respectively, considering or not the term that depends on the radius, and the flow rate for the initiation of the jet flow obtained by numerical simulation Qi_{sim}.

For the purpose of quantifying the influence of the radius on the flow rate for the initiation of the jet flow, the proposed methodology and formula were applied to three flip buckets with radius 0.2 m, 0.3 m, and 0.4 m, keeping the z parameter fixed with value of 0.054 m (Table 11). It was observed that, for the same z value, the flow rate for the initiation of the jet flow increased as the radius was reduced, or, equivalently, as the flip angle was increased (Figure 15). The increase in flow rate was 17.4% when the radius is halved, from 0.4 to 0.2 m. In accordance, the error due to neglecting the effect of the curvature of the flip bucket increases by decreasing the radius, in fact, the Relative Error increased from 12.97% to 25.87% when the radius was reduced by half, from 0.4 m to 0.2 m.

Table 11. Comparison using the Absolute Error and the Relative Error of the flow rate for the initiation of the jet flow, obtained by the proposed methodology and formula and considering whether or not the radius-dependent term, respectively Qi_p and Qi'_p, for z = 0.054 m.

R (m)	z (m)	α (°)	Qi_p (L/s)	Qi'_p (L/s)	$AE_{Q\text{-}Q'}$ (L/s)	$RE_{Q\text{-}Q'}$ (%)
0.4	0.054	30	2.93	2.55	0.38	12.97
0.3	0.054	34.78	3.09	2.55	0.54	17.50
0.2	0.054	42.95	3.44	2.55	0.89	25.87

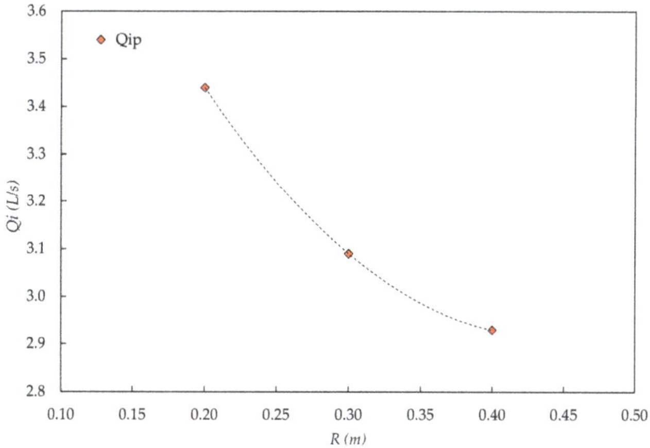

Figure 15. Flow rate for the initiation of the jet flow, obtained by the proposed methodology and formula for different values of R, Qi_p, for a constant value of z = 0.054 m.

It is interesting to compare the flow rate for the initiation of the jet flow obtained through the proposed methodology and formula, which includes the effect of the curvature of the flip bucket, with those obtained using the formulas previously proposed by various authors. The *flip bucket characteristic curve* is usually expressed as a relationship between parameter z/y_0 and the Froude number F_0, both at the lowest point of the flip bucket (Equations (1)–(6)). It can also be expressed as a relationship between y_0 and the unit flow rate for the initiation of the jet flow q (Table 12). This is useful for quantifying the flow rate for the initiation of the jet flow using the proposed *method of the characteristic curves*.

Table 12. Relationship between z/y_0 and the Froude number F_0 at the lowest point of the flip bucket, expressed as a function of the flow depth y_0 and the unit flow rate q.

Authors	Initiation Condition	End Condition
Abecasis-Quintela (Equations (1) and (2))	$y_0^3 - y_0\left(3\frac{q^{4/3}}{g^{2/3}} + 2z\frac{q^{2/3}}{g^{1/3}}\right) + 2\frac{q^2}{g} = 0$	$y_0^3 - z\,y_0^2 - y_0\left(3\frac{q^{4/3}}{g^{2/3}} + z\frac{q^{2/3}}{g^{1/3}}\right) + 2\frac{q^2}{g} = 0$
Muskatirovic-Batinic (Equations (3) and (4))	$\frac{y_0}{2}\left(\sqrt{1+\frac{8q^2}{y_0^3 g}} - 1\right) + \frac{q^2}{g y_0^2\left(1+\frac{4q^2}{g y_0^3} - \sqrt{1+\frac{8q^2}{g y_0^3}}\right)} - \frac{3q^{\frac{2}{3}}}{g^{\frac{1}{3}}} + z = 0$	$y_0^3 - y_0\left(3\frac{q^{4/3}}{g^{2/3}} + 2z\frac{q^{2/3}}{g^{1/3}} + z^2\right) + 2\frac{q^2}{g} = 0$
Heller-Hager-Minor (Equations (5) and (6))	$y_0^3 g - 2gy_0\left[\left(\frac{3}{2}\frac{q^{4/3}}{g^{2/3}}\right) + \left(z\,\frac{q^{2/3}}{g^{1/3}}\right) + \left(\frac{z^2}{2}\right)\right] + 2q^2 = 0$	$y_0^3 g - 2gy_0\left[\left(\frac{q^2}{g\left(\frac{q^{2/3}}{g^{1/3}}+z\right)}\right) + \left(\frac{q^{4/3}}{2g^{2/3}}\right) + \left(\frac{z^2}{2}\right) + \left(\frac{zq^{2/3}}{g^{1/3}}\right)\right] + 2q^2 = 0$

The *flip bucket characteristic curve* was built point by point, and the previously determined *chute characteristic curve* was used, and both were drawn for each geometrical configuration of the flip bucket, using the formula of different authors (Figure 16). The operating points were obtained at the intersection of the characteristic curves, and therefore the flow rate for the initiation of the jet flow and the flow depth at the lowest point of the flip bucket for that flow rate (Table 13; Figure 17). It is noted that the proposed formula provides the best prediction for the flow rate for the initiation of the jet flow in all cases, taking as reference the one obtained by numerical modeling, followed by the formula of Abecasis and Quintela [16,17]. The obtained results using the formulas of the rest of the authors differ significantly, showing great dispersion.

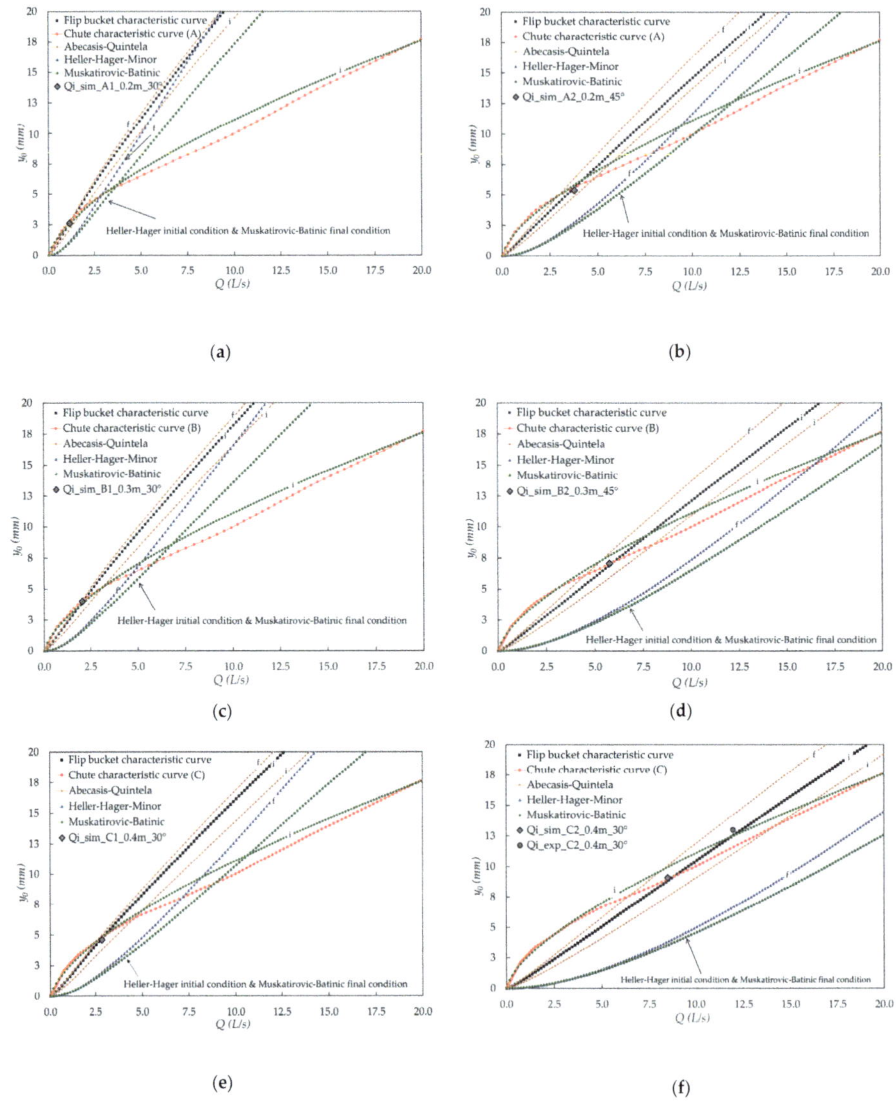

Figure 16. Comparison of the results obtained using the formulas of various authors. The *flip bucket characteristic curve* is represented according to the formula of different authors. For the previous formulas, it is indicated with "i" the initiation condition of the jet flow, and with "f" the end condition of the jet flow: (**a**) A1; (**b**) A2; (**c**) B1; (**d**) B2; (**e**) C1; (**f**) C2.

Table 13. Flow rate for the initiation of the jet flow Qi and flow depth at the lowest point of the flip bucket y_{0i} obtained using the formula of different authors, by means of the proposed formula and by numerical simulation Absolute Error and Relative Error, taking as the reference the flow rate obtained with the numerical simulation Qi_{sim}.

Tested Geometry	Author	Initial Condition	Qi (L/s)	y_{0i} (mm)	$AE_{Qi\,(p\text{-}sim)}$ (L/s)	$RE_{Qi\,(p\text{-}sim)}$ (%)
A1	Abecasis-Quintela	final	0.95	2.35	0.17	15.18
A1	Abecasis-Quintela	initial	1.89	3.86	0.77	40.74
A1	Muskaritovic-Batinic	final	3.69	5.56	2.57	69.65
A1	Muskaritovic-Batinic	initial	3.24	5.92	2.12	65.43
A1	Heller-Hager-Minor	final	2.93	4.88	1.81	61.77
A1	Heller-Hager-Minor	initial	3.69	5.56	2.57	69.65
A1	Proposed Model	initial	1.20	2.81	0.08	6.66
A1	Numerical Model	initial	1.12	2.61	-	-
A1	Experimental	initial	1.65	-	0.53	32.12
B1	Abecasis-Quintela	final	1.80	3.66	0.23	12.78
B1	Abecasis-Quintela	initial	3.18	5.17	1.15	36.16
B1	Muskaritovic-Batinic	final	6.02	7.21	3.99	66.28
B1	Muskaritovic-Batinic	initial	6.90	8.67	4.87	70.58
B1	Heller-Hager-Minor	final	4.74	6.29	2.71	57.17
B1	Heller-Hager-Minor	initial	6.02	7.21	3.99	66.28
B1	Proposed Model	initial	2.18	4.20	0.15	6.88
B1	Numeric Model	initial	2.03	4.01	-	-
B1	Experimental	initial	2.75	-	0.72	26.18
C1	Abecasis-Quintela	final	2.5	4.45	0.36	14.40
C1	Abecasis-Quintela	initial	4.73	6.46	1.87	39.53
C1	Muskaritovic-Batinic	final	9.11	9.38	6.25	68.61
C1	Muskaritovic-Batinic	initial	19.76	17.50	16.90	85.53
C1	Heller-Hager-Minor	final	7.36	8.18	4.50	61.14
C1	Heller-Hager-Minor	initial	9.11	9.38	6.25	68.61
C1	Proposed Model	initial	2.93	4.84	0.07	2.39
C1	Numeric Model	initial	2.86	4.61	-	-
C1	Experimental	initial	3.70	-	0.84	22.70
A2	Abecasis-Quintela	final	3.0	5.01	0.79	26.33
A2	Abecasis-Quintela	initial	4.91	6.50	1.12	22.81
A2	Muskaritovic-Batinic	final	10.20	10.14	6.41	62.84
A2	Muskaritovic-Batinic	initial	19.51	17.33	15.72	80.57
A2	Heller-Hager-Minor	final	8.09	8.61	4.30	53.15
A2	Heller-Hager-Minor	initial	10.20	10.14	6.41	62.84
A2	Proposed Model	initial	3.86	5.67	0.07	1.81
A2	Numeric Model	initial	3.79	5.38	-	-
A2	Experimental	initial	4.80	-	1.01	21.04
B2	Abecasis-Quintela	final	4.47	6.07	1.29	28.86
B2	Abecasis-Quintela	initial	8.18	6.68	2.42	29.58
B2	Muskaritovic-Batinic	final	23.68	20.42	17.92	75.58
B2	Muskaritovic-Batinic	initial	19.68	17.40	13.92	70.73
B2	Heller-Hager-Minor	final	16.35	14.99	10.59	64.77
B2	Heller-Hager-Minor	initial	23.68	20.42	17.92	75.68
B2	Proposed Model	initial	5.97	7.17	0.21	3.52
B2	Numeric Model	initial	5.76	7.06	-	-
B2	Experimental	initial	7.65	-	1.89	24.70
C2	Abecasis-Quintela	final	6.45	7.59	2.04	31.63
C2	Abecasis-Quintela	initial	13.95	13.08	5.46	39.14
C2	Muskaritovic-Batinic	final	50.10	39.73	41.61	83.05
C2	Muskaritovic-Batinic	initial	19.95	17.59	11.46	57.44
C2	Heller-Hager-Minor	final	29.25	24.57	20.76	70.97
C2	Heller-Hager-Minor	initial	50.10	39.73	41.61	83.05
C2	Proposed Model	initial	8.84	9.21	0.35	3.96
C2	Numeric Model	initial	8.49	9.08	-	-
C2	Experimental	initial	11.95	13.00	3.46	28.95

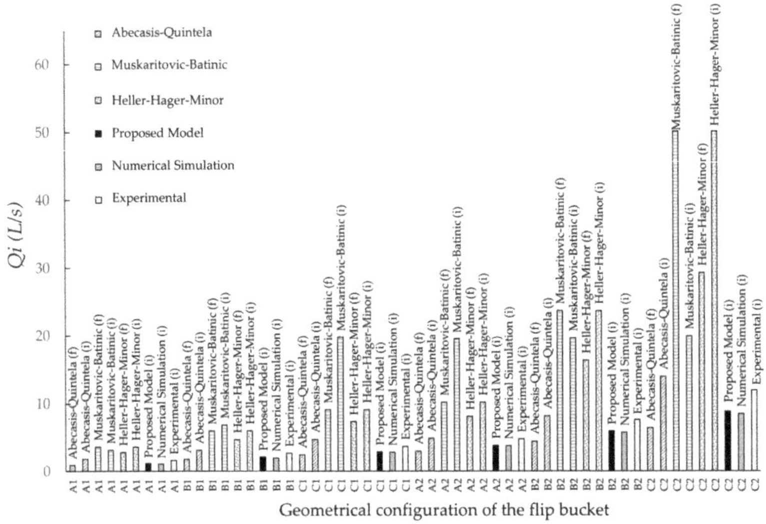

Figure 17. Flow rate for the initiation of the jet flow *Qi* obtained using the formula of various authors, the proposed formula, numerical modeling and physical models in the laboratory.

8. Conclusions

The flip bucket might be used as a stilling basin in the range of low discharges. In some cases, it might be advisable to delay the initiation of the jet flow. This way, the dissipation of most of the energy would occur inside the flip bucket, by means of the hydraulic jump, for low discharges, which are the most frequent working condition.

For the jet flow, the energy is fully dissipated through the impact with the riverbed, which implies lower erosion control for the lower range of discharges. If a pre-excavated stilling basin is considered, it should be larger, and therefore more expensive, if jet flow occurs for low discharges.

In order to minimize the scour basin, and increase the distance between the bucket toe and the impact point, the flow rate for initiation of the jet flow should be as high as possible. It implies a longer impact distance, something beneficial for the spillway, and for the dam if it is in the vicinity of the flip bucket.

Determining the position of the erosion basin downstream of a sky-jump spillway allows to control the impact area. Its shape and size depend on both the geometrical configuration of the flip bucket and the operating flow rates. Knowing the flow rate for the initiation of the jet flow is of interest, since it allows to establish the position of the impact area closest to the spillway, and usually to the dam itself.

This study proposes a methodology that allows to determine the flow rate for the initiation of the jet flows by the intersection of two curves, by analogy with a pumping system: The *chute characteristic curve* and the *flip bucket characteristic curve*. For the latter a new formula is proposed that incorporates the effect of the curvature of the flip bucket. The flow rate for the initiation of the jet flow obtained with the proposed methodology and formula are similar to the value obtained by numerical modeling, with differences less than 5%. Experimentally obtained flow rates are somewhat higher, possibly due to the influence of the aeration of the water stream.

Accurate determination of this flow rate is difficult, as small variations of any of the parameters involved might imply a relatively significant change in the flow rate for the initiation of the jet flow. This is explained graphically by the chute and *flip bucket characteristic curves*, which form very sharp angles at the intersection.

The curvature of the flip bucket was found to affect significantly the flow rate for the initiation of the jet flow, although it is not the most influential parameter. In the cases that were studied, the RE that is committed considering the complete proposed formula, with the term that includes the radius, does not exceed 8%, while if the proposed formula is considered by removing that term, the RE reaches 24%. Consequently, the inclusion of radius in the formulation is relevant, which justifies the interest of using the proposed new formula. It was observed that, being constant the depth of the flip bucket (z), the flow rate for the initiation of the jet flow increases as the radius is reduced or, equivalently, by increasing the throwing angle. The increase is 17.4% when the radius is halved, from 0.4 to 0.2 m. In line with this, the error made, without considering the effect of the curvature of the flip bucket, increases, by decreasing the radius, from a RE of 12.97% to 25.87% when the radius is reduced by half, from 0.4 m to 0.2 m.

The flow rate for initiation of the jet flow is determined during the design of the flip bucket, and we have shown that it depends not only on the depth of the flip bucket (z), but also on the bucket radius. If the designer desires a high flow rate for the initiation of the jet flow, a high lip angle and the minimum radius compatible with flow conditions should be specified at the design stage. The here proposed formula should be used for considering the effect of the bucket radius. The proposed method of the characteristic curves might be used for determining the flow rate for the initiation of the jet flow. It is a decision of the designer to choose where to dissipate the low and frequent discharge flows: in the riverbed or inside the flip bucket. The decision should me made taking into consideration the geological characterization of the riverbed and the flow depth in the impact area, and also the need and cost of implementing a pre-excavated basin.

It is necessary to carry out additional research on the influence of flow aeration on the value of the flow rate for the initiation of the jet flow.

Author Contributions: Conceptualization, M.Á.T.; methodology, M.Á.T., R.P., V.A.; validation, R.P.; formal analysis, R.P., V.A.; investigation, R.P., V.A.; resources, M.Á.T.; data curation, R.P.; writing—original draft preparation, R.P.; writing—review and editing, M.Á.T., V.A.; visualization, R.P.; supervision, M.Á.T.; project administration, M.Á.T. All authors have read and agreed to the published version of the manuscript.

Funding: This research received no external funding.

Acknowledgments: We are grateful to the members of the research group SERPA-Dam Safety Research for the support provided.

Conflicts of Interest: The authors declare no conflict of interest.

Abbreviations

The following abbreviations are used in this paper:

A	average flow area;
AR	absolute error;
α	flip angle;
$C_{1\varepsilon}, C_{2\varepsilon}$	k-ε turbulence model parameters;
$\sigma_k, \sigma_\varepsilon$	k-ε turbulence model parameters;
F	is the volume flow function;
f_{xi}, f_{xj}, f_{xk}	viscous acceleration;
F_0	Froude number;
g	gravity acceleration;
G_{xi}, G_{xj}, G_{xk}	body acceleration;
H	spillway total height;
La	horizontal distance from the upstream spillway vertical wall to the point where the straight part of the chute ends;
Lc	horizontal distance from the upstream spillway vertical wall and measurement point "p";

Lt	horizontal distance between the measurement point "p" and the flip bucket's lip;
MAE	mean absolute error;
MRE	mean relative error;
R	radius of curvature of flip bucket;
RE	relative error;
P	distance from the bottom of the flip bucket to the ground;
p	pressure;
P_k	production of turbulence kinetic energy;
Q	flow rate;
q	unit flow rate;
Qi	flow rate for the initiation of the jet flow;
Qi_{exp}	experimental flow rate for the initiation of the jet flow;
Qi_p	theoretical flow rate for the initiation of the jet flow;
Qi'_p	theoretical flow for the initiation of the jet flow not considering R;
Qi_{stm}	numerical flow rate for the initiation of the jet flow;
Qf	flow rate for the finishing of the jet flow;
y_c	critical flow depth;
y_0	flow depth;
y_{0i}	flow depth for the initiation of the jet flow;
y'_{0iexp}	experimental flow depth for the initiation of the jet flow with aerated upper area;
y_{0iexp}	experimental flow depth for the initiation of the jet flow without aerated upper area;
y_{0ip}	theoretical flow depth for the initiation of the jet flow;
$y_{0i\ sim}$	numerical flow depth for the initiation of the jet flow;
U	average velocity;
u_i, u_j, u_k	velocity component;
v_0	velocity;
v_c	critical velocity;
z	vertical height of the obstacle or depth of the flip bucket
x_i, x_j, x_k	Cartesian coordinates;
k	turbulence kinetic energy;
ε	rate of turbulence energy dissipation;
μ	dynamic viscosity;
μ_t	turbulent dynamic viscosity;
t	time;
ρ	fluid density.

References

1. Godon, R. Le barrage et l'usine hydro-électrique de Marèges sur la Dordogne. *Techniques des Travaux* **1936**, *12*, 101–110.
2. Rhone, T.J.; Peterka, A.J. Improved tunnel spillway flip buckets. *J. Hydraul. Div.* **1959**, *85*, 53–76.
3. Vischer, D.L.; Hager, W.H. *Dam Hydraulics*; Wiley: Chichester, UK; New York, NY, USA, 1998.
4. Rajan, B.H.; Shivashankara Rao, K.N. Design of trajectory buckets. *Water Energy Int.* **1980**, *37*, 63–76.
5. Bollaert, E.F.R.; Duarte, R.; Pfister, M.; Schleiss, A.J.; Mazvidza, D. Physical and numerical model study investigating plunge pool scour at Kariba Dam. In Proceedings of the 24th ICOLD Congress, Kyoto, Japan, 2–8 June 2012; pp. 241–248.
6. Peterka, A.J. *Hydraulic Design of Stilling Basins and Energy Dissipators*; U.S. Dept. of the Interior, Bureau of Reclamation, Technical Service Center: Denver, CO, USA, 1964; pp. 199–205.
7. Schleiss, A.J. Scour evaluation in space and time—The challenge of dam designers. In *Rock Scour due to Falling High-Velocity Jets*; A A Balkema Publishers: Rotterdam, The Netherlands, 2002; pp. 3–22.
8. Pfister, M.; Schleiss, A.J. Ski jumps, jet and plunge pools. In *Energy Dissipation in Hydraulic Structures*; Hubert Chanson; Taylor & Francis Group: London, UK, 2015; pp. 105–140.

9. Rouve, G. Some observation on flow over spillway flip buckets. In Proceedings of the 6th Symposium of Civil and Hydraulic Engineering Department, High Velocity Flows, Indian Institute of Science, Bangalore, India, 18–20 January 1967.
10. Baines, P.G. A unified description of two-layer flow over topography. *J. Fluid Mech.* **1984**, *146*, 127–167. [CrossRef]
11. Lawrence, G.A. Steady flow over an obstacle. *J. Hydraul. Eng. ACE* **1987**, *8*, 981–991. [CrossRef]
12. Pratt, L.J. A note on nonlinear flow over obstacles. *Geophys. Astrophys. Fluid Dyn.* **1983**, *24*, 63–68. [CrossRef]
13. Baines, P.G.; Whitehead, J.A. On multiple states in single-layer flows. *Phys. Fluids* **2003**, *15*, 298–307. [CrossRef]
14. Mehrotra, S.C. Hysteresis effect in one-and two fluid system. In Proceedings of the V Australian Conference on Hydraulics and Fluid Mechanics, University of Canterbury, Christchurch, New Zeland, 9–13 December 1974; Volume 2, pp. 452–461.
15. Austria, P.M. Catastrophe model for the forced hydraulic jump. *J. Hydraul. Res.* **1987**, *25*, 269–280. [CrossRef]
16. Abecasis, F.M.; Quintela, A.C. *Hysteresis in the Transition from Supercritical to Subcritical Flow*; memoria n.523; Laboratorio National de Engeharia Civil: Lisbon, Portugal, 1979.
17. Abecasis, F.M.; Quintela, A.C. Hysteresis in steady free-surface flow. *Water Power* **1964**, *4*, 147–151.
18. Muskatirovic, D.; Batinic, D. The influence of abrupt change of channel geometry on hydraulic regime characteristics. In Proceedings of the 17th IAHR Congress, Baden Baden, Germany, 15–19 August 1977; pp. 397–404.
19. Heller, V.; Hager, W.H.; Minor, H.E. Ski jump hydraulics. *J. Hydraul. Eng.* **2005**, *131*, 347–355. [CrossRef]
20. Arangoncillo, V. *Análisis del Funcionamiento con Pequeños Caudales, del Deflector de los Aliviaderos Trampolín*; Trabajo de Suficiencia Investigadora; Universidad Politécnica de Madrid: Madrid, Spain, 2011.
21. Falvey, H.T. *Engineering Monograph No.42: Cavitation in Chutes and Spillways*; US Department of the Interior, Bureau of Reclamation: Denver, CO, USA, 1990.
22. U.S. Bureau of Reclamation. *Design of Small Dams*; U. S. Government Printing Office: Washington, DC, USA, 1977.
23. Flow Science, Inc. *FLOW-3D User Manual Release 11.0.3*; Flow Science, Inc.: Santa Fe, NM, USA, 2014.
24. Hirt, C.H.; Sicilian, J.M. A porosity technique for the definition of obstacles in rectangular cell meshes. In *Proceedings 4th International Conference on Numerical Ship Hydrodynamics*; National Academy of Science: Washington, DC, USA, 1985; pp. 1–19.
25. Nichols, B.D.; Hirt, C.W. Methods for calculating multidimensional, transient free surface flows past bodies. In *Proceedings of 1st Int.Conf. Ship Hydrodynamics*; Schot, J.W., Salvesen, N., Eds.; Naval Ship Research and Development Center: Bethesda, MD, USA, 1975; pp. 253–277.
26. Nichols, B.D.; Hirt, C.W.; Hotchkiss, R.S. *Volume of Fluid (VOF) Method for the Dynamics of Free Boundaries*; Rep. LA-8355; Los Alamos Scientific Lab.: Los Alamos, NM, USA, 1980.
27. Hirt, C.W.; Nichols, B.D. Volume of fluid (VOF) method for the dynamics of free boundaries. *J. Comput. Phys.* **1981**, *39*, 201–225. [CrossRef]
28. Jakeman, A.J.; Letcher, R.A.; Norton, J.P. Ten iterative steps in development and evaluation of environmental models. *Environ. Modell. Softw.* **2006**, *21*, 602–614. [CrossRef]
29. Blocken, B.; Gualtieri, C. Ten iterative steps for model development and evaluation applied to Computational Fluid Dynamics for Environmental Fluid Mechanics. *Environ. Model. Softw.* **2012**, *33*, 1–22. [CrossRef]
30. Chanel, P.G. An Evaluation of Computational Fluid Dynamics for Spillway Modeling. Master' Thesis, University of Manitoba, Winnipeg, MB, Canada, 2008.
31. Yakhot, V.; Orszag, S.; Thangam, S.; Gatski, T.; Speziale, C. Development of turbulence models for shear flows by a double expansion technique. *Fluid Dyn.* **1992**, *4*, 1510–1520. [CrossRef]
32. Speziale, C.G.; Thangam, S. Analysis of an RNG based turbulence model for separated flows. *Int. J. Eng. Sci.* **1992**, *30*, 1379–1388. [CrossRef]
33. Pope, S.B. *Turbulent Flows*; Cambridge University Press: Cornell University, New York, NY, USA, 2000.

water

MDPI

Article

The Cross-Dike Failure Probability by Wave Overtopping over Grass-Covered and Damaged Dikes

Vera M. van Bergeijk [1,*], Vincent A. Verdonk [1,2], Jord J. Warmink [1] and Suzanne J. M. H. Hulscher [1]

1 Department of Marine and Fluvial Systems, University of Twente, Drienerlolaan 5, 7522 NB Enschede, The Netherlands; vincent.verdonk@anteagroup.com (V.A.V.); j.j.warmink@utwente.nl (J.J.W.); s.j.m.h.hulscher@utwente.nl (S.J.M.H.H.)
2 Antea Group, Tolhuisweg 57, 8443 DV Heerenveen, The Netherlands
* Correspondence: v.m.vanbergeijk@utwente.nl

Abstract: A probabilistic framework is developed to calculate the cross-dike failure probability by overtopping waves on grass-covered dikes. The cross-dike failure probability of dike profiles including transitions and damages can be computed to find the most likely location of failure and quantify the decrease in the failure probability when this location is strengthened. The erosion depth along the dike profile is calculated using probability distributions for the water level, wind speed and dike cover strength. Failure is defined as the exceedance of 20 cm erosion depth when the topsoil of the grass cover is eroded. The cross-dike failure probability shows that the landward toe is the most vulnerable location for wave overtopping. Herein, the quality of the grass cover significantly affects the failure probability up to a factor 1000. Next, the failure probability for different types of damages on the landward slope are calculated. In case of a damage where the grass cover is still intact and strong, the dike is most likely to fail at the landward toe due to high flow velocity and additional load due to the slope change. However, when the grass cover is also damaged, the probability of failure at the damage is between 4 and 125 times higher than for a regular dike profile.

Keywords: wave overtopping; erosion; levee; cover; probabilistic framework

Citation: van Bergeijk, V.M.; Verdonk, V.A.; Warmink, J.J.; Hulscher, S.J.M.H. The Cross-Dike Failure Probability by Wave Overtopping over Grass-Covered and Damaged Dikes. *Water* **2021**, *13*, 690. https://doi.org/10.3390/w13050690

Academic Editor: Miguel Á. Toledo

Received: 23 December 2020
Accepted: 26 February 2021
Published: 3 March 2021

1. Introduction

Coastal and fluvial areas are threatened by flooding by seas or rivers during a storm. Earthen dikes with a grass cover on top protect the hinterland against flooding and are one of the main flood defence structures in the Netherlands. Earthen dikes are also found among others in Western Europe [1], USA [2] and China [3]. However, many of these dikes need to be strengthened due to sea level rise and increase in peak river discharges as a result of climate change. This asks for cost-effective design solutions and accurate assessment tools.

Wave overtopping is one of the main reasons causing failure of grass-covered dikes. During a storm, high waves can overtop the dike although the water level is below the dike height and flow down the landward slope with significant erosive action [4,5]. The grass cover and the clay layer with the grass roots (Figure 1a) exert a crucial role in protecting earthen dikes from erosion [6–9]. The erosion resistance of the dike cover is determined by the topsoil, defined as the upper 20 cm of the grass cover (Figure 2), including the type of vegetation and its root system. The erosion resistance decreases in the subsoil layer underneath the topsoil where less roots are present; here, the erosion resistance is determined by the clay quality [4,10].

The probability of failure by wave overtopping is usually calculated based on the average overtopping discharge [11–13]. Failure is defined as the exceedance of a maximum allowable overtopping discharge that varies between 0.1 and 10 L/s/m. There are two disadvantages of using the overtopping discharge to describe failure. Firstly, the relation between the overtopping discharge and the resulting grass cover erosion is unclear. For example in the

Netherlands, failure is defined as an exceedance of 20 cm erosion depth [14]. An erosion depth of 20 cm can be caused by a wide range of overtopping discharges depending on hydraulic conditions and the strength of the dike cover [15,16]. Secondly, the methods based on the average overtopping discharge do not include a cross-dike component. Therefore, it is unknown where exactly the dike is most likely to fail.

Figure 1. (**a**) Dike cover erosion on the landward slope and toe during wave overtopping field tests in the Netherlands (Photo by Juan Pablo Aguilar Lopez). (**b**) A slope instability of a grass-covered clay dike in the Netherlands [17].

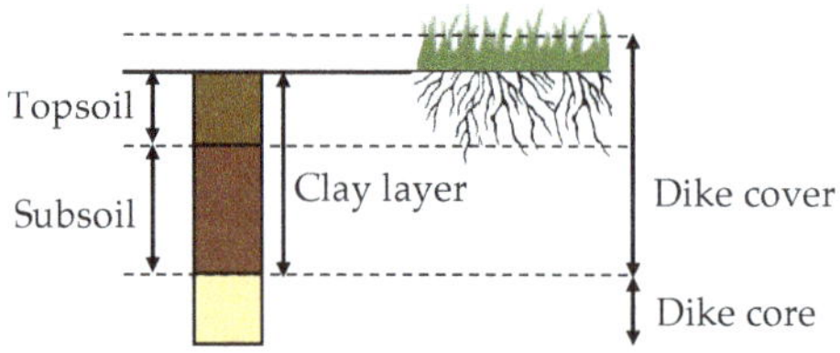

Figure 2. Schematization of the dike cover where the grass roots strengthen the topsoil of approximate 20 cm and the clay quality determines the strength of the subsoil.

Erosion models for wave overtopping are introduced to overcome these problems, for example the Cumulative Overload Method (COM) [18] or the analytical Grass-Erosion Model (GEM) [16,19]. The COM is based on the flow velocity and the cover strength is included in the model using a critical velocity. The damage number can be computed at multiple locations using influence factors for the flow acceleration and the effect of transitions [20]. The damage number for failure is empirically determined for the erosion of the topsoil, but the damage number for larger erosion depths is unknown. The GEM calculates the flow velocity and erosion depth along the dike profile and can easily be extended to larger erosion depths. The variation in flow velocity is calculated using the analytical formulas of Van Bergeijk et al. [21] and the effect of transitions can be included using a turbulence parameter. Although both methods are promising, they have solely been applied to compute failure during a storm and not yet for failure probabilities.

The cross-dike failure probability is computed by Aguilar-López et al. [15] for individual overtopping waves during a storm. This enables identification of the locations that are most likely to fail and might need additional strengthening measures. In the study of Aguilar-López et al. [15], Hoffmans' erosion model [4] is coupled to the numerical hydrodynamic model of Bomers et al. [22] to compute the erosion depth along the dike profile. However, due to high computational costs of the numerical model, probabilistic computations are only possible using an emulator [15]. This emulator was trained for a specific dike profile and needs to be trained again for other dike configurations. Moreover, the emulator is a black box and the link to the physical variables is not clear. Therefore, it is difficult to understand what the important variables are for dike cover failure. For example, the load is solely described by the average overtopping discharge that depends on many different parameters so the effect of the water level or wave height cannot be determined using this method.

A new framework for the failure probability by wave overtopping is developed based on the GEM [16]. This model couples the analytical formulas for the maximum flow velocity of Van Bergeijk et al. [21] with the erosion model of Hoffmans [4] to compute the erosion depth along the dike profile. The GEM has been used to compute the erosion depth of the upper cover layer during a storm [16], but needs to be extended for the computation of failure probabilities. The advantage of an analytical model compared to a numerical model is that analytical models are fast, thus failure probabilities can be computed without an emulator. Moreover, the analytical formulas contain physical parameters related to bottom friction, turbulence and cover strength, and therefore it is possible to find the effect of these physical variables on the failure probability. This new framework can be applied to multiple dike configurations, since the dike geometry and the hydraulic load are the only required boundary conditions.

Importantly, this framework can account for damages in the dike cover to obtain insights into the residual dike strength. Residual dike strength is defined as the ability of the flood defence to continue its water retaining function after it has failed according to the failure definition [23]. For wave overtopping, the residual dike strength is characterised by the clay cover underneath the grass cover once the topsoil is eroded. This clay layer still protects the dike core for overtopping and needs to be eroded completely before the dike breaches [9] (Figure 1a). However, more knowledge is required on how fast the erosion progresses once the topsoil is eroded before the residual strength of the clay layer can be considered and the failure definition for overtopping can be extended to larger erosion depths.

Moreover, the interaction between different failure mechanisms needs to be considered for some cases of residual dike strength. For example, a small slope instability does not result in flooding (Figure 1b) and progressive slope instabilities are necessary before the dike loses its water retaining function [23,24]. The effect of a small slope instability on other failure mechanisms needs to be determined before progressive slope instabilities can be included in safety assessments. A dike with a small slope instability is more vulnerable for wave overtopping due to the damaged dike cover and the formation of a vertical cliff at the instability that affects the hydraulic load [17]. The effect of a slope instability on the failure probability by wave overtopping is unknown.

The goal of this study is to calculate the effect of damages in the dike profile on the failure probability by wave overtopping using a new probabilistic framework. The main innovative component of the study is that the effect of transitions and damages on the failure probability can be computed to quantify the increase in failure probability caused by these weak spots in the dike profile. The landward toe is used as an example of a transitions in this study, where the slope change leads to an additional load on the cover. Additionally, the framework is applied to damaged dike profiles, such as a dike profile with an erosion hole or a small slope instability. These damages have two effects on the erosion by overtopping waves: (1) the load increases due to jet impact landward of the vertical cliff, and (2) the cover strength is reduced near the damage. Quantification of the failure probability at transitions and damages can help to improve local dike strengthening measures and maintenance strategies.

2. Framework for the Failure Probability by Wave Overtopping

A probabilistic framework is developed to calculate the failure probability by wave overtopping along a grass-covered dike crest and landward slope. First, the failure probability conditional to the water level $P_{f|h}$ is calculated (Figure 3). The hydraulic load distribution at the start of the dike crest consists of all overtopping wave volumes during the storm and depends on the water level h, the wind speed u_{10}, the fetch length and the geometry of the outer slope. The other required input variables are the dike geometry and the dike cover strength described by the critical velocity U_C. The GEM calculates the erosion depth along the dike profile of every overtopping wave based on the overtopping volume and the dike characteristics. The total erosion depth during the storm $d(x)$ is calculated at every cross-dike coordinate x by a summation of all overtopping waves.

The conditional failure probability is determined from the limit state function Z that expresses the difference between the strength and the load. The strength of dike cover is set to the upper cover layer of 20 cm and the load is described $d(x)$. Failure is defined as $Z < 0$ which happens when the erosion depth exceeds 20 cm according to the Dutch failure definition [14]. The probability of failure corresponds to the probability $P(Z < 0)$. A Monte Carlo analysis with 2×10^4 samples (see Appendix A for convergence of P_f) is performed where each sample corresponds to a storm event and therefore the failure probability in this study is the failure probability per storm event. The wind speed and critical velocity are sampled from their distribution for every storm event while keeping the water level constant, resulting in the failure probability conditional to the water level $P_{f|h}$.

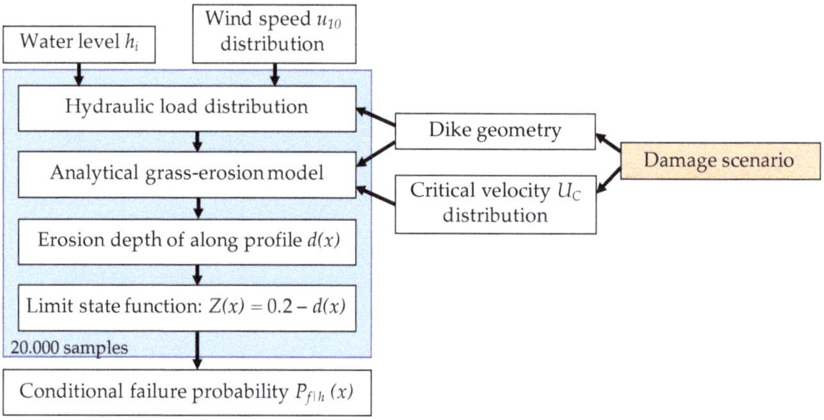

Figure 3. Schematization of the framework to calculate the failure probability conditional to the water level $P_{f|h}$.

The conditional failure probability $P_{f|h}$ is computed for several water levels to construct a fragility curve as illustrated in Figure 4. The conditional failure probabilities are significant for high water levels with a low probability. The probability of failure P_f is calculated by numerical integration of the conditional failure probability and the probability density function of the water level $f(h)$

$$P_f(x) = \int P_{f|h}(x) f(h) dh \tag{1}$$

The framework can be applied to damaged dike profiles such as an erosion hole or a slope instability. These damages influence the dike geometry and the strength of dike cover as indicated by the orange box in Figure 3.

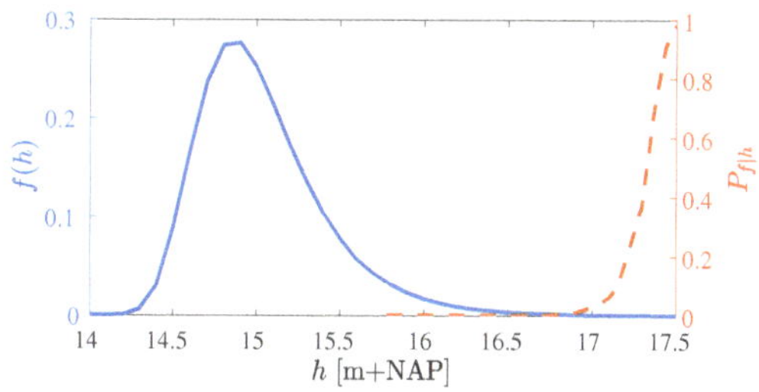

Figure 4. Example of the probability density function of the water level $f(h)$ (solid line) and a fragility curve (dashed line) showing the conditional failure probability $P_{f|h}$ as function of the water level h.

2.1. Hydraulic Load

The sampled water level and sampled wind speed are used together with the fetch length to calculate the significant wave height H_s and the wave period T_p at the outer toe of the dike using the Bretschneider equations. The return frequencies for the water levels are obtained from the Hydra-NL WBI 2017 software which includes the water levels and their return periods for every river dike section in the Netherlands based on an uncertainty analysis of numerical river models [25]. The wind statistics of Caires [26] are used to obtain the cumulative probability distribution for the wind speed u_{10} at a standard landscape roughness and a standard height of 10 m.

The storm duration was fixed for all simulations and set to 6 h based on previous research [17] and Dutch assessment requirements [27]. The 2% exceedance run-up height, average overtopping discharge and number of overtopping waves during the storm are calculated using the formulas in the EurOtop Manual [11]. These variables are used to generate the individual overtopping volumes during the storm according to the probability exceedance distribution as described by Van Bergeijk et al. [16]. The method for the hydraulic load is described in more detail in Verdonk [28].

2.2. Analytical Grass-Erosion Model

The erosion depth along the dike profile is calculated using the analytical grass-erosion model GEM where the analytical formulas for the overtopping flow velocities of Van Bergeijk et al. [21] are coupled to the erosion model of Hoffmans [4] that has been adapted to account for cross-dike variations in the load and strength so the effect of transitions and damages can be included [16,19]. Hoffmans' erosion model [4] describes the scour erosion of clay and grass covers by overtopping waves based on an erosion model for jet scour [29–32]. Scour erosion is the result of high flow velocities and locally increased turbulence, and therefore Jorissen and Vrijling [32] introduced a turbulence parameter ω to account for the effect of turbulence on the scour erosion. This turbulence parameter is related to the depth-averaged relative turbulence intensity r_0 as [7,32,33]

$$\omega = 1.5 + 5 \cdot r_0 \tag{2}$$

For wave overtopping flow, the depth-averaged relative turbulence intensity r_0 relates to the friction of the bed (see Appendix B for this relation) and the turbulence parameter in the erosion model accounts for the increase in the hydraulic load as the result of turbulence generated by bed friction [4,22]. Hoffmans et al. [7] estimated a range for the turbulence parameter on the slope ($\omega_{slope} = 2.00 - 2.75$) for wave overtopping on grass-covered dikes (Appendix B).

The total erosion depth during a storm event is calculated by first computing the flow velocity U along the dike profile of every overtopping wave i using analytical flow formulas that were derived from the 1D shallow water equations [21]. The overtopping volume is used as a boundary condition and the maximum flow velocity along the dike profile depends on the cover type and the dike geometry including the slope angle. Next, the total erosion depth along the dike profile $d(x)$ is calculated by summing over all overtopping waves N.

$$d(x) = \sum_{i}^{N} \left(\omega^2(x) U_i^2(x) - U_t^2 \right) T_0 C_E \quad \text{for } \omega(x) U_i(x) > U_t \tag{3}$$

with the threshold flow velocity U_t, the overtopping period T_0 and the inverse cover strength parameter C_E. Erosion of the grass cover starts once the hydraulic load-described by turbulence parameter and flow velocity (ωU)–exceeds the threshold flow velocity which is a factor 2.4 larger than the critical velocity U_C (Table 1). The turbulence parameter depends on the location along the dike profile based on three values for the crest, landward slope and landward toe (Table 1)

$$\omega(x) = \begin{cases} 2.35, & \text{Crest} \\ 2.00, & \text{Landward slope} \\ 2.75, & \text{Landward toe} \end{cases} \tag{4}$$

Table 1. The turbulence parameter ω and the threshold velocity U_t in the GEM are determined from calibration or measurements of wave overtopping field tests on grass-covered dikes in the Netherlands and Belgium.

Parameter	Relation	Method
Turbulence parameter	$\omega_{toe} = 2.75$	Calibration using the measured erosion depth at 7 field tests [19,34]
	$\omega_{crest} = 2.00$	Determined from measured pressure fluctuations at Millingen a/d Rijn [35]
	$\omega_{slope} = 2.35$	Determined from measured pressure fluctuations at Millingen a/d Rijn [35]
Threshold velocity	$U_t = 2.4 U_C$	Calibration using the measured erosion depth at 7 field tests [19,34]

The critical velocity and the inverse strength parameter depend on the quality of the dike cover which consists of grass or clay (Table 2). The uncertainty in the cover strength is simulated using a log-normal distribution for the critical velocity where the mean and the coefficient of variation are based on Aguilar-López et al. [15]. The inverse strength parameter is related to the erosion speed, which is larger for a poor grass cover compared to a good grass cover [36]. Once the topsoil of the grass cover is eroded (Figure 2), the clay layer underneath contains only a small amount of roots and the strength of the cover mainly depends on the cohesion of clay. The lower resistance against erosion results in a smaller critical velocity and a larger inverse strength parameter for clay compared to grass.

The GEM is the only model for the erosion of the topsoil layer by overtopping waves that takes cross-dike variations in load into account and can therefore be applied to transitions and damages. The GEM has been validated for a storm event by Van Bergeijk et al. [16] where they show that the model is able to accurately predict the erosion depth along the dike profile measured during wave overtopping field tests in the Netherlands and Belgium. Moreover, the erosion model and the grass-cover distributions in the framework have been validated by Bomers et al. [22] and Aguilar-López et al. [15].

Table 2. The critical velocity U_C and the inverse cover strength parameter C_E of the dike cover for three grass qualities and a good clay quality together with the coefficient of variation CV used for the distribution of the critical velocity.

		Grass			Clay	
		Good	Average	Poor	Good	Source
U_C	[m/s]	6.5	4	2.5	0.85	Aguilar-López et al. [15]
CV	[-]	0.3	0.3	0.3	0.1	Aguilar-López et al. [15]
C_E	[s/m]	1×10^{-6}	2×10^{-6}	3×10^{-6}	50×10^{-6}	Verheij et al. [36]

3. Methods

This framework is applied to a grass-covered river dike in the Netherlands. Firstly, the cross-dike failure probability is computed to identify vulnerable locations for wave overtopping failure. Secondly, a relation for the additional load due to jet impact near damages is derived and the framework is applied to damaged dike profiles. The failure probabilities for damaged spots are computed for a varying cover quality to simulate different types of damages. These failure probabilities of damaged spots are compared to the failure probability of the landward toe to determine the effects of damaged dike profiles on the failure probability.

3.1. Study Area

The framework is applied to a dike near Millingen a/d Rijn in the Netherlands close to the junction of the Rhine, the Pannerdensch canal, and the Waal (Figure 5a). This is a grass-covered dike with homogeneous clay core [22]. At this location, wave overtopping tests have been performed in 2013 to determine the erosion resistance of the grass cover for wave overtopping and the influence of an asphalt road on the erosion [37]. Aguilar-López et al. [15] used the same dike section enabling comparison of methods and failure probabilities. Additionally, the turbulence parameter was determined during the overtopping tests from the measured pressure fluctuations (Table 1).

The dike height of 17.93 m+NAP was determined using AHN viewer [38]. The dike geometry is characterised by an outer bed level of 9.4 m+NAP, a crest width of 4.20 m and a horizontal slope length of 17.20 m. Both the water side and landward slope have a steepness of $\cot(\varphi) = 3$ (Figure 5b). A friction coefficient f of 0.01 was used in the GEM for the grass cover. A smooth waterside slope was assumed with a reduction factor γ_f =1 for the run-up and overtopping equations [11].

The wind statistics of Caires [26] at the measurement station Deelden are used for this study area. Multiple simulations with different wind directions and their corresponding fetch length showed that the western wind direction (270 degree) with an effective fetch length of 2.785 km was dominant for wave overtopping [28].

The conditional failure probabilities are computed for 24 water levels varying between 15.78 m+NAP with a return period of 100 years and 17.64 m+NAP with a return period of 10^6 years, which is the maximum return period in the Hydra-NL software. The maximum water level of 17.64 m+NAP means that we do not compute the conditional failure probabilities for freeboards smaller than 0.3 m. However, this will not affect the total failure probability since the exceedance probabilities of these water levels are negligible for this location.

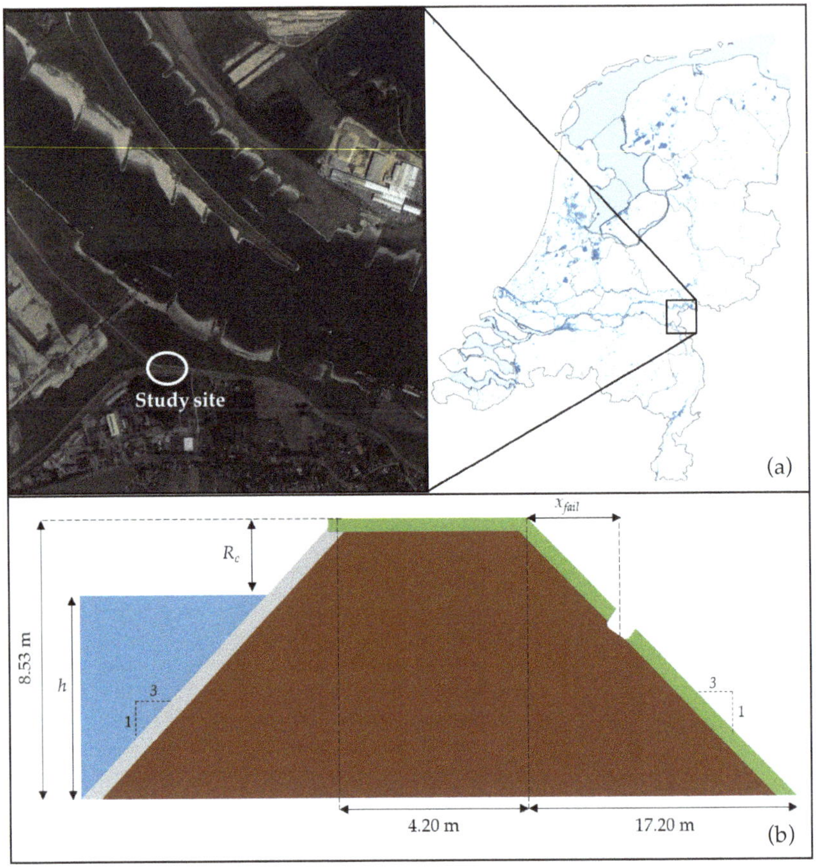

Figure 5. (**a**) Top view of the study site at Millingen a/d Rijn and the location in the Netherlands. Retrieved from Google Earth, earth.google.com and Oppervlaktewater in Nederland, www.clo.nl. (accessed on 13 November 2020). (**b**) The dike geometry with a smooth waterside slope and a grass-covered crest and landward slope including the water level h, the free crest height R_c and the location of failure x_{fail}.

3.2. Cross-Dike Failure Probability

The failure probability P_f is calculated along the dike profile every 0.5 m including the end of the crest and the landward slope to save computational costs. This spatial step of 0.5 m has no effect on the results since the analytical formulas in the framework are independent of the spatial step [21]. The 24 conditional failure probabilities $P_{f|h}$ are computed at every cross-dike location and numerically integrated using Equation (1) to obtain the cross-dike failure probability $P_f(x)$.

The effect of the grass cover quality on the failure probability is investigated by computing $P_f(x)$ for three grass qualities: poor, average and good (Table 2). The turbulence parameter ω is kept constant along the profile to solely identify the effect of the grass cover quality, where a value of $\omega = 2.00$ is used corresponding to the measured turbulence parameter on the slope (Table 1).

Next, $P_f(x)$ is also computed for a variation in the turbulence parameter along the dike profile as described by Equation (4) using an average grass quality. Comparison between the failure probability for a constant turbulence parameter and a varying turbulence parameter quantifies the underestimation in the failure probability when the turbulence parameter is not locally adapted for transitions.

3.3. Additional Load at Damaged Spots

A vertical cliff forms at damages on the landward slope such as an existing erosion hole or a slope instability [17,39] (Figures 1b and 6b). When the overtopping wave flows over this vertical cliff, a jet will form that impacts below the cliff (Figure 6a). The jet reattaches to the dike slope at a small distance landward of the cliff where the load on the cover increases due to the impact and a local increase in turbulence [4,40,41]. This additional load has not been quantified for wave overtopping erosion models.

The additional load on the cover is often expressed as the energy dissipation of the flow indicated by a friction factor, for example in the case of high-velocity air-water flows over a grass cover [42] or flow over stepped spillways [43]. The effect of bed friction is included in the GEM in the turbulence parameter, which originates from the local scour parameter that accounts for the effect of local turbulence on scour erosion created by both submerged and plunging jets [4,32,33]. The flow over a cliff at the damage shows similarities with both the flow over stepped spillways and a plunging jet and therefore the additional load by damages is included in the GEM using the turbulence parameter. The value of the turbulence parameter ω for overtopping flow over damaged spots is unknown and needs to be determined for this study. Since the energy dissipation depends on the cover type [42], we assume that the turbulence parameter for the damages depends on the critical velocity which is used in the model to simulate different cover types.

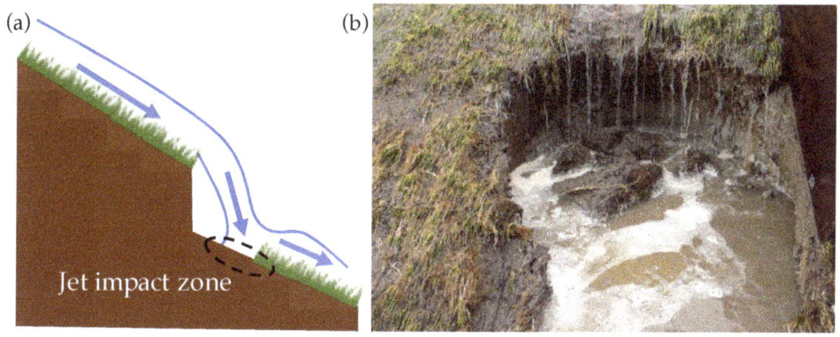

Figure 6. (**a**) Small cliff at a damaged spot leads to to a jet forming that impacts in the jet impact zone (white circle) resulting in an additional load. (**b**) Photo of damage during the overtopping test at St. Philipsland resulting in a small cliff [44].

The turbulence parameter for damaged spots is determined from the results of the wave overtopping tests on real grass-covered dikes in the Netherlands and Belgium in 2008–2012 (Table 3). During these tests, the dike cover eroded most at the landward toe due to high flow velocities [16], on the slope near weak spots such as molehills and bald spots [20] or on the upper slope near the crest due to wave impact [41]. The erosion in the latter two cases resulted in damages with a small cliff on the slope that led to failure of the dike cover (Figure 6b). Eight test sections (Table 3) were selected where the cover erosion on the slope resulted in the formation of a vertical cliff with a maximum depth of 20 cm. These damages to the grass cover were the result of animal burrows such as mice, moles and rabbits (Kattendijke 2, Wijmeers 1, Tielrodebroek 1 and 2) or an erosion hole formed by gradual scour erosion (Afsluidijk 2 and Wijmeers 3), bulging (Tholen 3) or roll-up (St. Philipsland). The test conditions and results are used to calibrate the turbulence parameter for small damages on the slope.

Table 3. The location of failure x_{fail} measured from the start of the landward slope for the eight test sections in the Netherlands and Belgium. The test conditions resulting in failure are described by the critical velocity U_C, the average overtopping discharge q and the simulated storm duration t_{storm}. The fractions for q indicates that the cover failed at a fraction of the storm duration for that specific discharge.

Test Section	x_{fail} [m]	U_C [m/s]	q [L/s/m]	t_{storm} [h]	Source
Afsluitdijk 2	2.8	4.0	1, 10	6	Bakker et al. [45]
Tielrodebroek 1	1.9	1.2	1, 10, 30($\frac{1}{3}$)	2	Peeters et al. [46]
Tielrodebroek 2	1.9	1.6	1, 10, 30($\frac{1}{6}$)	2	Peeters et al. [46]
Wijmeers 1	1.7	3.5	1, 5, 25	2	Pleijter et al. [47]
Wijmeers 3	1.3	3.0	25	2	Pleijter et al. [47]
Kattendijke 2	6.6	6.5	30, 50	6	Bakker et al. [44]
St. Philipsland	6.5	6.5	0.1, 1, 10, 30, 50	6	Bakker et al. [44]
Tholen 3	6.5	0.0	1, 5 ($\frac{2}{3}$)	6	Bakker et al. [48]

Firstly, the location of failure x_{fail} (Figure 5b) and the critical velocity of the test section are obtained from the reports (Table 3). Next, the test conditions are simulated in the GEM using the average overtopping discharge and the storm duration to find the combination of the turbulence parameter and the critical velocity that lead to failure ($d(x_{fail})$ = 20 cm) for each test section (Figure 7). This results in eight failure points for the eight test sections. A relationship between the turbulence parameters and the critical velocity is determined from a linear fit through these failure points. The inverse strength parameter depends on the cover type (Table 2) and during the calibration the cover types are distinguished as poor ($U_C \leq 3$ m/s), average (3 m/s $< U_C \leq 5$ m/s) and good ($U_C > 5$ m/s).

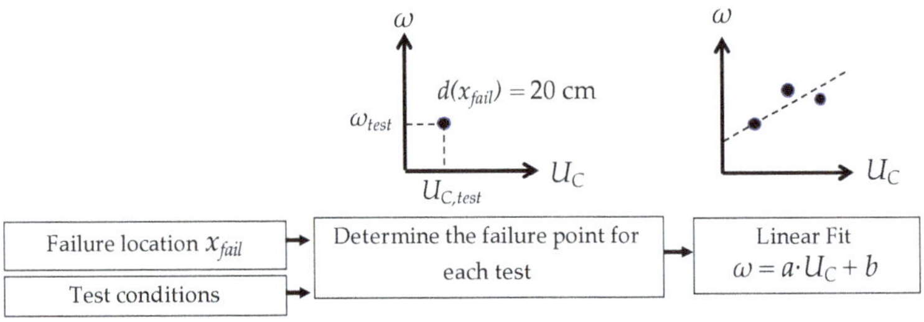

Figure 7. Schematization of the method for the calibration of the turbulence parameter ω at damaged spots as a function of the critical velocity U_C using the analytical grass-erosion model (GEM) and the overtopping tests.

3.4. Failure Probability of Damages on the Landward Slope

Damages on the landward slope have two effects on the cover erosion by overtopping waves. Firstly, a vertical cliff forms at the damaged spot leading to an additional load on the grass cover. This effect is included in the model using the turbulence parameter as discussed in the previous section. Secondly, the dike cover strength is reduced near the damage leading to a poorer grass quality or a bare clay cover in case of an erosion hole. The reduction in cover strength is modelled in the analytical grass model by changing the critical flow velocity and the inverse strength parameter from a good grass quality to an average or poor quality, or to clay for a damage where the grass cover is eroded completely (Table 2).

The failure probability for damaged spots $P_{f,damage}(x)$ is computed along the landward slope for four dike covers: bare clay, poor grass, average grass and good grass. The distributions for the critical velocity and the value of the inverse strength parameter in Table 2 are used to simulate the dike cover type. The mean critical velocity for each cover type is used to calculate the turbulence parameter near the damage using the calibrated relationship.

The ratio $\Delta P_f(x)$ is used to quantify the difference in failure probability for a undamaged and damaged dike profiles

$$\Delta P_f(x) = P_{f,damage}(x)/P_{f,toe} \tag{5}$$

with the failure probability at the landward toe $P_{f,toe}$. The landward toe is the location where a regular dike profile without any weak spots in the dike cover will fail due to high flow velocities at the end of the slope and additional load due to the slope change. The regular dike profile is modelled assuming a good grass cover and a turbulence parameter of 2.75 for the additional load at the landward toe (Table 1). In cases where the ratio is smaller than 1, the dike is most likely to fail at the landward toe and the damage is not the weakest location. Contrary, in cases where the ratio is larger than 1, the damaged spot is most likely to fail for overtopping.

4. Results

4.1. Cross-Dike Failure Probability

The cross-dike failure probability follows the variation in flow velocity along the profile and is maximum at the end of the slope (Figure 8). The flow velocity decreases over the crest ($x = 0 - 4.2$ m) due to bottom friction, increases over the slope ($x = 4.2 - 21.4$ m) until a balance is reached between the gravitational acceleration and the bottom friction followed by a decrease after the landward toe ($x > 21.4$ m) due to bottom friction. The same cross-dike variation is observed in the failure probability for a constant turbulence parameter, because the load variation in this case is solely determined by the flow velocity.

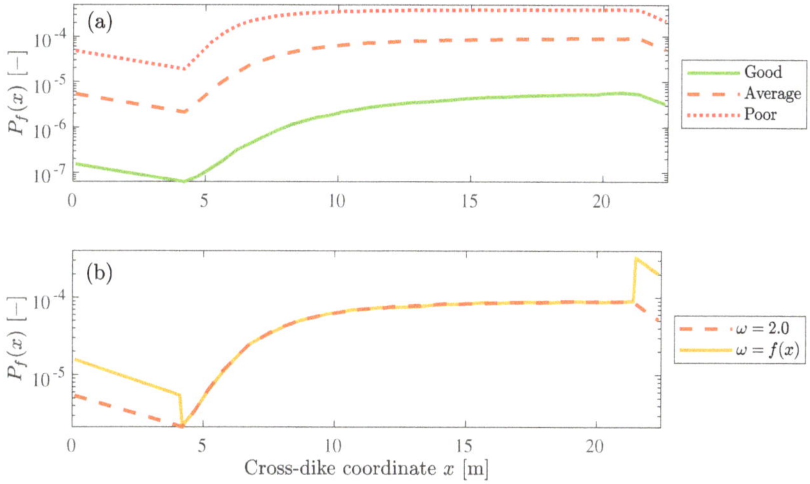

Figure 8. (**a**) The cross-dike failure probability $P_f(x)$ for three grass qualities and a constant turbulence parameter ω. (**b**) The cross-dike failure probability $P_f(x)$ for an average grass quality with a constant turbulence parameter ω and the turbulence parameter of Equation (4).

The grass quality has a large effect on the failure probability. The failure probability decreases with a factor up to 10^2 when the grass cover quality increases from average to good and increases with a factor 10 when the grass quality decreases from average to poor (Figure 8a). The difference between the failure probabilities is maximal at the end of the crest at a cross-dike distance x of 4.2 m. The turbulence parameter can locally increase the failure probability as can be seen on the crest and at the landward toe in Figure 8b. The failure probability for a varying turbulence parameter (that is $\omega(x)$ in Equation (4)) no longer follows the variation in flow velocity but increases at locations where the turbulence parameter is higher.

In all cases, the landward toe has the highest probability of failure. For an average grass quality, the landward toe is 35 times more likely to fail compared to the upper slope. This increases to a factor 150 when the turbulence parameter for the landward toe is used to describe the additional load due the change in slope. The failure probability at the landward toe increases from 9.1×10^{-5} for a constant turbulence parameter to 3.3×10^{-4} for the varying turbulence parameter. Therefore, the failure probability at the landward toe is underestimated by a factor 3.6 when the additional load at the landward toe is not taken into account and the turbulence parameter is kept constant along the dike profile.

4.2. Additional Load at Damaged Spots

The markers in Figure 9 show the combination of the turbulence parameter and critical velocity leading to failure of the dike cover for the eight test sections (Table 3). A linear fit through these failure points results in a relation for the turbulence parameter at damaged spots

$$\omega = 0.074\, U_C + 2.1 \tag{6}$$

with a root-mean-square error of 0.14 and a coefficient of determination R^2 of 0.56. The constant 0.074 has the units s/m to ensure that the turbulence parameter is dimensionless.

The calibrated relationship results in a turbulence parameter of 2.56, 2.38, 2.26 and 2.15 for a good grass cover, average grass cover, poor grass cover and clay cover, respectively, using the values for the critical velocity in Table 2. The calibrated turbulence parameter for additional load at damaged spots solely depends on the critical velocity, but the total load in the model is described by the turbulence parameter and the flow velocity. This means that the total load increases when the flow velocity increases and is therefore higher for damages on the lower slope compared to the upper slope.

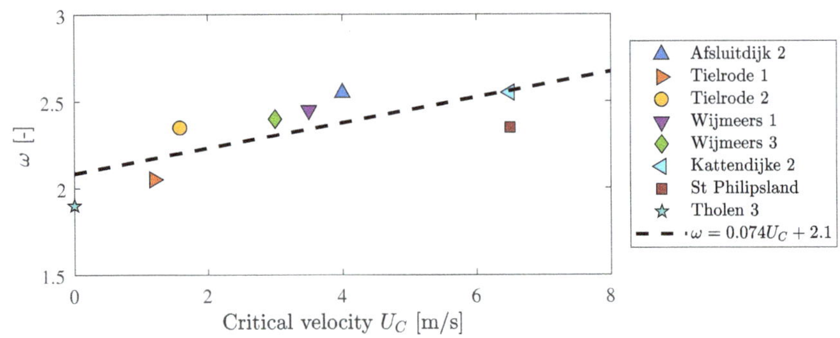

Figure 9. The relationship between the turbulence parameter ω and the critical velocity U_C based on a linear fit through the failure points of the eight test sections.

4.3. Failure Probabilities of Damaged Spots

The conditional failure probabilities $P_{f|h}$ and the total failure probabilities P_f are compared for failure at the landward toe and failure at damaged spots along the slope and for four cover qualities: clay, poor grass, average grass and good grass (Figure 10). The fragility curves show the conditional failure probability as function of the free crest height R_c and they do not reach 1 for the cases with a grass cover (Figure 10a). This is related to the crest height of 17.93 m+NAP and the maximum water level from the HydraNL software of 17.65 m+NAP, which means that no information is available for free crest heights smaller than 0.3 m where most of the wave overtopping occurs. However, these water levels have a return frequency smaller than 10^{-6} yr^{-1} and their contribution to the failure probability are therefore negligible. Significant wave overtopping occurs for the computed water levels with a maximum average overtopping discharge of 250 L/s/m and an average of all storm events of 20 L/s/m for a water level of 17.65 m+NAP. Wave overtopping tests on grass-covered dikes showed that good grass-covers can easily withstand an overtopping discharge of 20 L/s/m [20] which agrees with the small conditional failure probability for good grass covers.

The higher conditional failure probabilities for the clay cover are the result of a lower critical velocity and a larger inverse strength parameter. The lower critical velocity means that small waves with low velocities are able to erode the cover. Additionally, the waves erode more material due to the large inverse strength parameter so the erosion depth of 20 cm is already reached with little overtopping. This results in a fragility curve where the conditional failure probability reaches 1 at a free crest height of 0.6 m which corresponds to an average overtopping discharge of 3 L/s/m.

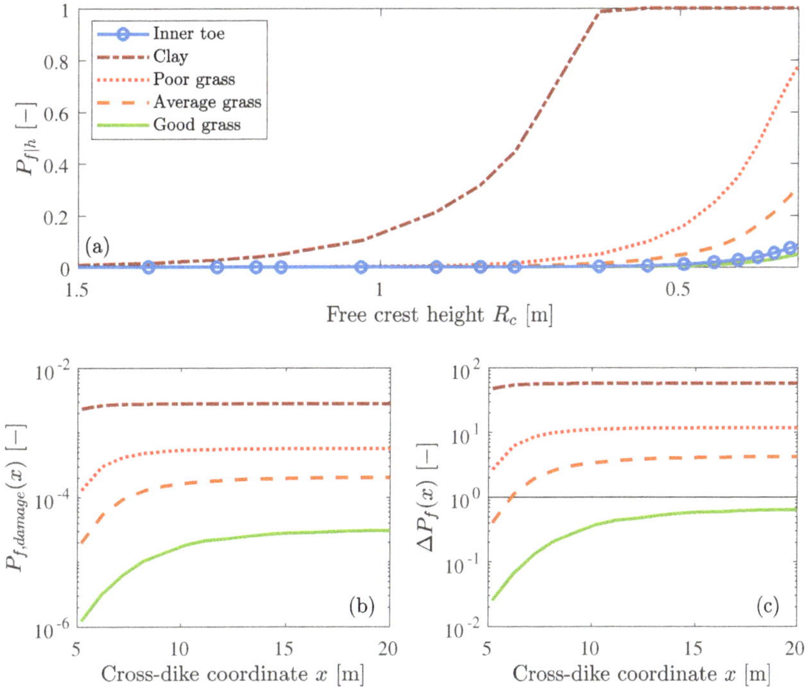

Figure 10. (**a**) The fragility curves showing the conditional failure probability $P_{f|h}$ as function of the free crest height R_c for failure at the landward toe and failure at a damage around 10 m from the crest for four cover qualities. (**b**) The cross-dike failure probability for a damaged spot on the slope $P_{f,damage}(x)$ for the four cover qualities. (**c**) The ratio $\Delta P_f(x)$ for the four cover qualities (Equation (5)).

The failure probabilities for a dike cover with a damage are compared to the failure probability at the landward toe using their ratio $\Delta P_f(x)$ (Equation (5)). The results show that a dike profile with a damaged spot is more likely to fail compared to an undamaged dike profile at the landward toe with a good grass cover, except when the grass cover at the damaged spot is still intact and of good quality. The ratio $\Delta P_f(x)$ is always smaller than 1 for damages with good grass cover (Figure 10c), which means that the dike is more likely to fail at the landward toe than at the damaged spot. An example of such a damage is a small slope instability where the grass cover is still intact, but the small cliff leads to an additional load at this location. The calibrated turbulence parameter for a good grass cover ($\omega = 2.56$) is smaller than the turbulence parameter at the landward toe ($\omega_{toe} = 2.75$)

leading to a lower failure probability at the damaged spot. Additionally, the load on the slope is lower compared to the landward toe because of a lower flow velocity.

The lower flow velocity on the slope is also the reason why the ratio is smaller than 1 for an average grass cover at the beginning of the slope, but the ratio increases along the slope to a ratio of 4. The ratio is especially large for damaged spots with a bare clay cover. The results show that a dike profile with only clay as cover is 60 times more likely to fail than a regular dike profile.

5. Discussion

In this study, we developed a framework to calculate the probability of dike cover failure by overtopping waves. The variation in the load along the dike profile is included in the GEM model using a turbulence parameter and the flow velocity along the profile. This developed framework has two innovative applications: it enables us to calculate (1) the cross-dike failure probability to identify vulnerable locations for dike cover failure, and (2) the failure probability of damaged dike profiles. Both applications are discussed together with the method to derive the additional load at the damage and a comparison with other studies.

5.1. The Cross-Dike Failure Probability

The cross-dike failure probability indicates that the landward toe is most vulnerable location for dike cover erosion due to the high flow velocities at the end of the slope. In case the turbulence parameter for the landward toe of Frankena [34] is used, the dike profile is 150 times more likely to fail at the landward toe compared to the upper slope.

The cross-dike variation in the hydraulic load is included in the model using the maximum flow velocity along the dike profile and the turbulence parameter. The turbulence parameter can be used to account for the additional load at damages and transitions and thereby the framework can also be applied to other irregularities in a grass-covered dike profile, such as transitions and multi-functional dikes [13,15,49]. This requires information on how these elements affect both the hydraulic load and the cover strength. Theoretical load and strength factors for transitions are derived by Van Hoven et al. [35] for geometric transitions, revetment transitions and vertical objects, but they recommended to validate these factors in more detail using prototype experiments. Warmink et al. [19] developed a method to calibrate these load factors and showed that the calibrated load factors are not comparable to the theoretical factors for a slope change. Detailed numerical models [5,22] can be used to determine the additional load at transitions and develop load factors such as the turbulence parameter that can be used in erosion models. Additionally, more measurements of the turbulence intensity along the profile are necessary to determine a relation for the cross-dike variation in load along damaged dike profiles and multi-functional dikes.

Kriebel [50] performed an sensitivity analysis on the GEM model for failure during a storm event and showed that the velocity on the crest and the critical velocity have the largest effect on the model results. However, the uncertainty in the velocity on the crest is small [11,50] and the uncertainty in the critical velocity is accounted for using a distribution. Therefore, [50] identified the inverse strength parameter and the turbulence parameter as the two parameters that have the most impact on the model results since literature reports a wide range of possible values of both parameters. The inverse strength parameter differs a factor 50 between grass and clay (Table 2) which is the main reason for the spread in the literature values. The values of Verheij et al. [36] are used in this study since the values were also used to calibrate the relation between the critical velocity and threshold velocity, where Frankena [34] showed that this combination of the threshold velocity and inverse strength parameter are able to accurately simulate the measured erosion depth during several wave overtopping tests on grass-covered dikes in the Netherlands and Belgium. The reported range of the turbulence parameter is 2.00–2.75 (Table 1) and the difference between the lower limit and upper limit of the turbulence parameter on the failure probability can be seen in Figure 8. The failure probability at the landward toe

($x = 21.4$ m) increase with a factor 3.5 when the turbulence parameter is increase from 2.00 to 2.75 in case of $w = f(x)$. This effect on the failure probability is smaller compared to the effect of grass quality where the failure probability increases by a factor 1000 from poor grass to good grass. The maximum flow velocity along the dike profile was not included in the uncertainty analysis since Van Bergeijk et al. [21] showed that the flow velocity can be accurately computed for a wide range of flood defences with their analytical formulas.

This study is limited to small damages with a maximum depth of 20 cm because the GEM solely describes scour erosion. This erosion model is able to accurately predict the erosion patterns during wave overtopping experiments on grass-covered dikes [16,22]. However, in case of larger damages with a higher cliff, other erosion mechanisms become dominant such as head-cut erosion [17,39]. These follow-up mechanisms need to be considered when the framework is extended to larger damages, which was outside of the scope of this study. Moreover, the erosion depth during a storm in the GEM is computed without a feedback between the erosion of previous waves on the flow of the next overtopping wave. Erosion holes lead to an increase in the hydraulic load and affect the overtopping flow and erosion downstream. The additional load is included in the model using the turbulence parameter, but the effects on the flow and erosion downstream are neglected in this study since we assumed that the dike would fail at the damage. The effect of erosion holes on the overtopping flow and erosion downstream is not fully understood at the moment and needs to be determined before these effects can be included in this framework. A possibility is to study the effect of erosion holes in detailed overtopping tests or with numerical models [5,22].

The framework is developed for river dikes where a constant water level during a storm is a good approximation. This approximation also holds for dikes at a lake, however, higher waves are expected at lake dikes due to a longer fetch length [51]. The framework is also applicable to sea dikes, but in this case the development of the hydraulic conditions-driven by a combination of the storm and the tides–needs to be included. Kriebel [50] showed how this framework can be extended to storms on sea dikes [52].

The framework is applied to a grass-covered river dike in the Netherlands and can be applied to flood defences globally. The water level and wind speed distribution are site specific and need to be adapted for other locations. The strength of the dike cover is described by a distribution for the critical velocity that is known for different clay and grass qualities [15,36]. Additionally, wave overtopping tests have been performed in Asia to determine the critical velocity of tropical grass species such as Bermuda, Carpet and Manilla [10,53] and therefore the framework can also be applied in these regions.

5.2. The Effect of Damages on the Failure Probability

The effect of damages on the landward slope depends on the type of damage and the remaining cover quality. In case of a damage where the grass cover is still intact and strong, the dike is most likely to fail at the landward toe. An example of such a damage is a small slope stability as depicted in Figure 1b where the grass cover is still intact. However, most damages to the dike profile result in weakening or removal of the grass cover. In these cases, the damaged dike is more than 4 (average grass), 12 (poor grass) or 60 (clay) times likely to fail compared to a regular profile. Damages to the grass cover are often the result of animal burrows such as mice, moles and rabbits [54] that reduce the cover strength. Therefore, these animal burrows are vulnerable locations for failure by wave overtopping. Other types of damages that are representative for this model study are erosion holes formed during wave overtopping by erosion mechanisms such as gradual scour erosion, bulging or roll-up [10]. These erosion holes often result in removal of the grass cover where only a bare clay cover remains to protect the dike core for a breach. For this study, we assumed that the clay layer was of good quality. In case of a poor clay quality [36], damages with only a clay cover have a failure probability that is 125 larger compared to a regular dike profile. This means that the clay quality of the cover layer is an important variable to determine the residual dike strength. The effect of the clay quality on the erosion rate needs to be studied in more detail before the failure definition can be extended to larger erosion depths.

The ratios between the failure probability of a damage and the landward toe can be used as a first estimates for damages on other grass-covered dikes. The dike geometry does not seem to have a large effect on the ratio, because the ratio becomes approximately constant on the lower slope. Simulations for a lake dike in the Netherlands showed similar ratios for damages with a grass cover and therefore these ratios are also applicable to other case studies with similar cover types as mentioned above. However, the simulations for damages with solely a clay cover shows that these have a higher failure probability for lake dikes due to the difference in hydraulic conditions [51]. Further research into damages with a clay cover needs to be done to determine in which cases these ratios can be applied.

The cross-dike failure probability for damaged dike profiles shows that a damage on the lower slope has a higher failure probability compared to a damage on the upper slope. In practice, a damage on the upper slope is more critical because less material needs to be eroded before the dike breaches since the upper slope is closer to the outer slope. In this study, this effect is not included because failure is defined as an erosion depth of 20 cm. However, when the failure definition moves towards a dike breach, this effect needs to be taken into account [9,17].

5.3. Additional Load at Damages

The additional load at damaged spots is simulated using the turbulence parameter, which was calibrated using the results of wave overtopping tests at eight grass-covered dike sections (Equation (6)). These test sections where damaged during the tests and a small cliff formed at these damaged spots resulting in the formation of a small jet that impacts landward of the damaged spot. Both sea and river dike sections at multiple locations in the Netherlands with different grass qualities (0 m/s $\leq U_C \leq$6.5 m/s) were used to calibrate a relation for the turbulence parameter at damaged locations. This relation is applicable to other cases with a similar cover type consisting of grass vegetation on a clay layer with a critical velocity in the range 0 m/s −6.5 m/s. The relation needs to be investigated further before it can be applied to other grass types or soil types, such as Bermuda grass or sand.

We increased the inverse strength parameter stepwise in the method for the calibration of the turbulence parameter, because we assumed that the inverse strength parameter is a cover characteristic and not a function of the critical velocity. This assumption does not affect the failure points, except for the failure point of Wijmeers 3 which is exactly on the boundary between poor and average grass with a critical velocity of 3 m/s. In the current method, Wijmeers 3 is classified as poor grass ($U_C \leq 3$ m/s) with $\omega_{test} = 2.4$ which would increase to $\omega_{test} = 2.7$ when classified as average grass (3 m/s $< U_C \leq 5$ m/s). Although the assumption for the inverse strength parameter leads to a significant increase for this failure point, the other failure points are not affected leading to a small change in the linear fit ($\omega = 0.071U_C + 2.1$) which only affects the second decimal of the calibrated turbulence parameter for each cover quality.

The calibrated turbulence parameter varies between 2.15 and 2.56 for the different cover types, which means that the turbulence parameter for damages is within the range 2.00–2.75 reported by Hoffmans [4] and smaller than the turbulence parameter for the transition at the landward toe (Table 1). The turbulence parameter is related to the Darcy-Weisbach friction factor f_{WD} used to express the energy dissipation of the flow (Appendix B). Scheres et al. [42] derived a Darcy-Weisbach friction factor of 0.19 for high-velocity air-water flows over a grass cover corresponding to a turbulence parameter of 2.42, which is close to the calibrated turbulence parameter for average grass ($\omega = 2.38$). The cliff near the damage shows similarities with stepped spillways. Felder and Chanson [43] determined $0.1 \leq f_{DW} \leq 0.4$ for flow over stepped spillways which corresponds to $2.16 \leq \omega \leq 2.83$, which coincides with the range of the calibrated turbulence parameter.

The calibrated turbulence parameter for additional load at damaged spots depends on the critical velocity since the amount of bed turbulence depends on the cover type, which is included in the model using the critical velocity. The turbulence parameter determined for small damages with a maximum vertical cliff height of 20 cm. For higher cliffs, the height of the cliff as well as the impinging angle will affect the load in the impact zone [39]. Analytical formulas for jet impact show that the normal stress [41] and pressure [55] of the jet in the impact zone increases with the impinging angle. Additionally, wave impact simulations on grass covers using a jet show that the impact pressure increases with height [56,57]. Therefore, the effect of the cliff height and the impinging angle on the load needs to be investigated further when the model approach is to be extended for larger damages leading to higher cliffs.

5.4. Comparison to Other Studies

Aguilar-López et al. [15] calculated the cross-dike failure probability for the same dike section that we analysed using a different hydrodynamic model [22] in combination with an emulator. However, instead of using the water level and wind speed as stochastic variables, the failure probability for an average overtopping discharge using the dike cover strength as only stochastic variable. The failure probabilities of Aguilar-López et al. [15] are of the same magnitude but slight smaller compared to our results (Table 4). However, only the crest and upper slope were included in the hydrodynamic model used by Aguilar-López et al. [15] with the highest failure probability at the end of the slope. Therefore, it is likely that the failure probabilities of Aguilar-López et al. [15] increase when the model is extended to the landward toe where the load is highest and are closer to the probabilities in this study.

Table 4. Comparison between the model results and the results of Aguilar-López et al. [15] for the maximum failure probability for the grass-covered dike profile of Millingen a/d Rijn without any damages.

	Poor	Average	Good
Our framework	3.9×10^{-4}	9.0×10^{-5}	5.8×10^{-6}
Aguilar-López et al. [15]	8.2×10^{-5}	5×10^{-5}	$\leq 10^{-6}$

Marijnissen et al. [13] computed the failure probability for multi-functional dikes using solely the average overtopping discharge as failure criterion. The failure probabilities by wave overtopping are not reported, but the total failure probability by wave overtopping, piping and macro-stability combined is in the same range as the failure probabilities computed in this study (10^{-6}–10^{-2}). The fragility curves for wave overtopping show a steep curve where the conditional failure probability is always 1 for small free-crest boards [13]. For comparison, the fragility curves for macro-stability and piping usually increase from 0 to 1 over a few meters of water depth, while for overtopping the increase from 0 to 1 is over less than one meter water depth. The fragility curves in this study show a steep curve similar to Marijnissen et al. [13] with an increase in the conditional failure probability over less than one meter water depth.

In our framework, each sample corresponds to a storm event and therefore the failure probability per storm event is computed similar to the failure probabilities of Aguilar-López et al. [15] and Marijnissen et al. [13]. However, annual failure probabilities are often required for the safety assessment of dikes. The distributions for the water level and wind speed are currently per storm event and need to be transferred to a distribution per year to change the failure probability per storm event to an annual failure probability. An explanation of this method is provided by Vuik et al. [12] where it is important to take the correlation between the water level and the wind speed into account.

The vulnerability of dike profiles with a slope instability for wave overtopping was also investigated by Van Hoven [17]. The head-cut erosion model [39] was applied to an overtopping test on bare clay at Delftzijl and an overflow test at a slope instability at Bergambacht in the Netherlands. Van Hoven [17] concluded that an additional crest width of 1.5 m is necessary after an instability to be safe for overtopping with an average overtopping discharge of 1 L/s/m. The residual strength for higher overtopping discharges was not determined due to uncertainties in the head-cut model, although no significant erosion was observed for both Delfzijl and Bergambacht [17]. In this study, the failure probability for damages with solely a clay cover resulted in a conditional failure probability of 1 around a free crest height of 0.6 m in which case the average overtopping discharge is approximately 3 L/s/m. However, in cases where a grass cover is still present at the damage, the failure probability is much lower and the dike cover is able to withstand higher overtopping discharges. Van Hoven [17] did not consider the first phase of erosion, namely the failure of the vegetal cover protection where the topsoil eroded by the overtopping waves. Since most of the cover strength is in the topsoil, it is important to take this initiation phase into account in order to determine the maximum allowable overtopping discharges for grass-covered dikes with a slope instability.

6. Conclusions

We have developed a framework to calculate the probability of dike cover failure by overtopping waves along the dike profile. This framework enables to determine the weakest location along the dike profile for dike cover erosion. The effect of transitions and damages on the hydraulic load are included in this framework using the turbulence parameter. A relationship for the turbulence parameter at damages is calibrated and used to determine the vulnerability of grass damages for wave overtopping.

The landward toe is identified as the most vulnerable location for wave overtopping for a regular grass-covered dike profile without any damages. The quality of the grass cover has a large influence on the failure probability and can increase the failure probability with a factor 1000. Furthermore, a formulation for the varying turbulence parameter shows that transitions and damages can locally increase the failure probability and are therefore vulnerable locations.

The vulnerability of grass damages for wave overtopping depends on the type and quality of the dike cover at the damaged location. When the grass cover remains intact and has a good quality, the landward toe is the most likely location to fail. However, damages that lead to a reduction of the cover strength or removal of the topsoil result in higher failure probabilities compared to failure at the landward toe. These damages are more than 10 times as likely to fail for a poor grass quality and more than 100 times more vulnerable in case of a clay cover.

The large variation in failure probability between grass and clay covers shows the importance of including the initiation phase of cover erosion in overtopping calculations. More knowledge on the erosion process of different types of grass and clay is necessary to understand the erosion resistance and speed of the different types of cover layers. This knowledge is required before the strength of the entire cover layer can be included in the safety assessment of grass-covered dikes and the failure definition for wave overtopping can be extended to larger erosion depths. If the dike strength of the entire cover layer can be used properly in erosion models, the failure calculations for overtopping will become less conservative resulting in more cost-effective design solutions and more accurate assessment tools.

Author Contributions: Conceptualization, V.M.v.B., V.A.V., J.J.W. and S.J.M.H.H.; methodology, V.M.v.B., V.A.V.; software, V.M.v.B., V.A.V.; validation, V.M.v.B., V.A.V.; writing—original draft preparation, V.M.v.B.; writing—review and editing, V.A.V., J.J.W. and S.J.M.H.H.; visualization, V.M.v.B., V.A.V.; supervision,V.M.v.B., J.J.W. and S.J.M.H.H.; project administration, J.J.W. and S.J.M.H.H.; funding acquisition, J.J.W. and S.J.M.H.H. All authors have read and agreed to the published version of the manuscript.

Funding: This research was funded by the Netherlands Organisation for Scientific Research (NWO), research programme All-Risk with project number P15-21.

Acknowledgments: We would like to thank Matthijs Gensen, Guido Remmerswaal, Mark van der Krogt and Joost Pol for their valuable input on failure probabilities and slope-instabilities.

Conflicts of Interest: The authors declare no conflict of interest.

Appendix A. Convergence of the Failure Probability

Figure A1 shows the convergence of the failure probability for a constant turbulence parameter of 2.0 and an average grass quality. For a small number of samples, the samples are not uniformly distributed in the domain and are therefore do not correctly consider the stochastic nature of the variables resulting in a large variation in the failure probability as function of the number of samples. From 10^4 samples onward, the failure probability convergences and the failure probability shows a small variation as function of the number of samples. The failure probability for 2×10^4 samples only differs 0.3% from the failure probability for 10^6 samples. Therefore, the 2×10^4 samples used in this study are sufficient to obtain convergence of the failure probability.

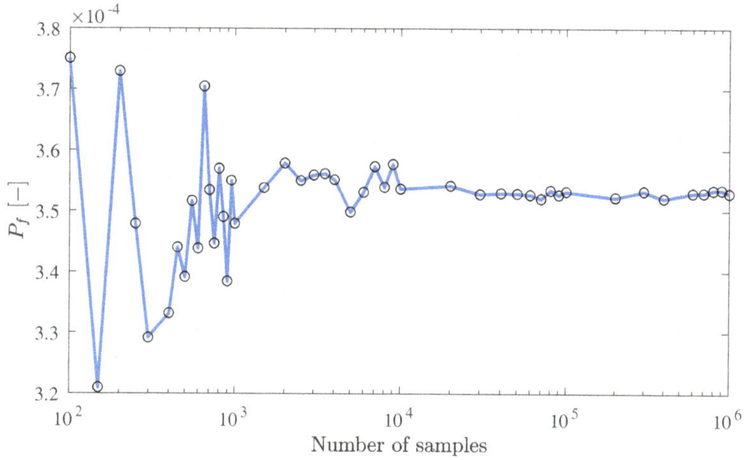

Figure A1. Convergence of the failure probability P_f for a constant turbulence parameter of 2.0 and an average grass quality.

Appendix B. The Depth-Averaged Relative Turbulence Intensity

Hoffmans [4] derived a formula for the depth-averaged relative turbulence intensity r_0 under uniform flow conditions, which was used to estimate the range of the turbulence ω for overtopping (Section 2.2).

The depth-averaged relative turbulence intensity r_0 is defined as

$$r_0 = \frac{\sqrt{k_0}}{U} = \alpha_0 \frac{u_*}{U_0} \tag{A1}$$

with the bed shear velocity u_*, the depth-averaged flow velocity U_0 and the constant $\alpha_0 = 1.2$. The depth averaged turbulence energy k_0 is defined as

$$k_0 = \frac{1}{h} \int_0^h \frac{1}{2} \left(u_{rms'}^2 + v_{rms'}^2 + w_{rms'}^2 \right) dz \tag{A2}$$

with the root mean square values of the fluctuating flow velocities in the streamwise $u_{rms'}^2$, transverse $v_{rms'}^2$ and normal $w_{rms'}^2$ directions. Under uniform flow conditions, $k_0 = (\alpha_0 u_*)^2$

resulting in Equation (A1). In uniform flow, the bed shear velocity is related to the Chezy coefficient C as

$$u_* = \frac{\sqrt{g U_0}}{C} \tag{A3}$$

The Chezy formula for overtopping flow is written as

$$C = \frac{U_0}{\sqrt{h*(1-\eta_a)S_b}} \tag{A4}$$

with the layer thickness $h*$ defined as the water depth of the wave on the crest and landward slope, the air content η_a and the dike slope parameter $S_b = 1/\cot(\varphi)$. Combining Equations (A1), (A3) and (A4), the formula for r_0 becomes

$$r_0 = \frac{\alpha_0\sqrt{g}}{C} = \frac{1.2\sqrt{gh*(1-\eta_a)/\cot(\varphi)}}{U_0} \tag{A5}$$

For an overtopping wave, the flow velocity and layer thickness are maximum at the front of the wave [5,18] resulting in

$$r_0 = \frac{1.2\sqrt{gh_m(1-\eta_a)/\cot(\varphi)}}{U_m} \tag{A6}$$

with the maximum layer thickness h_m and the maximum flow velocity U_m. Hoffmans [4] used measurements of h_m, U_m and η_a during wave overtopping experiments on sea dikes to calculate values for r_0 using Equation (A6). The calculated r_0 varied between 0.10 and 0.25 for volumes in range of 400–5500 l/m corresponding to a turbulence parameter between 2.00 and 2.75. No clear relation between r_0 and the overtopping volume was found and therefore the same value of r_0 is used for all overtopping volumes.

The Darcy-Weisbach friction factor f_{DW} is related to the Chezy coefficient as

$$f_{DW} = \frac{8g}{C^2} \tag{A7}$$

resulting in the following relation between r_0 and f_{DW}

$$r_0 = 0.42\sqrt{f_{DW}} \tag{A8}$$

References

1. Scheres, B.; Schüttrumpf, H. Enhancing the ecological value of sea dikes. *Water* **2019**, *11*, 1617. [CrossRef]
2. Hughes, S.A. *Adaptation of the Levee Erosional Equivalence Method for the Hurricane Storm Damage Risk Reduction System (HSDRRS)*; Technical Report; U.S. Army Corps of Engineers: Vicksburg, MS, USA, 2011.
3. Zhang, Y.; Chen, G.; Hu, J.; Chen, X.; Yang, W.; Tao, A.; Zheng, J. Experimental study on mechanism of sea-dike failure due to wave overtopping. *Appl. Ocean Res.* **2017**, *68*, 171–181. [CrossRef]
4. Hoffmans, G.J.C.M. *The Influence of Turbulence on Soil Erosion*; Eburon Uitgeverij BV: Delft, The Netherlands, 2012.
5. Van Bergeijk, V.M.; Warmink, J.J.; Hulscher, S.J.M.H. Modelling the Wave Overtopping Flow over the Crest and the Landward Slope of Grass-Covered Flood Defences. *J. Mar. Sci. Eng.* **2020**, *8*, 489. [CrossRef]
6. Mazzoleni, M.; Dottori, F.; Brandimarte, L.; Tekle, S.; Martina, M.L. Effects of levee cover strength on flood mapping in the case of levee breach due to overtopping. *Hydrol. Sci. J.* **2017**, *62*, 892–910. [CrossRef]
7. Hoffmans, G.; Akkerman, G.J.; Verheij, H.; Van Hoven, A.; Van Der Meer, J. The erodibility of grassed inner dike slopes against wave overtopping. *Coast. Eng.* **2009**, 3224–3236.
8. Thieu Quang, T.; Oumeraci, H. Numerical modelling of wave overtopping-induced erosion of grassed inner sea-dike slopes. *Nat. Hazards* **2012**, *63*, 417–447. [CrossRef]
9. Oumeraci, H.; D'Eliso, C.; Kortenhaus, A. *Breaching of Coastal Dikes: State of the Art*; LWI Report Uumber: 910; TU Braunschweig: Braunschweig, Germany, 2005.
10. Le, H.T.; Verhagen, H.J.; Vrijling, J.K. Damage to grass dikes due to wave overtopping. *Nat. Hazards* **2017**, *86*, 849–875. [CrossRef]

11. Van der Meer, J.W.; Allsop, N.W.H.; Bruce, T.; De Rouck, J.; Kortenhaus, A.; Pullen, T.; Schüttrumpf, H.; Troch, P.; Zanuttigh, B. *Manual on Wave Overtopping of Sea Defences and Related Structures. An Overtopping Manual Largely Based on European Research, but for Worldwide Application*; EurOtop, 2018. Available online: www.overtopping-manual.com (accessed on 1 July 2020).
12. Vuik, V.; van Vuren, S.; Borsje, B.W.; van Wesenbeeck, B.K.; Jonkman, S.N. Assessing safety of nature-based flood defenses: Dealing with extremes and uncertainties. *Coast. Eng.* **2018**, *139*, 47–64. [CrossRef]
13. Marijnissen, R.; Kok, M.; Kroeze, C.; van Loon-Steensma, J. Re-evaluating safety risks of multifunctional dikes with a probabilistic risk framework. *Nat. Hazards Earth Syst. Sci.* **2019**, *19*, 737–756. [CrossRef]
14. t Hart, R.; de Bruijn, H.; de Vries, G. *Fenomenologische Beschrijving, Faalmechanismen WTI (in Dutch)*; Technical Report; Rijkswaterstaat Water, Verkeer en Leefomgeving: Lelystad, The Netherlands, 2016.
15. Aguilar-López, J.P.; Warmink, J.J.; Bomers, A.; Schielen, R.M.J.; Hulscher, S.J. Failure of grass covered flood defences with roads on top due to wave overtopping: A probabilistic assessment method. *J. Mar. Sci. Eng.* **2018**, *6*, 74. [CrossRef]
16. Van Bergeijk, V.M.; Warmink, J.J.; Frankena, M.; Hulscher, S.J.M.H. Modelling Dike Cover Erosion by Overtopping Waves: The Effects of Transitions. In *Coastal Structures 2019*; Goseberg, N., Schlurmann, T., Eds.; Bundesanstalt für Wasserbau: Karlsruhe, Germany, 2019; pp. 1097–1106. [CrossRef]
17. Van Hoven, A. *Residual Dike Strength after Macro-Instability, WTI2017*; Technical Report; Deltares: Delft, The Netherlands, 2014.
18. Van der Meer, J.W.; Hardeman, B.; Steendam, G.J.; Schüttrumpf, H.; Verheij, H. Flow depths and velocities at crest and landward slope of a dike, in theory and with the wave overtopping simulator. *Coast. Eng. Proc.* **2010**, *1*, 10. [CrossRef]
19. Warmink, J.J.; Van Bergeijk, V.M.; Frankena, M.; Van Steeg, P.; Hulscher, S.J.M.H. Modelling Transitions in Grass Covers to Quantify Wave Overtopping Erosion. *Coast. Eng. Proc.* **2020**, 1–8. [CrossRef]
20. Steendam, G.J.; Van Hoven, A.; Van der Meer, J.W.; Hoffmans, G. Wave Overtopping Simulator Tests on Transitions and Obstacles At Grass-covered Slopes of Dikes. *Coast. Eng. Proc.* **2014**, *1*, 34. [CrossRef]
21. Van Bergeijk, V.M.; Warmink, J.J.; Van Gent, M.R.A.; Hulscher, S.J.M.H. An analytical model of wave overtopping flow velocities on dike crests and landward slopes. *Coast. Eng.* **2019**, *149*, 28–38. [CrossRef]
22. Bomers, A.; Aguilar-López, J.P.; Warmink, J.J.; Hulscher, S.J.M.H. Modelling effects of an asphalt road at a dike crest on dike cover erosion onset during wave overtopping. *Nat. Hazards* **2018**, *93*, 1–30. [CrossRef]
23. Remmerswaal, G.; Hicks, M.A.; Vardon, P.J. Influence of Residual Dyke Strength on Dyke Reliability Using the Random Material Point Method. In Proceedings of the 7th International Symposiumon Geotechnical Safety and Risk (ISGSR), Taipei, Taiwan, 11–13 December 2019; pp. 775–780. [CrossRef]
24. Van Der Krogt, M.G.; Schweckendiek, T.; Kok, M. Do all dike instabilities cause flooding? In Proceedings of the 13th International Conference on Applications of Statistics and Probability in Civil Engineering, ICASP 2019, Seoul, Korea, 26 May 2019. [CrossRef]
25. Duits, M.T.; Kuijper, B. *Hydra-NL–Systeemdocumentatie–Versie 2.4. HKV-Rapport PR3598*; Technical Report; HKV: Lelystad, The Netherlands, 2018 .
26. Caires, S. *Extreme Wind Statistics for the Hydraulic Boundary Conditions for the Dutch Primary Water Defences. SBW-Belastingen: Phase 2 of Subproject "Wind Modelling"; Technical Report, Deltares, 1200264-005*; Deltares: Delft, The Netherlands, 2009.
27. Rijkswaterstaat. *Regeling Veiligheid Primaire Waterkeringen 2017: Bijlage II Voorschriften Bepaling Hydraulische Belasting Primaire Waterkeringen (in Dutch)*; Rijkswaterstaat: Utrecht, The Netherlands, 2016.
28. Verdonk, V.A. The Impact of Overtopping on the Failure Probability of Slipped river dikes. Master's Thesis, University of Twente, Water Engineering and Management, Enschede, The Netherlands, 2020.
29. Breusers, H.N.C. *Conformity and Time Scale in Two-Dimensional Local Scour*; Technical Report; Delft Hydraulic Laboratory: Delft, The Netherlands, 1966.
30. Breusers, H.N.C. Time Scale of Two Dimensional Local Scour. In Proceedings of the 12th Congress IAHR, Fort Collins, CO, USA, 11–14 September 1967; Volume 3, pp. 275–282.
31. Van der Meulen, T.; Vinjé, J.J. Three-dimensional local scour in non-cohesive sediments. In Proceedings of the 16th IAHR-Congress, Sao Paulo, Brazil, 27 July–1 August 1975.
32. Jorissen, R.E.; Vrijling, J.K. Local scour downstream hydraulic constructions. In Proceedings of the 23rd IAHR Conference Ottawa, Ottawa, ON, Canada, 21–25 August 1989 .
33. Hoffmans, G.; Verheij, H. Jet scour. *Proc. Inst. Civ. Eng. Marit. Eng.* **2011**, *164*, 185–193. [CrossRef]
34. Frankena, M. Modelling the Influence of Transitions in Dikes on Grass Cover Erosion by Wave Overtopping. Master's Thesis, University of Twente, Water Engineering and Management, Enschede, The Netherlands, 2019.
35. Van Hoven, A.; Verheij, H.; Hoffmans, G.; Van der Meer, J. *Evaluation and Model Development: Grass Erosion Test at the Rhine Dike*; Technical Report; Deltares: Delft, The Netherlands, 2013.
36. Verheij, H.J.; Meijer, D.G.; Kruse, G.A.M.; Smith, G.M.; Vesseur, M. *Investigation of the Strength of a Grass Cover Upon River Dikes*; Technical Report; Deltares: Delft, The Netherlands, 1995.
37. Bakker, J.; Melis, R.; Mom, R.; Bakker. *Factual Report: Overslagproeven Rivierenland*; Technical Report; Infram: Maarn, The Netherlands, 2013.
38. Actueel Hoogtebestand Nederland. Available online: https://www.ahn.nl/ahn-viewer (accessed on 1 September 2020).
39. Natural Resources Conservation Service. Chapter 51 Earth Spillway Erosion Model and Chapter 52 Field Procedures Guide for the Headcut Erodability Index. In *National Engineering Handbook Part 628 Dams*, 210-vi-NEH ed.; Natural Resources Conservation Service: Washington, DC, USA, 1997.

40. Valk, A. Wave Overtopping Impact of Water Jets on Grassed Inner Slope Transitions. Master's Thesis, Technical University of Delft, Civil Engineering and Geosciences, Delft, The Netherlands, 2009.
41. Ponsioen, L.; van Damme, M.; Hofland, B.; Peeters, P. Relating grass failure on the landside slope to wave overtopping induced excess normal stresses. *Coast. Eng.* **2019**, *148*, 49–56. [CrossRef]
42. Scheres, B.; Schüttrumpf, H.; Felder, S. Flow Resistance and Energy Dissipation in Supercritical Air-Water Flows Down Vegetated Chutes. *Water Resour. Res.* **2020**, *56*, 1–18. [CrossRef]
43. Felder, S.; Chanson, H. Simple Design Criterion for Residual Energy on Embankment Dam Stepped Spillways. *J. Hydraul. Eng.* **2015**, *142*, 04015062. [CrossRef]
44. Bakker, J.; Mom, R.; Steendam, G. *Golfoverslagproeven Zeeuwse Zeedijken-Factual Report (in Dutch)*; Technical Rport; Infram: Maarn, The Netherlands, 2008.
45. Bakker, J.; Mom, R.; Steendam, G. *Factual Report: Overslagproeven en Afschuifproef Afsluitdijk (in Dutch)*; Technical Report; Infram: Maarn, The Netherlands, 2009.
46. Peeters, P.; de Vos, L.; Vandevoorde, B.; Taverniers, E.; Mostaert, F. *Stabiliteit van de Grasmat bij Golfoverslag: Golfoverslagproeven Tielrodebroek (in Dutch)*; Technical Report; Waterbouwkundig Laboratorium, INBO en afdeling Geotechniek: Antwerp, Belgium, 2012.
47. Pleijter, G.; van Steeg, P.; ter Horst, W.; van Hoek, M.; Pol, J. *Data Analyse Golfoverslag en Overloop-Proeven Proeflocatie Wijmeers (in Dutch)*; Technical Report; HKV Lijn in Water & Deltares: Delft, The Netherlands, 2018.
48. Bakker, J.; Mom, R.; Steendam, G.; Van der Meer, J.W. *Factual Report: Overslagproeven en Oploopproef Tholen (in Dutch)*; Technical Report; Infram: Maarn, The Netherlands, 2011.
49. Van Bergeijk, V.M.; Warmink, J.J.; Hulscher, S.J.M.H. Modelling of Wave Overtopping Flow over Complex Dike Geometries: Case Study of the Afsluitdijk. *Coast. Eng. Proc.* **2020**, 1–9. [CrossRef]
50. Kriebel, M. Quantification of Grass Erosion Due to Wave Overtopping at the Afsluitdijk. Master's Thesis, University of Twente, Water Engineering and Management, Enschede, The Netherlands, 2019.
51. Van Bergeijk, V.M.; Verdonk, V.A.; Warmink, J.J.; Hulscher, S.J.M.H. The Vulnerability of Damaged River and Lake Dikes to Wave Overtopping. In Proceedings of the FLOODrisk 2020-4th European Conference on Flood Risk Management, Budapest, Hungary, 21–25 June 2021.
52. Warmink, J.J.; Van Bergeijk, V.M.; Kriebel, M.; Trul, H.; Kuiper, C.; Hulscher, S.J.M.H. The failure probability of grass erosion due to wave overtopping. A case study of the Afsluitdijk, The Netherlands. In Proceedings of the FLOODrisk 2020-4th European Conference on Flood Risk Management, Budapest, Hungary, 21–25 June 2020.
53. Van der Meer, J.W.; Steendam, G.J.; Mosca, C.A.; Guzzo, L.B.; Takata, K.; Cheong, N.S.; Eng, C.K.; Lj, L.A.; Ling, G.P.; Wee, C. Wave Overtopping Test to Determine Tropical Grass Species and Topsoils for Polder Dikes in a Tropical Country. *Coast. Eng. Proc.* **2020**, 1–15. [CrossRef]
54. Klerk, W.J.; Roscoe, K.L.; Tijssen, A.; Nicolai, R.P.; Sap, J.; Schins, F. Risk based inspection of flood defence dams: An application to grass revetments. In Proceedings of the 6th International Symposium on Life-Cycle Civil Engineering, IALCCE 2018, Ghent, Belgium, 28–31 October 2018.
55. Beltaos, S. Oblique Impingement of Circular Turbulent Jets. *J. Hydraul. Res.* **1976**, *14*, 17–36. [CrossRef]
56. Stanczak, G. Breaching of Sea Dikes Initiated from the Seaside by Breaking Wave Impacts. Ph.D. Thesis, University of Braunschweig, Braunschweig, Germany, University of Florence, Florence, Italy, 2008.
57. Scheres, B.; Schüttrumpf, H. Investigating the Erosion Resistance of Different Vegetated Surfaces for Ecological Enhancement of Sea Dikes. *J. Mar. Sci. Eng.* **2020**, *8*, 519. [CrossRef]

water

MDPI

Article

Fragility Curves for Slope Stability of Geogrid Reinforced River Levees

Nicola Rossi *, Mario Bačić, Meho Saša Kovačević and Lovorka Librić

Faculty of Civil Engineering, University of Zagreb, 10000 Zagreb, Croatia; mbacic@grad.hr (M.B.); msk@grad.hr (M.S.K.); llibric@grad.hr (L.L.)
* Correspondence: nrossi@grad.hr; Tel.: +385-1-4639-491

Abstract: When constructing flood protection structures such as river levees, oftentimes due to various factors engineers must design composite structures, i.e., reinforced earthen structures which comply with all the stability criteria. The most common way of reinforcing such structures is the usage of geosynthetics, or mostly geogrids when talking about stability. Since geosynthetics are man-made materials produced in a controlled environment and go through quality control measures, their characteristics contain a negligible amount of uncertainty compared to natural soils. However, geosynthetic handling, their installation in the levee, and their long-term degradation can all have significant effects of variable magnitude on geosynthetic characteristics. These effects and their variability can be considered as random variables, which can then be used in probabilistic analyses together with soil properties. To investigate the effects of the geogrid's resistance variability on slope stability compared to soil properties variability, probabilistic analyses are conducted on a river levee in northern Croatia. It is found that the geogrid's variability generally has very little effect on the total uncertainty compared to the friction angle's variability, but out of the three geogrid layers used the top grid has the most influence.

Keywords: fragility curves; river levees; geogrid reinforcement; First Order Reliability Method (FORM); Surface Response Method (SRM); slope stability

Citation: Rossi, N.; Bačić, M.; Kovačević, M.S.; Librić, L. Fragility Curves for Slope Stability of Geogrid Reinforced River Levees. *Water* **2021**, *13*, 2615. https://doi.org/10.3390/w13192615

Academic Editors: Miguel Á. Toledo and Rafael Morán

Received: 9 September 2021
Accepted: 16 September 2021
Published: 23 September 2021

1. Introduction

River levees for flood protection are structures usually made from earthfill material, and their cross section can be made up of multiple distinct parts, which serve specific purposes in the protection from high waters. However, as Wang et al. [1] noted, levees cannot completely exclude flood disasters, and living behind a levee poses unique flood risks since levees are designed to reduce the impact of a flood event at a certain scale.

Their stability is mostly affected by the material used for the levee body, the foundation material, and is also a function of the water level on the riverside. Often, due to cadastral parcels owned by the investor, stability cannot be ensured for required crown heights corresponding to defined return periods of flood events by using conventional solutions due to the need of building steep slopes to fit the levee into the parcel width. This issue is commonly solved by introducing ground reinforcement techniques that allow for steeper slopes. One common technique in such structures is the reinforced fill built by placing geosynthetic layers during the construction or reconstruction of a levee. The use of geosynthetic materials generally in reinforced earth structures started to increase after 1971 when the first geotextile reinforced wall was constructed in France, and their beneficial effect was noticed. At a later date, around 1980, geogrids were developed [2]. Nowadays, geosynthetics are widely used in various fields of geotechnical engineering, such as shallow footing to increase bearing capacity and decrease settlement [3–5], retaining walls [6–8], and road construction [9,10]. When used in levees, their benefit has also been shown in decreasing settlement of levees on soft soil [11] and increasing slope stability [12], or both. Their effects have been studied under undrained [13,14], partially drained [14], and drained [15] conditions, during and after embankment construction. Hird and Kwok [13] studied the stress distribution in the geosynthetic element depending on its stiffness, and the strength and stiffness of the embankment material. As the levees can be made from various materials, Balakrishnan and Viswanadham [16] studied the tensile load-strain behaviour of geogrids embedded in different soil types and under variable normal stress. Other ground reinforcement methods can also be combined with geosynthetics. For example, Zheng et al. [17] have used stone columns in conjunction with geosynthetics to achieve stable embankments on soft soil and have studied their interaction.

Studies have shown the stability benefit of using geosynthetics to ensure embankment stability, as well as the economic advantages, with the help of limit equilibrium based methods [18–20] and numerical methods [14,21,22] in 2D and 3D [23], physical models [24], as well as various other methods mentioned by Tandjiria et al. [25]. In practice, the most used method is the limit equilibrium due to its simplicity, despite all the limitations and assumptions, which has shown good performance in real-life problems [25].

The introduction of geosynthetics for stability, mostly geogrids, is significant not only because it means a higher stability, but also because it is a reinforcement element which can be made from various materials (polyester, polypropylene, polyethylene, polyamide, polyester, and polyvinyl chloride) [26], and whose characteristics can be controlled during their production, which in turn means a higher reliability in their parameters' values and less variability. Nevertheless, some variability within geogrid parameters can still arise from various sources, namely biases regarding strength reduction factors, which consider installation damage, creep, and durability.

As Rowe and Soderman [18] stated, geosynthetics can fail by two mechanisms, either on the soil-reinforcement interface, or internally as the rupture of the reinforcement element itself. To resist the tensile rupture of the element, the resistance is straightforwardly calculated by using the material's parameters and the cross section. To resist pull out, multiple effects are in place, whose relative contribution to the total pull-out resistance effect has been studied by various authors [24,27,28]. When such elements are placed within a levee, a few failure modes can be expected, namely internal, external, and compound [29], as shown in Figure 1. Internal stability refers to slip surfaces which pass entirely through the reinforcement layers, which means that the reinforcement failed either by tensile rupture, or by pull out. External failure refers to deeper slip surfaces which go around all the reinforcement layers. The compound failure is the most common type, where the slip surface goes around and through various reinforcement layers. On top of those mentioned failure modes, if the spacing between neighbouring reinforcement layers is too big and secondary reinforcement is not provided, failure can initiate by soil sliding between those layers, which then leads to a global failure. Thus, geogrid reinforced slope sections usually consist of primary or principal, and secondary or intermediate, geogrid layers [2,29–31]. Failure of the slope can also occur without the need of reinforcement failure, i.e., if the reinforcement is a low stiffness geosynthetic whose failure strain is much larger than the strain at which the slope fails, then the whole slope might fail without reaching any of the previously defined geosynthetic failure mechanisms [18]. Which failure mechanism will occur in a levee highly depends on the cross section of the levee and whether it is a newly constructed levee or a reconstructed one, because these parameters will dictate the placement of geogrids.

Even though levees are characterized by a number of failure mechanisms [32,33], and that about half of earth embankment failures occur as a result of processes related to piping [34], this study considers only the slope stability of a reconstructed and additionally reinforced river levee. The primary purpose of this study is to investigate the sensitivity of reinforced levees to rising water levels and uncertainties in geotechnical materials, while also promoting the usage of probabilistic analyses which can take those uncertainties into consideration. Thus, probabilistic analyses are conducted with the objective of quantifying the effects of uncertainties related to geogrid reinforcement on the slope stability of levees, and to construct fragility curves which show the probability of failure of the levee for any water level. Such probabilistic slope stability analyses can be conducted using numerous methods [35–43]. In this study, the limit equilibrium method is adopted due to its simplicity and wide usage in geotechnical practice, while results are further processed with programmed probabilistic methods to find the probabilities of unwanted behavior of the levee subjected to various water levels with steady state conditions. The statistical techniques and probabilistic methods used in this study are the Surface Response Method (SRM) and the First Order Reliability Method (FORM), which have been programmed with MATLAB. The variability values of random variables used for probabilistic calculations are selected as reported in literature. Since the considered sources of variability of geogrids include the long-term degradation, and no seismic event is considered, the conditions considered for the whole levee are drained. The case study is described in Section 3.

Fragility curves, which will be constructed as a result of this study, are curves showing the conditional probability of an unwanted behaviour occurring as a result of increasing the design event intensity. Their usage in civil engineering began in, at least, 1980 with the work of Kennedy et al. [44] related to the safety of a nuclear power plant. Later, their usage in flood protection started in 1991 with a USACE Policy Guidance Memorandum [45], followed by a further explanation in a 1993 USACE Engineer Technical Letter [46], as reported by [47]. Today, their usage in slope stability is widespread [35,39,47], with regards to seismic events, rainfall, and rising water level.

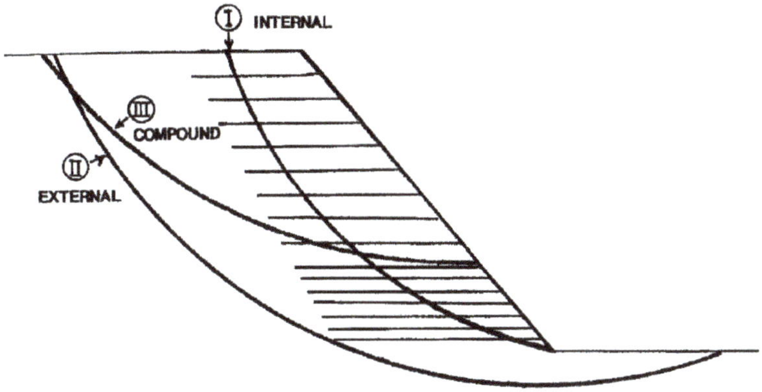

Figure 1. Failure modes of reinforced slopes [2].

2. Methodology

In this study, the Hasofer-Lind method is employed [48], also known as the First Order Reliability Method (FORM), together with the surface response method (SRM) for approximatively calculating the reliability of the flood protection embankment. The SRM is a statistical technique used to approximate the response of a model to input variables by using a suitable function when the true relationship is unknown. The approximation is done by fitting the selected function to the original model evaluated at multiple sample points, i.e., the coefficients of the function are determined by an error minimization technique. It is chosen as a relatively simple tool to complement the FORM by defining the required performance function. In this study it is used to construct an n-dimensional surface which approximates the response of the levee, where n is the number of random variables, based on known function values and on regression analysis. The surface used in this study is a quadratic function defined by a second-order polynomial, as shown in Equation (4). The coefficients of the function are obtained by minimizing the error between the original and approximated functions [49]. After that, the probability of failure is obtained through FORM optimization. The FORM is an upgrade to the First Order Second Moment (FOSM) method with its geometrical interpretation of the reliability index, which is invariant to the performance function format. To employ it, the first step is to convert all random variables to independent variables in the standard normal space with zero mean and unit standard deviation, and the performance function needs to be known. It offers a solution which defines the reliability index (β) as the shortest distance from the failure function (defined by SRM and the performance function) to the origin of the standard variable space, which is the mean of the joint probability distribution, and is the most efficient method for estimating p_f for problems involving one dominant failure mode [50]. Rackwitz [51] noted that, for 90% of all application, the FORM fulfils all practical needs, and its numerical accuracy is usually more than sufficient. Since all the random variables are normally distributed and independent, the transformation to the standard normal space is simply done by Equation (1) [52].

$$\overline{x_i} = \frac{x_i - \mu}{\sigma} \quad (1)$$

where $\overline{x_i}$ is the standard normal variable value, μ the mean value of the original variable, and σ the standard deviation of the original variable.

The first step in the analyses is to determine the number of random variables to be used, and their respective statistics. The mechanisms of failure of the geogrids are the rupture of the elements, or the pull out of the grid from the soil. Regarding the tensile strength, as there are three rows of geogrids reinforcing the body, the ultimate tensile strength of each of them is simulated as an independent random variable. The interaction between reinforcement and soil depends on various factors, including grid parameters such as roughness, grid opening dimensions, thickness of transverse ribs and deformability characteristics, as well as soil parameters such as friction angle, grain size distribution, particle shape, density, water content, cohesion, and stiffness [2]. For the pull-out parameters in this study, the soil-grid interface friction is taken as a fraction of the soil internal friction angle, while the cohesion is ignored. As the soil-grid interface friction depends on the friction angle of the material which covers the grid, the internal friction angle of that material is also taken as a random variable. Thus, a total of 4 random variables are considered (Table 3). Since there is no face anchorage, the sliding of the soil on the soil-grid interface can happen on either end of the grid, i.e., inside the body or at the face. Throughout the analyses, a specific soil-grid friction ratio is kept constant to investigate the behaviour at various ratios. The analyses are performed for three design cases with different interface friction, named here as SIF (small interface friction), MIF (mean interface friction) and HIF (high interface friction). Sia and Dixon [53] analysed the variability of interface strength parameters between soil and geotextiles or geomembranes in coarse- and fine-grained soils. In this paper, the ratio is held constant as a deterministic parameter. For the MIF case, a contact friction angle of 2/3 φ is used, which is the recommended conservative value for geosynthetics [2]. This is closely in agreement with values obtained by Yu and Bathurst [54] who used a reduction factor applied to the tangent of backfill friction angle of 0.5–0.8, with the best agreement between pull out tests and numerical model results being 0.67 or 2/3. Other studies propose different values, e.g., Ferreira et al. [12] define the interface friction angle as 6/7 φ, while Jewell [55] takes a factor of 0.8 as the "direct sliding coefficient" as a value to "safely encompass most practical cases". In this study, the factor 0.67 is used as the mean value but applied directly to the friction angle instead of its tangent (which is equivalent to applying a factor of 0.63 to the tangent). For the other two cases, SIF and HIF, interface friction ratios of 0.5 and 1 are used, respectively.

Next, an arbitrary number of different deterministic slope stability analyses are conducted for each water level by varying the random variables' values. In this study, this is achieved with the help of Latin Hypercube simulations, which varied the grids' strength for each chosen friction angle. All the variables' values are then transformed into the standard normal variable space by Equation (1), and the resulting safety factor is corrected accordingly with the appropriate performance function, as follows. The performance function is defined for two cases and shown in Equations (2) and (3), one for failure condition where $F_S = 1$ (ULS), and one for an arbitrary safety factor value of $F_S = 1.5$. The "probability of failure" calculated for the second case actually refers to the probability of reaching the defined threshold. When the performance functions defined by the left and middle terms in Equations (2) and (3) are equated to zero, this becomes the limit state function which defines failure or unwanted behaviour. Deterministic slope stability analyses, as well as steady seepage analyses, are conducted using Slide2 v9.009, Rocscience Inc., Toronto, ON, Canada.

$$g(\overline{x}) = F_s - 1 = 0 \tag{2}$$

$$g(\overline{x}) = F_s - 1.5 = 0 \tag{3}$$

Such defined groups consisting of standard normal variables' values and the respective performance function values for each water level are fitted with a polynomial shown in Equation (4) [56].

$$g'(\overline{x}) = c + \sum_{i=1}^{N} b_i x_i + \sum_{i=1}^{N}\sum_{j=1}^{N} a_{ij} x_i x_j \tag{4}$$

The $g'(\overline{x})$ symbolizes an approximation of the real performance function, where "c", "b", and "a" are its coefficients, N is the number of random variables, and "x" the random variables' values. The fitting is done in MATLAB by minimizing the sum of squared residuals with the *lsqcurvefit* function, where the value or the performance function and the random variables are known. The results of such minimization are the coefficients "c", "b", and "a" for a polynomial, which approximates the performance function in the vicinity of the design point (1 or 1.5). Now that the coefficients are known, a constrained optimization (minimization) is run. What we are searching for is the minimal value of the Euclidean norm of the standard normal variables which satisfies the condition that $g(\overline{x}) = 0$. This is done by minimizing the vector of standard normal variables x_i with the constraint $g(\overline{x}) = 0$, by using the MATLAB function *fmincon*.

$$\beta = \sqrt{\mathbf{x_i}' \cdot \mathbf{x_i}} = \min \tag{5}$$

The result is, (1) the reliability index β defined as the shortest value of the radius vector $\mathbf{x_i}$ which defines the limit state function, and (2) the standard normal variables' values $\mathbf{x_i}$ which give the previously defined distance. After a few iterations, these values converge towards the true limit state function. When the difference between two iterations becomes negligible, the procedure stops. This is usually achieved within 2–4 iterations for this study. Each iteration contains new deterministic slope stability analyses with new random variables' values, which resulted from previous iterations. To calculate the probability of failure from the resulting reliability index, the cumulative standard normal distribution is calculated for the reliability index with inversed sign. As the optimization needs a set of starting values, they are varied for the same calculation to check for the robustness of the result and for local minima. Another quality check is made by plotting surfaces in a 3-dimensional space by ignoring 2 of the random variables. To accept the result, not only is a small change in consecutive iterations needed, but also the quality of regression between the real performance function and the approximated one, as shown in Equation (6) [56], needs to be ≥ 0.95.

$$R^2 = 1 - \frac{\sum_{i=1}^{r}[g(\overline{x}) - g'(\overline{x})]^2}{\sum_{i=1}^{r}[E[g(\overline{x})] - g(\overline{x})]^2} \rightarrow 1 \tag{6}$$

where $E[g(\overline{x})]$ is the expected value of the performance function, simply taken as the arithmetic mean of all the deterministic performance function values. On top of that, the mean square error (MSE) is also calculated by Equation (7), and the results varied between 5×10^{-6} and 2×10^{-19}.

The whole process is repeated for various water levels, and fragility curves are constructed.

$$MSE = \frac{1}{n}\sum_{i=1}^{n}[g(\overline{x}) - g'(\overline{x})]^2 \rightarrow 0 \tag{7}$$

The previously discussed methodology for the development of fragility curves for levee stability, summarized in Figure 2, requires the proper water pressure state to be established. For the design situations which include water level up to the top of levee crown, numerical analyses include commonly used boundary conditions of properly defined hydraulic heads on the riverside and landside. However, as Librić et al. [57] found in their study, the overtopping of the case study levee has a high risk exposure compared to other risks, thus overflow is also considered in this study. Overtopping (i.e., overflow) is usually a result of a high-water event (surge), or it can occur due to wave overtopping. The combined effect of surge and wave overflow is discussed by many authors [58–60]. However, for the river levee considered, only a surge type of overflow is relevant. When the water level rises higher than the crown, the stress analyses are supplemented with the additional boundary shear stress along the crown and landside slope. Additionally, a trapezoidal stress is applied over the crown during overflow to simulate water pressure, corresponding to the height of water on the upstream side, and to the water height on the downstream side calculated by Equation (10). Given that this aspect goes beyond standard analyses, a cautious evaluation of these shear stresses is required. The boundaries of both analyses are defined far from the levee region, enough to not affect the results. Seepage analyses require only hydraulic boundary conditions, which in this case consist of constant or varying hydraulic head values applied on lateral boundaries and on the top boundary of the model, up to the required height. During surge overflow, the water velocity increases down the slope until a terminal velocity is reached at equilibrium between water momentum and slope frictional resistance, after which the flow becomes steady and the velocity can be calculated by the following equation:

$$v_0 = \left[\frac{\sqrt{\sin\theta}}{n}\right]^{3/5} \cdot q_0^{2/5}\ (\mathrm{m/s}) \tag{8}$$

where v_0 (m/s) is the steady flow velocity, θ (°) is the landside slope angle, n (-) is the Manning's coefficient, and q_0 (m^2/s) is the steady discharge [58]. For supercritical flow, which develops on the landside slope—as shown in Figure 3—Hewlett et al. [61] proposed a value of Manning's coefficient of n = 0.02, relevant for slopes of 1:3.

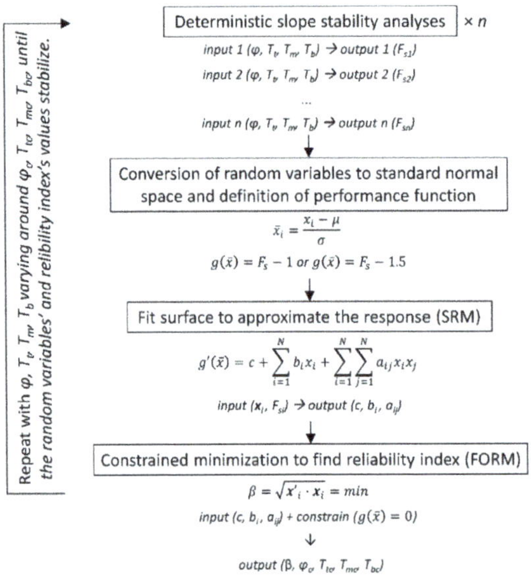

Figure 2. Flowchart of the applied methodology.

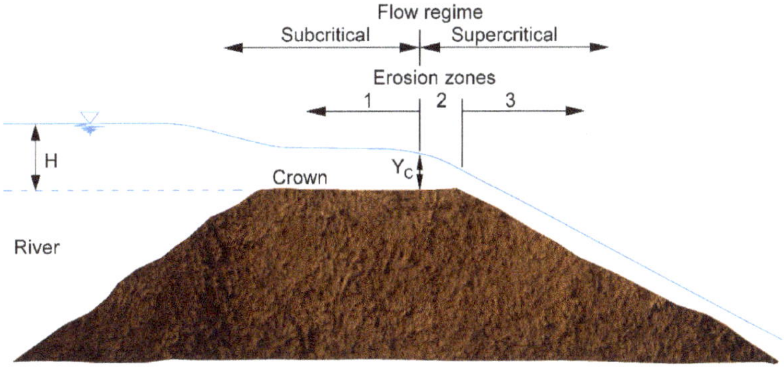

Figure 3. Flow regimes during overflow of a dam, re-drawn from [62].

The discharge over the levee crown can be calculated using the equation for flow over a broad-crown weir, which gives slightly conservative results due to not taking into consideration frictional losses [63]:

$$q = \left(\frac{2}{3}\right)^{3/2} \cdot \sqrt{g} \cdot h_1^{3/2}\ (m^2/s) \tag{9}$$

where g (m/s^2) is the gravitational acceleration and h_1 (m) is the upstream head (elevation over the levee crown). If a steady flow is assumed, the discharge is constant along the slope. Therefore, the height of water perpendicular to the slope in the steady, uniform flow area for unit length of the levee can be calculated from Equations (8) and (9) as:

$$h = \frac{q}{v_0}\ (m) \tag{10}$$

Finally, when steady, uniform, flow is reached, the shear stress resulted from surge overflow, is equal to:

$$\tau_0 = \gamma_w \cdot h \cdot \sin\theta\ (kPa) \tag{11}$$

where γ_w (kN/m^3) is the unit weight of water. Equation (11) conservatively overestimates results, since the resulting pressure is a little bit higher than the pressure in area above the steady flow [60]. Shear stresses calculated this way are applied along the crown and landside slope, as shown in Figure 4 for the case study numerical model.

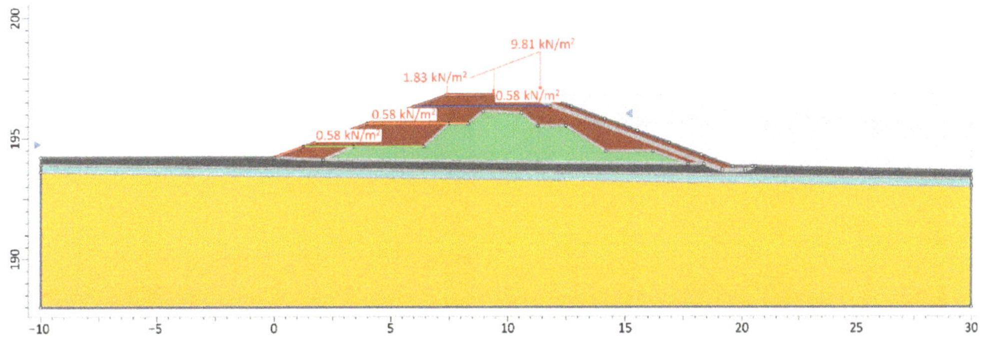

Figure 4. Numerical model for analyses of case study levee.

Stability analyses aim to find the critical slip surface by using a population-based stochastic algorithm, Cuckoo Search, which searches for non-circular slip surfaces, together with Monte Carlo optimization to potentially find even more critical surfaces [64]. All the slope stability analyses are deterministic with values of the four random variables previously described varied over appropriate ranges, while probabilistic analyses are conducted after the results of deterministic analyses are obtained. The variation is manually performed for the friction angle, while for the geogrids it is performed with the help of Latin Hypercube simulations. It was initially conducted over a range of $\pm 3\sigma$ with steps of 1σ to detect the approximate location of the design point and was then corrected to smaller steps closely spaced around the design point.

3. Case Study

River Drava, with the overall length of 710 km, flows from Italy to eastern Croatia where it merges with Danube, and is historically known for major flood events [65], where prominent events occurred in last several years. For this case study, a reach of a 3.7 km long section of flood protection embankment running from Otok Virje to Brezje on the Drava River in Croatia is analysed. The reach lies on sediments from the Holocene period. They are mostly sediments of the first alluvial terraces of Drava, composed of large amounts of sand and gravel, which at places surpass 100 m in depth. Closer to the surface, layers of silty material can be found.

In 2012, a water level of 1000-year return period was measured in the Drava River, which caused the overflow of the embankment over a length of more than 1 km, and breaching over a length of 50 m, causing huge damages to the surrounding area. Since the original embankment was built in 1968 with the design high water level from 1965 [66], a reconstruction of the existing embankment is required for raising its crown height to new design water levels. The new required height corresponds to the new 100-year return period water level +1 m, which is between a few cm and 1 m above the old crown. Raising the height also implies a widening of the embankment cross section, which can be accomplished in three ways: by keeping the existing embankment on the landward side of the new one (i.e., reconstructing towards the river side), keeping the existing embankment on the river side (i.e., reconstructing towards the landward side), and by coinciding the existing and new axes (i.e., reconstructing on both sides). The selected reach for this case study is defined by the reconstruction direction and subsoil stratigraphy—the reconstruction on both sides is chosen. A situational view of the embankment section on the Drava River is shown in Figure 5.

To prove the stability of the newly reconstructed embankment in all the relevant design situations, calculations are made using deterministic limit equilibrium analyses, and all according to valid norms for geotechnical design, i.e., *EN 1997-1:2012 Eurocode 7: Geotechnical design—Part 1: General rules* and its respective Croatian national annex for static design situations, *EN 1998-1:2011 Eurocode 8: Design of structures for earthquake resistance—Part 1: General rules, seismic actions and rules for buildings* and its respective national annex for seismic design. The analyses resulted in the deterministic safety factors shown in Table 1. It can be seen that safety factor values for all design situations are acceptable.

Table 1. Deterministic safety factors for the cross section of interest in various design situations.

Design Situation						Safety Factor [1]
Reconstruction on both sides of the existing levee	Low water	Riverside	Static + traffic		Drained	1.79
			Seismic	475-year RP	Undrained	1.47
		Landside	Static + traffic		Drained	2.18
			Seismic	475-year RP	Undrained	1.48
	High water (100-year RP)	Landside	Static + traffic		Drained	1.72
			Seismic	475-year RP	Undrained	1.49
	Water at crown height	Landside	Static + traffic		Drained	1.66
	RDD	Riverside	Static		Drained	1.21

[1] Analyses are made using EC7, DA3, thus the minimum required safety factor is 1.

The reconstruction of the levee is made with well graded gravel (GW by USCS classification). Since gravel is highly permeable, GCL membranes are used to make sure the free water surface stays inside the levee body during high water events. Suzuki et al. [67] performed field and laboratory tests with various types of GLC to find their effect on the stability of the embankments. However, since the GCL in this study is located on the riverside of the levee, while the stability of the landside is analysed, their effect is not relevant for this study. Other than that, the body is further strengthened using TENAX TT 045 GS, HDPE uniaxial geogrids. The embankment's cross section used in calculations is shown in Figure 6. Geogrids are placed on 0.7 and 0.9 m distance from one another to fit the height of the embankment, while the maximum suggested height for reinforced slopes as per [31] is 1 m due to local face stability. This way, local face instabilities are partly mitigated. Instabilities on the landside may also be initiated by surface erosion during overflow. The resistance against such action can be increased by placing a reinforcing layer of standard geosynthetics or other specific products [68] such as biopolymers [69,70] over the slope, but in this case, there is no such additional protection. The same applies for the riverside slope where surface erosion might be caused by the flow of the river and during the rapid decrease of water level in the river (RDD).

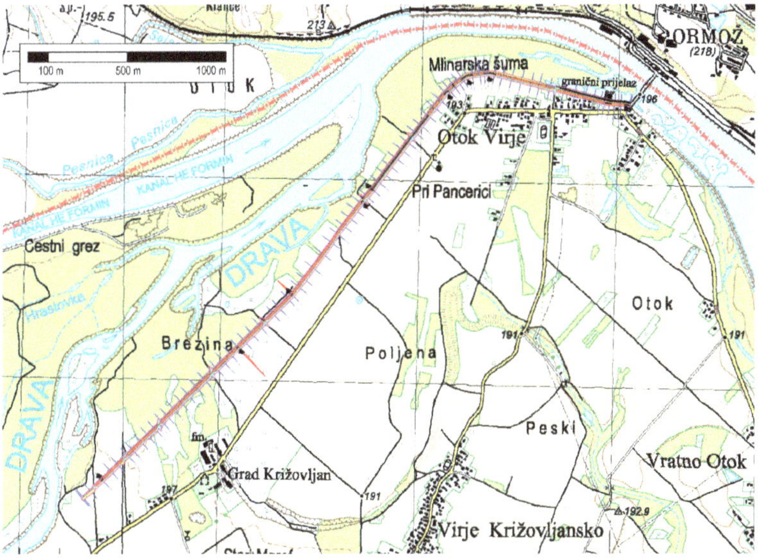

Figure 5. Situational view of levee section.

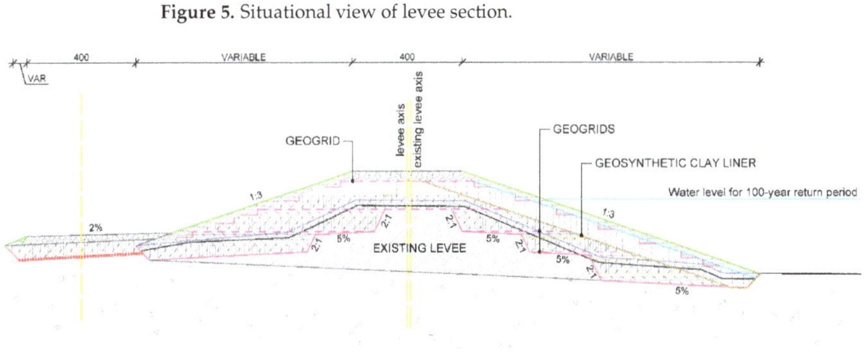

Figure 6. Levee cross section.

Variability of Materials' Parameters

Deterministic parameters for all soils are carefully chosen from the available laboratory and field data conducted by the authors of this paper. The mean values of geogrid parameters are taken from the manufacturer's specification sheet. Statistics for the parameters assigned as random variables are chosen from literature. Table 2 shows the design values of deterministic parameters for each material, while Table 3 shows the statistics of the random variables. As all three grids inside the embankment are the same, their statistics are also the same, but each grid is modelled as an independent random variable.

Table 2. Deterministic values of parameters.

Material	USCS Symbol	φ_d (°)	c_d (kPa)	γ_d (kN/m^3)	k (m/s)
Reconstruction material—GW	GW	Random	0	20	2.5×10^{-2}
Existing body	SM	25.1	1.6	19	1.4×10^{-5}
Thin surface layer	MI	18.8	3.3	19	5×10^{-6}
Second thin layer	SP-SM	25.6	0	19	4.7×10^{-4}
Foundation soil	GP-GM	28.4	0	19	8.6×10^{-4}
GCL [1]		-	-	-	1×10^{-7}

[1] GCL is only relevant in seepage modelling.

Table 3. Statistics of random variables.

Material	Tensile Strength (kN/m)		Friction Angle (°)	
	Mean	CoV	Mean	CoV
Geogrids	19.06	0.122	-	-
Reconstruction material—GW	-	-	35	0.1
Distribution	Normal		Normal	

The statistical parameters for the geogrids are determined as follows. The mean value is taken from the manufacturer's specification sheet where the characteristic value is divided by a series of factors, namely the factor for installation damage (RF_{ID}), creep (RF_C), and degradation due to chemical and/or biological processes (RF_D), to obtain a design tensile strength of 18.5 kN/m. To transform the manufacturer's proposed long term design strength into a random variable, it is first multiplied with the mean values of bias factors for installation damage ($\mu_{X_{ID}}$), creep (μ_{X_C}) and durability (μ_{X_D}), whose statistics are determined from literature [71–74] to obtain the mean, while the CoVs of different bias factors are together taken as the CoV for tensile strength (Equation (12)). Theoretically, this is valid for uncorrelated log-normal random variables, but with small CoVs it is sufficiently accurate for uncorrelated normal random variables [75]. Since the durability factor is mostly project-specific, it is taken with an arbitrary CoV = 0.1 [72]. Chosen statistics for all factor's bias values are shown in Table 4. The chosen geogrids are made from HDPE (High Density Polyethylene), which showed the lowest mean and CoV of the bias factors values, and their statistics are found to be independent of soil type [72]. As the random variables of the geogrids are normally distributed [72] and uncorrelated, the simple conversion to standard normal variables as shown in Equation (1) can be employed.

$$CoV = \sqrt{{CoV_{X_{ID}}}^2 + {CoV_{X_C}}^2 + {CoV_{X_D}}^2} \quad (12)$$

Table 4. Statistics for reduction factor's bias values.

Statistic	X_{ID}	X_C	X_D
Mean	1.03	1	1
CoV	0.06	0.036	0.1

Since partial factors in various design approaches are calibrated using reliability analyses [76], the mean friction angle of the reconstruction material is left at its characteristic value and the variability is applied to it, while all the other deterministic values are factored using partial factors from Eurocode 1997 DA3.

4. Results and Discussion

During slope stability calculations at higher water levels, small variations in random variables' values resulted in shallow and deep sliding surfaces with highly different safety factors, such as those shown in Figure 7. In such cases, the shallow and deep surfaces are separated, and two probabilities of failure are calculated, one pertaining to the shallow sliding and the other to deep sliding. Fragility curves are then constructed for two limit states defined by safety factors 1 and 1.5 (LSF10 and LSF15 respectively), for both types of failures. Figure 8 shows the resulting fragility curves for the two limit states, for varying water levels from the toe to the crown of the levee (located at 196.8 m.a.s.l.), and over to simulate surge overflow. The water level is increased until almost certain failure is ensured. However, Rackwitz [51] noted that FORM works well only for sufficiently large reliability indices, which he defined as $\beta > 1$, as otherwise it might not be the best linearization point [77], which in this case corresponds to water level of around 200.5 m.a.s.l. for LSF10, and 196 m.a.s.l. for LSF15. The curves used to fit the points are represented by general sigmoid function with the following equation [78]:

$$f(x) = p_{f,min} + (p_{f,max} - p_{f,min}) / \left(1 + 10^{[(\overline{H} - x)k]}\right) \quad (13)$$

where $p_{f,min}$ and $p_{f,max}$ are minimum and maximum values of the function respectively, $\overline{H}$ the mean hydraulic head, and k is the slope of the curve at the mean value. The curve is fitted to the points using least-squares. By using such a function, a curve can be defined even by not having the whole range of points from zero to one probability of failure. As can be seen from the figures, for the cases where the limit state function is defined by $F_S = 1.5$ (LSF15), the probabilities of failure occur over the whole range, and certain limit-state-behaviour with probability of one is already reached at the crown water level. For the limit state function defined by $F_S = 1$ (LSF10), the maximum calculated probabilities of failures reach from around 65% to as low as 10%, while the rest of the curve is based just on the fitting to those smaller values.

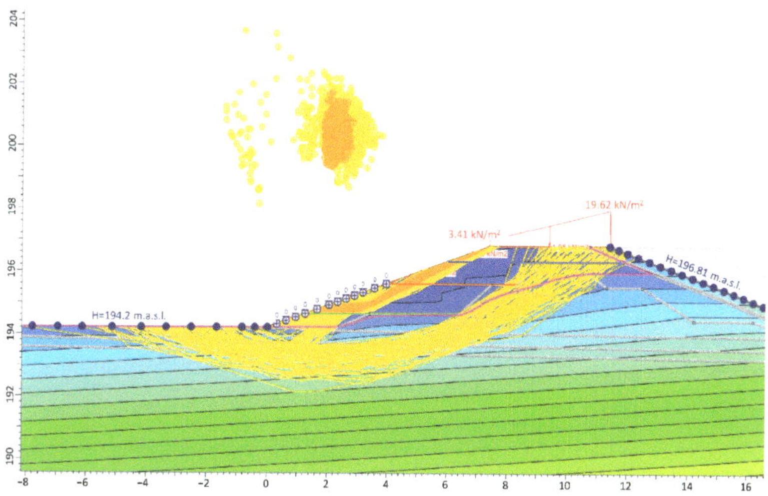

Figure 7. Shallow and deep surfaces on a characteristic slope stability analysis.

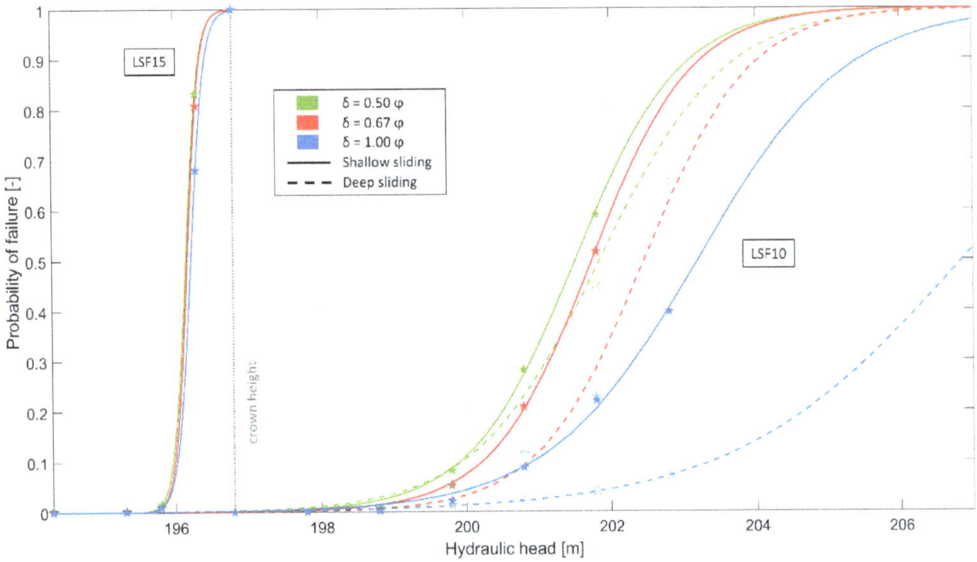

Figure 8. Fragility curves for LSF10 and LSF15, for shallow and deep surfaces, and all interface friction angles.

From the diagrams it is seen that with increasing interface friction, the distinction between shallow and deep surfaces starts to show earlier, i.e., at lower water levels, for LSF10. For $\delta/\varphi = 0.5$, the distinction occurs at $p_f > 0.3$, for $\delta/\varphi = 0.67$ at $p_f > 0.05$, and for $\delta/\varphi = 1$ at $p_f > 0.002$. Regardless, this effect never reached water levels as low as the levee crown for this case. For the LSF15, the distinction occurs at lower water levels rather than higher and is not so large, which is the reason why it is not noticeable on the normal scale. Also, the interface friction in this regard does not have any noticeable effect in this case. At water levels where there is no distinction between deep and shallow surfaces, this happens because of two reasons. One is that all the failures occurred as either deep or shallow failures, and the other is that there is little distinction between safety factors of deep and shallow surfaces. The first reason indicates that the curves are constructed for the stability of the slope regardless of the failure mode, as long as all the surfaces followed the same mode. Only when different modes appear, the curves become separated.

Even though the diagrams for both limit states seem to merge at lower water levels, obviously this is not the case because the diagrams refer to different limit states which cannot be achieved with the same strength parameters for a specific water level. Thus, Figure 9 shows the diagram in a logarithmic scale to see the difference at lower water levels. The LSF15 points can still be approximated as relatively good by the same sigmoid function. For LSF15 the probability of failure starts to noticeably increase only after the water level reaches circa 60% of the levee height. On the other hand, the LSF10 points have a worse fit which is caused by the fact that the p_f stays almost the same for water levels between 0 and the levee crown, and start to substantially change only for the surge overflow. Thus, to fit the LSF10 points, the first point referring to a no-water situation is ignored. The reason for the constant probability of failure is that the parameters needed to achieve the defined limit state are such that they produce small slip surfaces on which water has no effect in this case. This means that regardless of the water level up to the crown, the p_f of the levee stays the same. It can be noted from the figures that deeper failure surfaces are generally less likely to occur during failure than shallower surfaces. This is certainly conditioned by the fact that the levee body through which the deeper surfaces pass has been modelled as a deterministic material. An interesting thing to note for the minimum friction LSF15 case is that the points which refer to deep surfaces show a slight decrease of p_f with the increase of the water level at the beginning of the curve. This means that for the lower water level there is a higher probability that the levee will fail overall, than there is for the higher water level that the levee will fail through deep sliding. The reason for such behaviour is because, for the smaller water level, no deep surfaces are found. For other interface friction angles, these two values are quite similar.

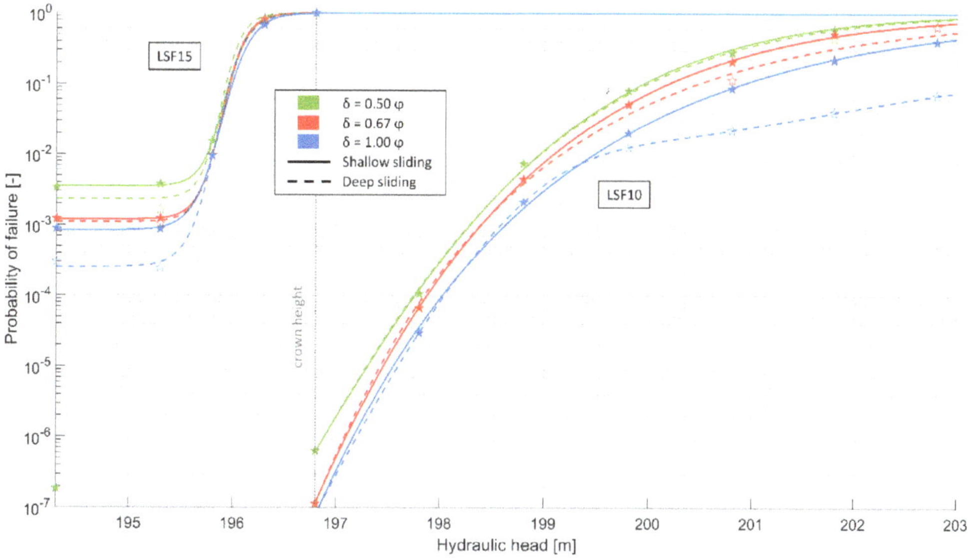

Figure 9. Diagrams shown in logarithmic scale.

While interpreting the results of computed conditional probabilities of failure, it must be kept in mind that they are calculated based on only one point closest to the origin, defined by the reliability index. This implies the linearization of the limit state function for integration below the standard normal joint distribution, which can lead to an overestimation or underestimation of the real probability, depending on the true shape of the limit state function [52]. To investigate their shapes, 2D representations of the limit state functions for mean interface friction, and for LSF10 and LSF15, are shown in Figure 10, where the black lines represent LSFs for soil internal friction and top grid strength as random variables, red lines LSFs for soil internal friction and middle grid strength as random variables, and blue lines LSFs for soil internal friction and bottom grid strength as random variables. The rightmost curves correspond to higher water levels, decreasing towards the leftmost curves. The circles in Figure 10 represent the standard normal joint distribution, i.e., each circle corresponds to one standard deviation. From the figures it can be seen that LSFs are slightly curved in either direction, without any notable trend, thus giving mixed results in terms of conservativeness. However, the effect of linearization is not expected to be high in most cases due to the curvatures being relatively mild. Similar shapes are noted for all interface friction angle values and are not shown here.

To better investigate the effect of increasing reliability with increase of interface friction ratio, defined as δ/φ, graphs are plotted in Figure 11 showing these trends for LSF10 with the normalized reliability index on the vertical axis. The normalized reliability index is simply the reliability index for the mean interface friction ratio for each respective water level subtracted from the reliability indices at other friction ratios ($\beta_n = \beta - \beta_{0.67}$). This way all the curves are translated over the vertical axis for better comparison. From Figure 11 two characteristics are noticed, one being that the trend is approximately linear, changing from a power law for the lower water levels (i.e., higher β, lower p_f) to a positive parabola for the higher water levels (i.e., lower β, higher p_f). The second relates to the increase of steepness of the curve from lower to higher water levels, which is more pronounced on the higher friction ratio than on the lower. Also, the same increasing effect is seen for deeper versus shallow surfaces for the same water level (the two highest curves). For the LSF15 case (not shown on figures), even though some differences apparently exist between p_fs for different friction ratios, there are no visible trends, except for the deep sliding curve being steeper than the corresponding shallow sliding curves.

To analyse the sensitivity of the slope stability to each of the defined random variables, 2D sections of the response surfaces through the design point are plotted on Figure 12. For each graph, one variable of interest is varied in the vicinity of the design point, while the other random variables are held at their respective design point values. The horizontal axes on the graphs are normalized such that the design point value is at zero and show the number of standard deviations away from the point. In other words, the curves are shifted from values obtained through Equation (1) to align all the design points at zero. This helps comparing the trends of the response surface at various water levels. From the response surfaces, it is found that two main trends exist, namely parabolic (positive or negative) and linear. This is of course constrained by the function used to approximate the response surface (Equation (4)), which is a quadratic polynomial. It is intuitive that the increase of any strength or resistance parameter's value causes the stability of a slope to increase by increasing the safety factor. However, some curves shown in Figure 12 seem to contradict this statement as they are parabolas which have maxima and minima at, or close to, the design point. This is just an apparent problem caused only by the chosen approximation function and does not show any inconsistencies considering the friction angle because all the maxima are found on the right side of the design point, while all the minima are found on the left side, and the curves are fitted to the data by their increasing parts. This mostly occurred when smaller friction angles did not have a distinction between deep and shallow sliding, but higher friction angles did have it. In those cases, a sudden increase in safety factor occurs and a parabolic response surface cannot be generated with the whole range of data for deep sliding. This means that to achieve a good fit of the data to a parabola, one needs to discard all but the closest sample points on either side of the design point, while only keeping all the points on the opposite side. An example of such situation is shown in Figure 13. It is obvious that a higher order function should be used to approximate such data. It could be argued that if only a narrow range of data around the design point is used, then there would be no need to approximate the data using higher order function. While this may be true in some cases, in other cases the range which would be needed to avoid higher order functions is relatively narrow, and would complicate the analysis to find values only inside that narrow range.

For the geogrids' tensile strengths, the trends are generally constant, which means they do not affect the safety factor. However, there are multiple increasing curves, and others which actually do show a decrease of the safety factor with a strength increase. Regarding the latter, it should be noted that the range of safety factors on Figure 12 for the grids is from 0.95 to 1.05, and thus such trends cannot be deemed as true trends. They can instead be attributed to data scatter in both deep and shallow surfaces, as well as to the shape limitations of the selected approximation function. This data scatter occurs mostly for the bottom and middle grid layers, and only at the higher end of friction angles. In the same region of friction angles, the top grid's strength start showing a linear to parabolic trend, as shown in Figure 14. This kind of data, however, also did not cause any inconsistencies with results, as the design point tensile strengths are practically at the mean values for most cases (Figure 15).

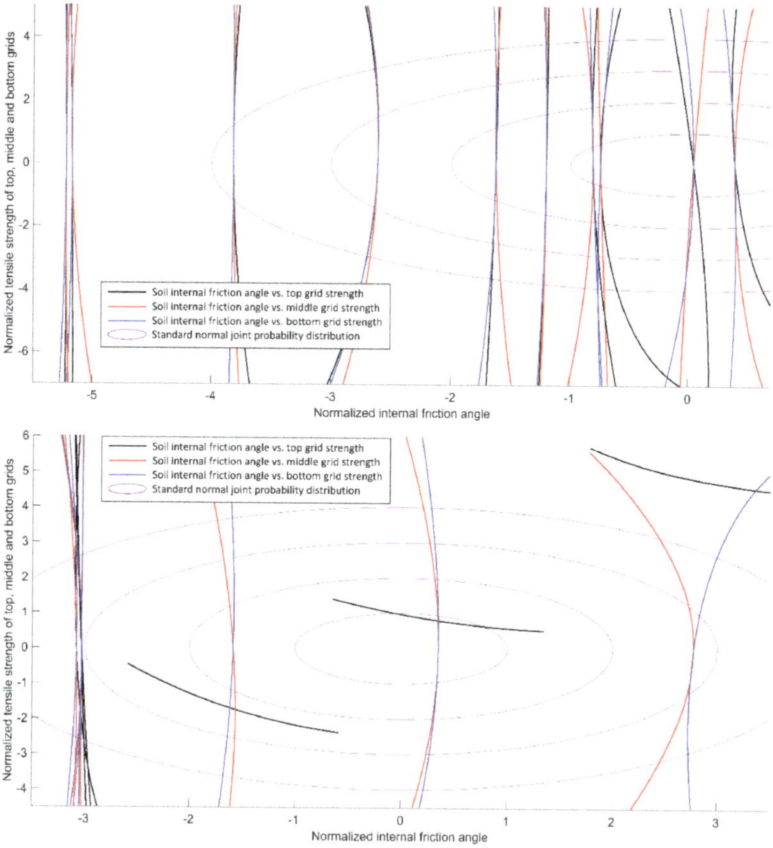

Figure 10. Limit state functions for LSF10 (**top**) and LSF15 (**bottom**).

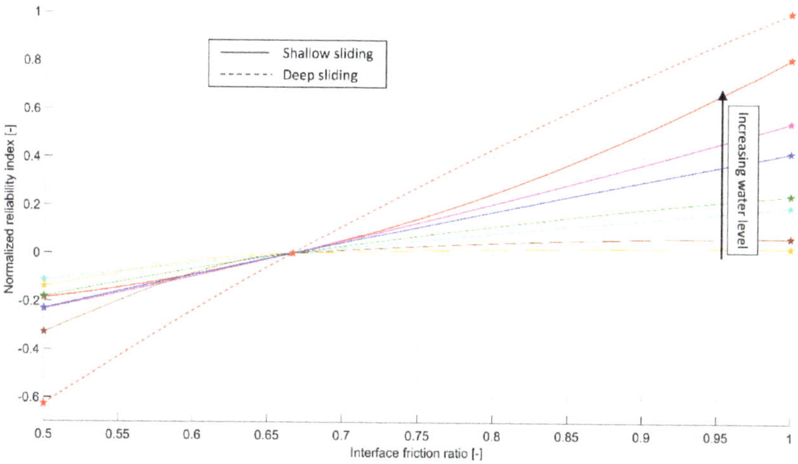

Figure 11. Trend of reliability index increase with increase of interface friction ratio for LSF10.

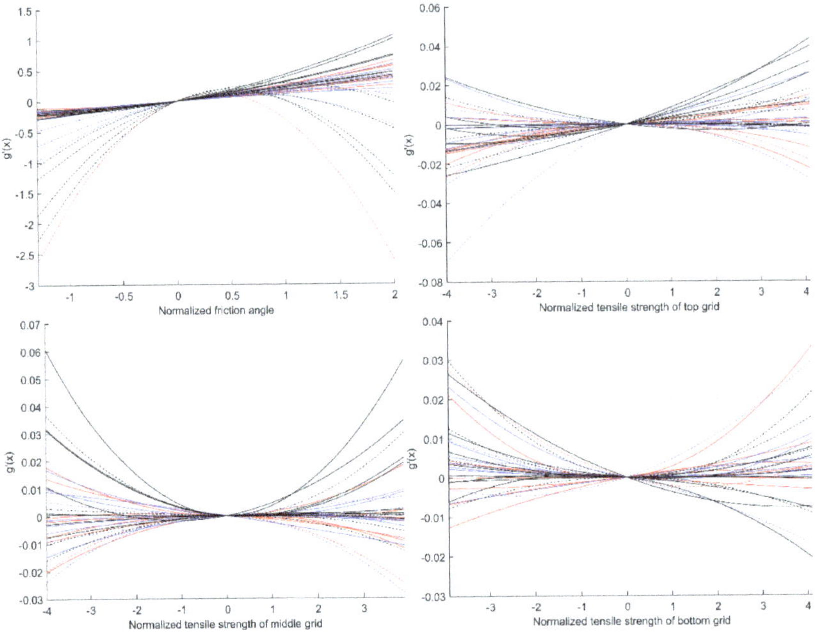

Figure 12. Trends of the response surface around the design point for LSF10 for low (red lines), mean (blue lines) and high interface friction angle (black lines), with full lines representing response for shallow sliding, and dashed lines for deep sliding.

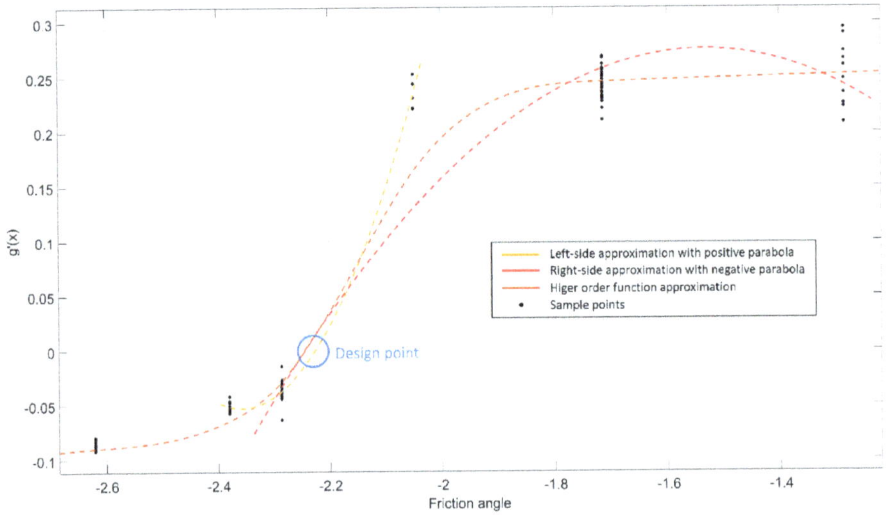

Figure 13. Situation where a cubic equation would be more appropriate (on the horizontal axis an interval of 0.1 = 0.35°).

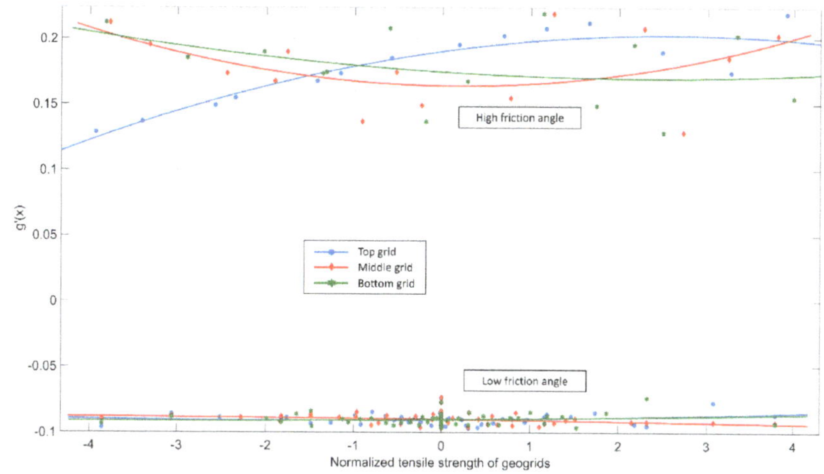

Figure 14. Effect of tensile strength of all geogrid layers on the performance function for mean interface friction angle, for two different values of reconstruction material friction angle.

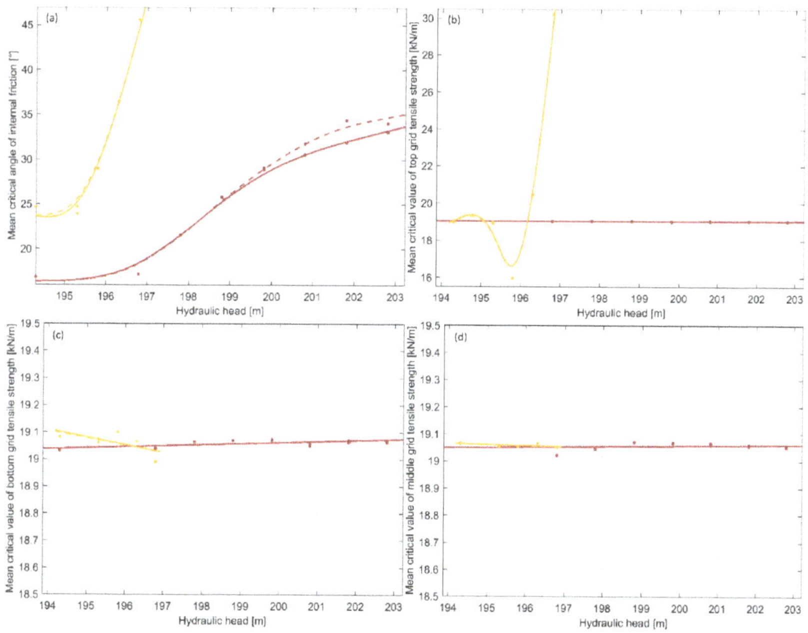

Figure 15. Critical values of random variables for various water levels; (**a**) friction angle (top left), (**b**) top grid (top right), (**c**) middle grid (bottom left), (**d**) bottom grid (bottom right); dark lines represent LSF10, light lines LSF15, full lines shallow sliding, and dashed lines deep sliding.

With each increment in water level there is change in probability of failure/reliability index, which is caused by the different critical values of random variables needed to reach a specific LSF at that specific water level. Even though the critical values differ when the same water level is evaluated with different interface friction ratios, because the difference is not large, Figure 15 shows the mean critical value of each random variable. Figure 15 is a representation of Equation (5), where the total reliability index can also be calculated for each water level by knowing the corresponding critical values of each random variable. Even though Figure 15b shows a decrease in critical tensile strength for the LSF15 at hydraulic head near 196 m.a.s.l. compared to lower heads, the reliability index still decreases (Figures 8 and 9) due to an increase in critical friction angle. On the other hand, the critical value for the LSF10 shows practically no change with the increase of water level. This is also true for both LSFs for the middle and bottom grids.

To investigate the relative contribution of the uncertainty of each random variable to the total uncertainty, the direction cosines (or sensitivity factors) are calculated as the ratio of each random variable's critical value to the reliability index. The squared sensitivity factors give us the values of interest [79]. For all interface friction angles (low, mean, high) for LSF10, the contribution of the internal friction angle is >99.84%, with a mean value of 99.97%. The small remainder (<0.15% of total uncertainty) belongs to all three layers of geogrids, with almost 3/4 of that belonging to the top grid, and the rest somewhat evenly distributed between the middle and bottom grids. For LSF15, the relative contribution of the internal friction angle decreases with the increase of water level, from almost 100% to 17%, while the rest is attributed to the top grid, as shown in Figure 16. The middle and bottom grid's contribution stayed close to zero at all times.

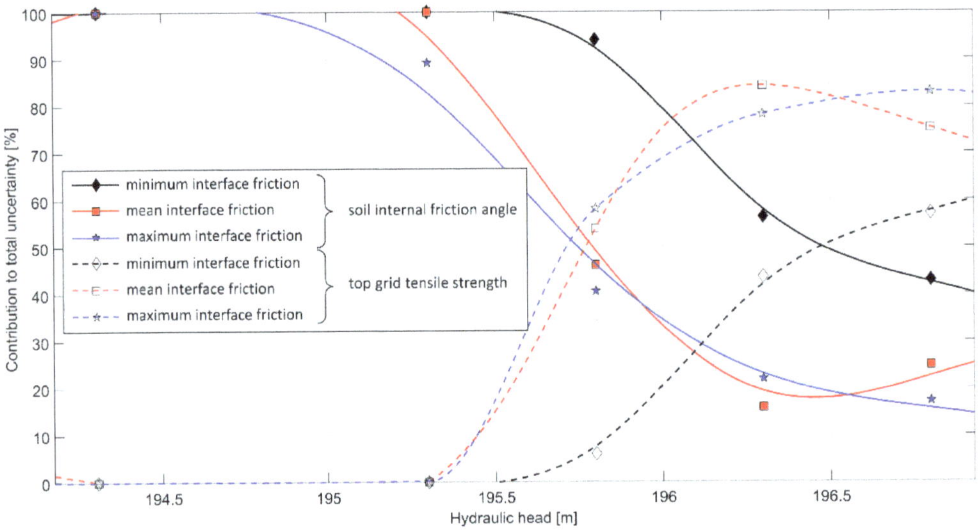

Figure 16. Relative contribution of internal friction's and top grid's uncertainty to total uncertainty for LSF15, for all three interface friction angles (full lines refer to internal friction angle, dashed lines refer to top grid).

5. Conclusions

It should be noted that conclusions drawn from this study are only valid for systems similar to the analyzed levee, where local face stability in ensured between geogrid layers, and where there is no anchorage on the front face, which would increase stability even more at the cost of additional material for the anchorage. From the presented results, a few conclusions can be drawn:

- Close to the ULS, at higher water levels (in this case $p_f > 0.002$), small variations in random variables cause deep and shallow sliding surfaces with highly different safety factors and p_fs. With the increase of interface friction angle, the water level at which this distinction becomes visible decreases. Farther from the ULS, this effect occurs at low water levels rather than higher, and the interface friction doesn't have any noticeable effect on the occurrence. Shallow sliding is shown to be more likely to occur for both limit states.

- Linearization of the performance function required for calculation of the probability of failure with the FORM does not influence the results greatly as the curvatures of the limit state functions are generally not large.
- The increase of reliability with increase in interface friction ratio is approximately linear in proximity of the ultimate limit state, with higher steepness for higher water levels. Also, deep surfaces seem to have a steeper curve than shallow surfaces. This is not true farther away from the ULS.
- Constant, linear, and parabolic trends, and those of higher order, are found for the performance function dependency to the reconstruction material friction angle and to the geogrid layers. The order of the function tends to increase with water level, i.e., with probability of failure. The higher order trends occur mostly for deep sliding when a sudden increase of safety factor occurs as a result of small increase in friction angle of the levee body. For this reason, quadratic functions should be used with care, and perhaps a function with an inflection point (e.g., cubic function) should be employed in some cases.
- The internal friction angle contributes almost completely to the total uncertainty when close to the ULS (the contribution of the grids is negligible). However, it seems that geogrids placed near the top contribute the most out of all the geogrids. The contribution of the internal friction angle seems to diminish going farther away from the ULS (e.g., LSF15), and it transfers to the grid placed near the top, while the other grids' contribution remains negligible. This importance of the top grid, however, needs to be considered carefully, because in this study the top grid is the only one that goes from one slope of the levee to the other, while the middle and bottom grids are only placed on one side. The relative contribution might be different in case all grids are the same length.
- The reason for the grids' extremely low contribution to the total uncertainty lies with the small variability of their tensile strengths. But as seen in the LSF15 case, as the required critical tensile strength reaches the actual tensile strength, their variability has more effect on the stability, which indicates a way of determining the required strength for each grid layer. Moreover, since the increase of both the soil friction angle and the soil-grid interface friction individually tend to generate deeper surfaces, it is implied that a balance between these parameters can be found. Both procedures would lead to a balanced reinforced slope design with regard to geogrid rupture strength and geogrid pull out.

At this time, deterministic analyses are still the dominant type of analyses when it comes to designing levees for flood protection in terms of slope stability. These calculations require safety factors >1 to be acceptable by definition of the safety factor. However, it is rarely deemed acceptable to reach safety factors around one, and higher values are often targeted. This study deals with slope stability of a geogrid reinforce levee in a probabilistic manner, and shows trends in the response of the levee and behaviour of the reinforcement components for two target safety factors of one and 1.5. Even though some trends are not as clear as others, and/or they are not quantitatively defined, and for which further investigations are needed, the identified trends might still serve as guidelines for design and can be loosely interpolated (where applicable) to increase understanding of the system behaviour for a targeted safety level.

Author Contributions: Conceptualization, N.R. and M.S.K.; Formal analysis, N.R.; Investigation, N.R., M.B., M.S.K. and L.L.; Methodology, N.R. and M.S.K.; Software, N.R.; Supervision, M.S.K.; Validation, N.R., M.B. and M.S.K.; Visualization, N.R.; Writing—original draft, N.R. and M.B.; Writing—review and editing, M.S.K. and L.L. All authors have read and agreed to the published version of the manuscript.

Funding: This research was funded by European Union Civil Protection Mechanism, under UCPM-2019-PP-AG call, Grant Agreement Number 874421, oVERFLOw project (Vulnerability assessment of embankments and bridges exposed to flooding hazard).

Conflicts of Interest: The authors declare no conflict of interest. The funders had no role in the design of the study; in the collection, analyses, or interpretation of data; in the writing of the manuscript, or in the decision to publish the results.

References

1. Wang, X.; Wang, L.; Zhang, T. Geometry-Based Assessment of Levee Stability and Overtopping Using Airborne LiDAR Altimetry: A Case Study in the Pearl River Delta, Southern China. *Water* **2020**, *12*, 403. [CrossRef]
2. Elias, V.; Christopher, B.R.; Berg, R.R. *Mechanically Stabilized Earth Walls and Reinforced Soil Slopes Design and Construction Guidelines*; National Highway Institute, Federal Highway Administration, U.S. Department of Transportation: Washington, DC, USA, 2001.
3. Avesani Neto, J.O.; Bueno, B.S.; Futai, M.M. A Bearing Capacity Calculation Method for Soil Reinforced with a Geocell. *Geosynth. Int.* **2013**, *20*, 129–142. [CrossRef]
4. Kolay, P.K.; Kumar, S.; Tiwari, D. Improvement of Bearing Capacity of Shallow Foundation on Geogrid Reinforced Silty Clay and Sand. Available online: https://www.hindawi.com/journals/jcen/2013/293809/ (accessed on 16 November 2020).
5. Wang, J.-Q.; Zhang, L.-L.; Xue, J.-F.; Tang, Y. Load-Settlement Response of Shallow Square Footings on Geogrid-Reinforced Sand under Cyclic Loading. *Geotext. Geomembr.* **2018**, *46*, 586–596. [CrossRef]
6. Barani, O.R.; Bahrami, M.; Sadrnejad, S.A. A New Finite Element for Back Analysis of a Geogrid Reinforced Soil Retaining Wall Failure. *Int. J. Civ. Eng.* **2018**, *16*, 435–441. [CrossRef]
7. Wang, J.-Q.; Xu, L.-J.; Xue, J.-F.; Tang, Y. Laboratory Study on Geogrid Reinforced Soil Wall with Modular Facing under Cyclic Strip Loading. *Arab. J. Geosci.* **2020**, *13*, 398. [CrossRef]
8. Wang, H.; Yang, G.; Wang, Z.; Liu, W. Static Structural Behavior of Geogrid Reinforced Soil Retaining Walls with a Deformation Buffer Zone. *Geotext. Geomembr.* **2020**, *48*, 374–379. [CrossRef]
9. Abu-Farsakh, M.Y.; Akond, I.; Chen, Q. Evaluating the Performance of Geosynthetic-Reinforced Unpaved Roads Using Plate Load Tests. *Int. J. Pavement Eng.* **2016**, *17*, 901–912. [CrossRef]
10. Cuelho, E.V.; Perkins, S.W. Geosynthetic Subgrade Stabilization—Field Testing and Design Method Calibration. *Transp. Geotech.* **2017**, *10*, 22–34. [CrossRef]
11. Varuso, R.J.; Grieshaber, J.B.; Nataraj, M.S. Geosynthetic Reinforced Levee Test Section on Soft Normally Consolidated Clays. *Geotext. Geomembr.* **2005**, *23*, 362–383. [CrossRef]
12. Ferreira, F.B.; Topa Gomes, A.; Vieira, C.S.; Lopes, M.L. Reliability Analysis of Geosynthetic-Reinforced Steep Slopes. *Geosynth. Int.* **2016**, *23*, 301–315. [CrossRef]
13. Hird, C.C.; Kwok, C.M. Finite Element Studies of Interface Behaviour in Reinforced Embankments of Soft Ground. *Comput. Geotech.* **1989**, *8*, 111–131. [CrossRef]
14. Rowe, R.K.; Li, A.L. Reinforced Embankments over Soft Foundations under Undrained and Partially Drained Conditions. *Geotext. Geomembr.* **1999**, *17*, 129–146. [CrossRef]
15. Rowe, R.K.; Gnanendran, C.T.; Landva, A.O.; Valsangkar, A.J. Calculated and Observed Behaviour of a Reinforced Embankment over Soft Compressible Soil. *Can. Geotech. J.* **1996**. [CrossRef]
16. Balakrishnan, S.; Viswanadham, B.V.S. Evaluation of Tensile Load-Strain Characteristics of Geogrids through in-Soil Tensile Tests. *Geotext. Geomembr.* **2017**, *45*, 35–44. [CrossRef]
17. Zheng, G.; Yu, X.; Zhou, H.; Wang, S.; Zhao, J.; He, X.; Yang, X. Stability Analysis of Stone Column-Supported and Geosynthetic-Reinforced Embankments on Soft Ground. *Geotext. Geomembr.* **2020**, *48*, 349–356. [CrossRef]
18. Rowe, R.K.; Soderman, K.L. An Approximate Method for Estimating the Stability of Geotextile-Reinforced Embankments. *Can. Geotech. J.* **1985**. [CrossRef]
19. Deng, M. Reliability-Based Optimization Design of Geosynthetic Reinforced Embankment Slopes. Ph.D. Thesis, Missouri S&T, Rolla, MO, USA, 2015.
20. Sharma, P.; Mouli, B.; Jakka, R.S.; Sawant, V.A. Economical Design of Reinforced Slope Using Geosynthetics. *Geotech. Geol. Eng.* **2020**, *38*, 1631–1637. [CrossRef]
21. Rowe, R.K.; Li, A.L. Geosynthetic-Reinforced Embankments over Soft Foundations. *Geosynth. Int.* **2005**, *12*, 50–85. [CrossRef]
22. Derghoum, R.; Meksaouine, M. Coupled Finite Element Modelling of Geosynthetic Reinforced Embankment Slope on Soft Soils Considering Small and Large Displacement Analyses. *Arab. J. Sci. Eng.* **2019**, *44*, 4555–4573. [CrossRef]
23. Avesani Neto, J.O.; Bueno, B.S.; Futai, M.M. Evaluation of a Calculation Method for Embankments Reinforced with Geocells over Soft Soils Using Finite-Element Analysis. *Geosynth. Int.* **2015**, *22*, 439–451. [CrossRef]
24. Mehrjardi, G.T.; Ghanbari, A.; Mehdizadeh, H. Experimental Study on the Behaviour of Geogrid-Reinforced Slopes with Respect to Aggregate Size. *Geotext. Geomembr.* **2016**, *44*, 862–871. [CrossRef]
25. Tandjiria, V.; Low, B.K.; Teh, C.I. Effect of Reinforcement Force Distribution on Stability of Embankments. *Geotext. Geomembr.* **2002**, *20*, 423–443. [CrossRef]
26. Mulabdić, M.; Kaluđer, J.; Minažek, K.; Matijević, J. *Priručnik za Primjenu Geosintetika u Nasipima za Obranu od Poplava (Manual for Application of Geosynthetics in Flood Protection Embankments)*; Faculty of Civil Engineering Osijek: Osijek, Croatia, 2016.
27. Moraci, N.; Cardile, G.; Gioffrè, D.; Mandaglio, M.C.; Calvarano, L.S.; Carbone, L. Soil Geosynthetic Interaction: Design Parameters from Experimental and Theoretical Analysis. *Transp. Infrastruct. Geotech.* **2014**, *1*, 165–227. [CrossRef]

28. Skejic, A.; Medic, S.; Ivšić, T. Numerical Investigations of Interaction between Geogrid/Wire Fabric Reinforcement and Cohesionless Fill in Pull-out Test. *Gradevinar* **2020**, *72*, 237–252. [CrossRef]
29. Strata Systems, Inc. *Reinforced Soil Slopes and Embankments*; Strata Systems, Inc.: Cumming, GA, USA, 2010.
30. Yamanouchi, T.; Fukuda, N. Design and Observation of Steep Reinforced Embankments. In Proceedings of the Third International Conference on Case Histories in Geotechnical Engineering, St. Louis, MO, USA, 1–6 June 1993.
31. GEO. *Guide to Reinforced Fill Structure and Slope Design*; Geotechnical Engineering Office, Civil Engineering and Development Department, HKSAR Government: Hong Kong, China, 2017; p. 218.
32. Wolff, T.F. Reliability of levee systems. In *Reliability-Based Design in Geotechnical Engineering*; Phoon, K.-K., Ed.; Taylor & Francis: Abingdon, UK, 2008; pp. 448–496.
33. Kirca, V.S.O.; Kilci, R.E. Mechanism of Steady and Unsteady Piping in Coastal and Hydraulic Structures with a Sloped Face. *Water* **2018**, *10*, 1757. [CrossRef]
34. Bujakowski, F.; Falkowski, T. Hydrogeological Analysis Supported by Remote Sensing Methods as a Tool for Assessing the Safety of Embankments (Case Study from Vistula River Valley, Poland). *Water* **2019**, *11*, 266. [CrossRef]
35. Martinović, K.; Reale, C.; Gavin, K. Fragility Curves for Rainfall-Induced Shallow Landslides on Transport Networks. *Can. Geotech. J.* **2017**, *55*, 852–861. [CrossRef]
36. Li, Y.; Qian, C.; Fu, Z.; Li, Z. On Two Approaches to Slope Stability Reliability Assessments Using the Random Finite Element Method. *Appl. Sci.* **2019**, *9*, 4421. [CrossRef]
37. MacKillop, K.; Fenton, G.; Mosher, D.; Latour, V.; Mitchelmore, P. Assessing Submarine Slope Stability through Deterministic and Probabilistic Approaches: A Case Study on the West-Central Scotia Slope. *Geosciences* **2019**, *9*, 18. [CrossRef]
38. Far, M.S.; Huang, H. A Hybrid Monte Carlo-Simulated Annealing Approach for Reliability Analysis of Slope Stability Considering the Uncertainty in Water Table Level. *Procedia Struct. Integr.* **2019**, *22*, 345–352. [CrossRef]
39. Hu, H.; Huang, Y.; Chen, Z. Seismic Fragility Functions for Slope Stability Analysis with Multiple Vulnerability States. *Environ. Earth Sci.* **2019**, *78*, 690. [CrossRef]
40. Pan, Q.; Qu, X.; Wang, X. Probabilistic Seismic Stability of Three-Dimensional Slopes by Pseudo-Dynamic Approach. *J. Cent. South Univ.* **2019**, *26*, 1687–1695. [CrossRef]
41. Wang, L.; Wu, C.; Gu, X.; Liu, H.; Mei, G.; Zhang, W. Probabilistic Stability Analysis of Earth Dam Slope under Transient Seepage Using Multivariate Adaptive Regression Splines. *Bull. Eng. Geol. Environ.* **2020**, *79*, 2763–2775. [CrossRef]
42. Zhang, S.; Li, Y.; Li, J.; Liu, L. Reliability Analysis of Layered Soil Slopes Considering Different Spatial Autocorrelation Structures. *Appl. Sci.* **2020**, *10*, 4029. [CrossRef]
43. Rossi, N.; Bačić, M.; Kovačević, M.S.; Librić, L. Development of Fragility Curves for Piping and Slope Stability of River Levees. *Water* **2021**, *13*, 738. [CrossRef]
44. Kennedy, R.P.; Cornell, C.A.; Campbell, R.D.; Kaplan, S.; Perla, H.F. Probabilistic Seismic Safety Study of an Existing Nuclear Power Plant. *Nucl. Eng. Des.* **1980**, *59*, 315–338. [CrossRef]
45. DePoto, W.; Gindi, I. *Hydrology Manual*; Los Angeles County Department of Public Works: Los Angeles, CA, USA, 1991.
46. DePoto, W.; Gindi, I. *Hydrology Manual*; Los Angeles County Department of Public Works: Los Angeles, CA, USA, 1993.
47. Jasim, F.H.; Vahedifard, F. Fragility Curves of Earthen Levees under Extreme Precipitation. In Proceedings of the Geotechnical Frontiers 2017 GSP 278, Orlando, FL, USA, 12–15 March 2017; American Society of Civil Engineers: Orlando, FL, USA, 2017; pp. 353–362.
48. Hasofer, A.M.; Lind, M.C. An Exact and Invariant First Order Reliability Format. *J. Eng. Mech.* **1974**, *100*, 111–121.
49. Singh Arora, J. Additional Topics on Optimum Design. In *Introduction to Optimum Design*; Academic Press: Cambridge, MA, USA, 2017; pp. 795–849. ISBN 978-0-12-800806-5.
50. Phoon, K.-K. Numerical recipes for reliability analysis—A primer. In *Reliability-Based Design in Geotechnical Engineering*; Phoon, K.-K., Ed.; Taylor & Francis: Abingdon, UK, 2008; pp. 1–75.
51. Rackwitz, R. Reliability Analysis—A Review and Some Perspectives. *Struct. Saf.* **2001**, *23*, 365–395. [CrossRef]
52. Baecher, G.B.; Christian, J.T. The Hasofer-Lind Approach (FORM). In *Reliability and Statistics in Geotechnical Engineering*; Wiley: Chichester, UK, 2003; pp. 377–397.
53. Sia, A.H.I.; Dixon, N. Distribution and Variability of Interface Shear Strength and Derived Parameters. *Geotext. Geomembr.* **2007**, *25*, 139–154. [CrossRef]
54. Yu, Y.; Bathurst, R.J. Influence of Selection of Soil and Interface Properties on Numerical Results of Two Soil–Geosynthetic Interaction Problems. *Int. J. Geomech.* **2017**, *17*, 04016136. [CrossRef]
55. Jewell, R.A. Application of Revised Design Charts for Steep Reinforced Slopes. *Geotext. Geomembr.* **1991**, *10*, 203–233. [CrossRef]
56. Pendola, M.; Mohamed, A.; Lemaire, M.; Hornet, P. Combination of Finite Element and Reliability Methods in Nonlinear Fracture Mechanics. *Reliab. Eng. Syst. Saf.* **2000**, *70*, 15–27. [CrossRef]
57. Librić, L.; Kovačević, M.S.; Ivoš, G. Determining of Risk Ranking for Otok Virje – Brezje Levee Reconstruction. In Proceedings of the ICONHIC2019, Chania, Greece, 23–26 June 2019.
58. Hughes, S.A.; Nadal, N.C. Laboratory Study of Combined Wave Overtopping and Storm Surge Overflow of a Levee. *Coast. Eng.* **2009**, *56*, 244–259. [CrossRef]
59. Hughes, S.A.; Shaw, J.M.; Howard, I.L. Earthen Levee Shear Stress Estimates for Combined Wave Overtopping and Surge Overflow. *J. Waterw. Port Coast. Ocean. Eng.* **2012**, *138*, 267–273. [CrossRef]

60. Xu, Y.; Li, L.; Amini, F. Slope Stability of Earthen Levee Strengthened by Roller-Compacted Concrete under Hurricane Overtopping Flow Conditions. *Geomech. Geoengin.* **2013**, *8*, 76–85. [CrossRef]
61. Hewlett, H.W.M.; Boorman, L.A.; Bramley, M.E. *Design of Reinforced Grass Waterways*; Construction and Industry Research and Information Association: London, UK, 1987; p. 116.
62. Powledge, G.R.; Ralston, D.C.; Miller, P.; Chen, Y.H.; Clopper, P.E.; Temple, D.M. Mechanics of Overflow Erosion on Embankments. II: Hydraulic and Design Considerations. *J. Hydraul. Eng.* **1989**, *115*, 1056–1075. [CrossRef]
63. Hughes, S.A. *Combined Wave and Surge Overtopping of Levees: Flow Hydrodynamics and Articulated Concrete Mat Stability*; USACE, Engineer Research and Developement Center, Coastal and Hydraulics Laboratory: Vicksburg, MS, USA, 2008.
64. Wu, A. *Locating General Failure Surfaces in Slope Analysis via Cuckoo Search*; Rocscience Inc.: Toronto, ON, Canada, 2012.
65. Krušelj, Ž. Katastrofalne Poplave u Koprivničkoj i Đurđevačkoj Podravini 1965., 1966. i 1972. Godine. *Podrav. Časopis Za Multidiscip. Istraživanja* **2017**, *16*, 5–35.
66. Radni Obilazak Gradilišta Vodoopskrbnog Sustava Općine Cestica i Nasipa Otok Virje—Brezje | Hrvatske Vode. Available online: https://www.voda.hr/hr/radni-obilazak-gradilista-vodoopskrbnog-sustava-opcine-cestica-nasipa-otok-virje-brezje (accessed on 22 February 2021).
67. Suzuki, M.; Koyama, A.; Kochi, Y.; Urabe, T. Interface Shear Strength between Geosynthetic Clay Liner and Covering Soil on the Embankment of an Irrigation Pond and Stability Evaluation of Its Widened Sections. *Soils Found.* **2017**, *57*, 301–314. [CrossRef]
68. Vaníček, I.; Jirásko, D.; Vaníček, M. *Modern Earth Structures for Transport Engineering*; Taylor & Francis: London, UK, 2020; ISBN 978-0-367-20834-9.
69. Ko, D.; Kang, J. Experimental Studies on the Stability Assessment of a Levee Using Reinforced Soil Based on a Biopolymer. *Water* **2018**, *10*, 1059. [CrossRef]
70. Ko, D.; Kang, J. Biopolymer-Reinforced Levee for Breach Development Retardation and Enhanced Erosion Control. *Water* **2020**, *12*, 1070. [CrossRef]
71. Hufenus, R.; Rüegger, R.; Flum, D.; Sterba, I.J. Strength Reduction Factors Due to Installation Damage of Reinforcing Geosynthetics. *Geotext. Geomembr.* **2005**, *23*, 401–424. [CrossRef]
72. Bathurst, R.J.; Huang, B.; Allen, T.M. Analysis of Installation Damage Tests for LRFD Calibration of Reinforced Soil Structures. *Geotext. Geomembr.* **2011**, *29*, 323–334. [CrossRef]
73. Bathurst, R.J.; Huang, B.-Q.; Allen, T.M. Interpretation of Laboratory Creep Testing for Reliability-Based Analysis and Load and Resistance Factor Design (LRFD) Calibration. *Geosynth. Int.* **2012**, *19*, 39–53. [CrossRef]
74. Miyata, Y.; Bathurst, R.J. Reliability Analysis of Geogrid Installation Damage Test Data in Japan. *Soils Found.* **2015**, *55*, 393–403. [CrossRef]
75. Ang, A.H.-S.; Tang, W.H. *Probability Concepts in Engineering Planning and Design, Volume 1: Basic Principles*; Wiley: New York, NY, USA, 1975; Volume 1.
76. Orr, T.L.L.; Breysse, D. Eurocode 7 and reliability-based design. In *Reliability-Based Design in Geotechnical Engineering*; Taylor & Francis: Abingdon, UK, 2008.
77. Phoon, K.-K.; Nadim, F. Modeling Non-Gaussian Random Vectors for FORM: State-of-the-Art Review. In Proceedings of the International Workshop on Risk Assessment in Site Characterization and Geotechnical Design, Indian Institute of Science, Bangalore, India, 26 November 2004; pp. 55–85.
78. R P Sigm_fit. MATLAB Central File Exchange. 2016. Available online: Https://www.Mathworks.Com/Matlabcentral/Fileexchange/42641-Sigm_fit) (accessed on 24 November 2020).
79. CISM Courses and Lectures; Griffiths, D.V.; Fenton, G.A. (Eds.) *Probabilistic Methods in Geotechnical Engineering*; Springer: New York, NY, USA, 2007.

water MDPI

Article

Development of Fragility Curves for Piping and Slope Stability of River Levees

Nicola Rossi *, Mario Bačić, Meho Saša Kovačević and Lovorka Librić

Faculty of Civil Engineering, University of Zagreb, 10000 Zagreb, Croatia; mbacic@grad.hr (M.B.); msk@grad.hr (M.S.K.); llibric@grad.hr (L.L.)
* Correspondence: nrossi@grad.hr; Tel.: +385-1-4639-491

Abstract: The design code Eurocode 7 relies on semi-probabilistic calculation procedures, through utilization of the soil parameters obtained by in situ and laboratory tests, or by the means of transformation models. To reach a prescribed safety margin, the inherent soil parameter variability is accounted for through the application of partial factors to either soil parameters directly or to the resistance. However, considering several sources of geotechnical uncertainty, including the inherent soil variability, measurement error and transformation uncertainty, full probabilistic analyses should be implemented to directly consider the site-specific variability. This paper presents the procedure of developing fragility curves for levee slope stability and piping as failure mechanisms that lead to larger breaches, where a direct influence of the flood event intensity on the probability of failure is calculated. A range of fragility curve sets is presented, considering the variability of levee material properties and varying durations of the flood event, thus providing crucial insight into the vulnerability of the levee exposed to rising water levels. The procedure is applied to the River Drava levee, a site which has shown a continuous trend of increased water levels in recent years.

Keywords: levee; slope stability; piping; overtopping; fragility curves; Monte Carlo simulation

Citation: Rossi, N.; Bačić, M.; Kovačević, M.S.; Librić, L. Development of Fragility Curves for Piping and Slope Stability of River Levees. *Water* **2021**, *13*, 738. https://doi.org/10.3390/w13050738

Academic Editors: Jihn-Sung Lai and Miguel Á. Toledo

Received: 7 January 2021
Accepted: 3 March 2021
Published: 9 March 2021

1. Introduction

As earthen structures constructed for the purpose of flood defence, the levees should be verified for several potential failure modes. According to Wolff [1], these include overtopping, slope stability, external erosion, underseepage and through-seepage, with the latter two being considered as internal erosion mechanisms. These failure modes are conditioned by the levee's geometrical configuration, its material properties, and overall hydraulic conditions of the site. Based on the examined breach characteristics of hundreds of failures, Özer et al. [2] identified the external erosion as the most frequent for levees, while failures due to internal erosion and instability are less frequent but lead to larger breaches, and as such are emphasized within this study. Of all the internal erosion mechanisms, backward erosion piping is considered to be the primary failure mechanism for levees [3], and even accounts for one-third of all piping failures that occurred in the last century [4]. Various design situations such as rainfall, high water level, seismic peak ground acceleration, etc., can be triggering factors for one or more failure mechanisms, directly or indirectly. Extensive studies have been conducted with various approaches regarding slope stability, Figure 1a, with respect to rainfall [5,6], high water levels [7,8], and peak ground accelerations [9,10], as well as combinations of various events [11]. Regarding the piping failure, Figure 1b, and depending on the mechanism of soil particle removal (e.g., removal of particles by water forces, chemical dispersion of clays, migration of fine material through coarse matrix, etc.), various modes are identified, all pertaining to internal erosion under or through the levee [12–15].

The analysis of different levee failure mechanisms within the Eurocode 7 design code [16] is based on the use of recommended, singular values of partial safety factors

(PSFs), with a defined combination of PSFs for action and resistivity (material), depending on the adopted calculation approach for a specific design situation. The code, however, prescribes constant values of PSFs for limit states, with no variation depending on the nature or the duration of the design situation and no recommendation regarding the target reliability values. Du Thinh [17] notes that, during the design process, an engineer must select a set of characteristic values and the corresponding PSFs, hoping to obtain in the end a design that satisfies a prescribed reliability level. On the other hand, the design code Eurocode 0 [18] provides minimum values for the reliability index for three consequence classes, but these are only defined for buildings, not for geotechnical structures. Some other design codes, such as those in [19], acknowledge the uncertain nature of soils, by defining target values of p_f and equivalent reliability indices for three consequence levels, based on random finite-element analyses.

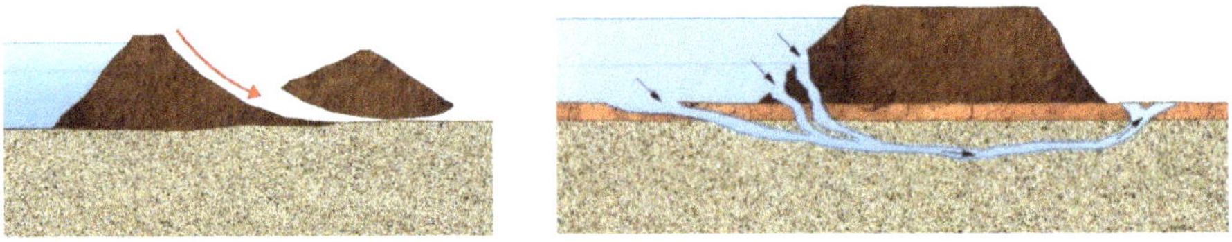

(a) (b)

Figure 1. Levee failure mechanisms analysed in the study: slope instability (**a**) and internal erosion (**b**).

Concerning the soil related uncertainties, Phoon and Kulhawy [20] identified three main sources of geotechnical uncertainties: (1) inherent soil variability, which describes the variation of properties from one spatial location to another, (2) measurement error, which implies the scatter of measurements on presumably homogeneous soil volumes, and (3) transformation uncertainty, where, in the process of model characterization, which includes linking the on-site and laboratory test results to the design parameters, some degree of uncertainty is introduced. By implementing a Eurocode 7 semi-probabilistic approach, which utilizes statistical methods to select characteristic values of geotechnical parameters, both spatial correlations between the same parameter sat different sampling points and cross-correlations between different parameters at the same sampling point are neglected [21]. Further, depending on the associated failure mechanisms of levees, different material parameters will control the limit states and different models are necessary to predict the resistance, and thus no uniform reliability level can be obtained with the load and resistance factor design method [22]. The degree of uncertainty involved in calculation of levees is especially high for slope stability [23] and piping mechanisms [24]. Even though the geotechnical community has been more progressive in the implementation of different probability-based methods in analyses of levees, understanding of levee failure mechanisms is still limited [25], while their behaviour during critical conditions mostly remains uncertain.

This paper contributes to the efforts of levee vulnerability evaluations, through the demonstration of a methodology for calculation of fragility curves for relevant failure mechanisms of slope stability and piping. Among the many available probabilistic methods [26], this study adopts the Monte Carlo simulation to determine the levee probability of failure when the hydraulic head rises on the riverside up to the levee crown and over, to simulate overflowing. Even though this method takes the most time to run due to its slow convergence, which is its largest disadvantage, it gives the most accurate results when a sufficient number of runs is chosen. For relatively simple calculations such as the limit equilibrium formulation, the number of runs and computation time to solve the problem are acceptable. Additionally, given that it is applicable to both linear and particularly to nonlinear problems [26,27], with many random variables which may be differently distributed [26], this method is used in this study. The method does not identify the relative contribution of each random variable to the safety factor, as some other methods do (e.g., FOSM), but for this purpose sensitivity analyses were conducted. The demonstrated methodology, applied to the river levee in Croatia, results in sets of fragility curves, which can then be used in risk assessment and categorization of levees [28,29], based on calculated probabilities of failures. This provides a support for the decision making process regarding the optimization of resources for levee reconstructions or maintenance [30]. Furthermore, future design protocols and monitoring activities of levees can be enhanced [31].

2. Methodology for the Development of Fragility Curves

To assess the vulnerability of the levee exposed to raising water levels up to and over the levee crown, with respect to identified failure mechanisms of landside slope stability and piping in the foundation soil, a series of numerical simulations were conducted. Within these simulations, the water level on the riverside is raised until the levee is sure to fail ($p_f \approx 1$). Overtopping (i.e., overflow) is usually a result of a high-water event (surge) or it can occur due to wave overtopping. The combined effect of surge and wave overflow is discussed by many authors [32–34]. However, if a river levee is considered, only a surge type of overflow is relevant. The results of numerical simulations feed into the proposed methodology of fragility curve development, giving insight into the probability of failure relative to the design event intensity. Sensitivity analyses indicate the influence of certain parameters on the fragility curves' shapes—i.e., on the variation of the failure probability.

2.1. Slope Stability Evaluation

Slope stability analyses can generally be conducted by limit equilibrium (LEM) and numerical methods incorporated in many commercially available programs, where each method has its own pros and cons [35]. As one of the oldest methods for slope stability calculations, the LEM has been significantly modified, from the introduction of the circular sliding surface [36] to its enhanced versions [37,38], and is still one of the most used methods for slope stability analyses.

Opposite to the deterministic approach which searches and pinpoints a critical slip surface with the lowest factor of safety, many probabilistic studies, such as multivariate adaptive regression splines analysis (Wang et al. [8]), utilize slope stability methods to identify a slip surface with the highest probability of failure. To give an overview of different possibilities and the results they yield, Akbas and Huvaj [39] compared results of probabilistic slope stability analyses by using LEM with integrated Latin Hypercube and Monte Carlo simulation, and by a numerical finite element method (FEM) with integrated Rosenblueth's point estimate method, as well as random finite element method analyses. They found that FEM analyses resulted in higher probabilities of failure.

This study utilizes LEM and Monte Carlo simulations to conduct series of probabilistic slope stability analyses. The initial total stress state and the pore pressure distribution from steady and transient seepage analyses are modelled separately for each water level increment using FEM, with triangle mesh element sizes of 0.2 m in the body, and 0.5 m elsewhere. The results of both analysis types are tested with different element shapes and sizes, and the resulting distributions are unchanged. Both steady state and transient two-dimensional seepage analyses are governed by a partial differential equation (Equation (1)), where the term on the right-hand side is equal to zero for the steady state.

$$\frac{\partial}{\partial x}\left(k_x \frac{\partial H}{\partial x}\right) + \frac{\partial}{\partial y}\left(k_y \frac{\partial H}{\partial y}\right) + Q = \frac{\partial \theta}{\partial t} \tag{1}$$

where H (m) is the total head, k_i (m/s) is the hydraulic conductivity in the i direction, Q $(m^3/s/m^2)$ is the applied boundary flux, and θ $(-)$ is the volumetric water content.

The stress states as well as the water pressures feed into the LEM for slope stability calculation, as shown on the diagram in Figure 2.

A reason for generating a stress state separately is to yield more realistic results by defining a stress state with stress concentrations closer to the levee toe, instead of calculating it as the product of unit weight and depth. The advantages of defining the stress state in this way do not come to fore with a low angle of levee slope where a low stress concentration can be expected; however, a significant difference is evident in the case of levee overflow where shear stress can be applied on the surface of the slope and the stress state adjusted accordingly. Thus, for consistency reasons, all the analyses are conducted using this procedure. Another benefit of separate generation of the stress state is a drastic reduction in calculation time, since the LEM utilizes an iterative procedure to find the interslice forces and thus requires multiple calculations to find the safety factor for just one slip surface. With the imported stress state, the stresses are already defined so the safety factor can be immediately calculated for a trial slip, which is significant considering the number of runs required to conduct a Monte Carlo simulation.

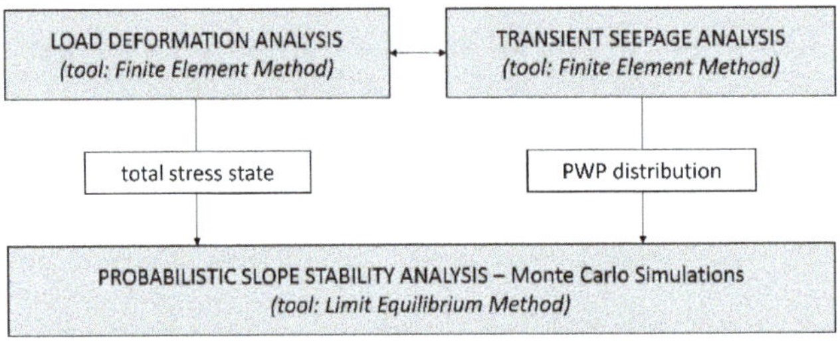

Figure 2. Workflow of probabilistic analysis of levee slope stability.

Since the intention of this study is to inspect the probability of failure of the levee due to the water rising, the fragility curves were constructed by incrementally increasing the water level from the levee toe to the levee crown, and over for the case of surge overflow, by calculating the pore pressure distribution from seepage analyses. The overflow was simulated by applying the equivalent shear stress, caused by water flow, along the crown and landside slope, while keeping the water level at the crown height for the free water surface generation through the levee body. When the complete fragility curve was obtained, based on the input of parameters with probabilistic distribution, a sensitivity analysis followed. This included varying the values of parameters, while keeping the other parameters at their means, to assess their influence on the shape of the curve and stability of the levee. The investigated parameters are the statistics of the strength parameters, the permeability, and the duration of the flood. For the latter, additional transient flow analyses were conducted with various water level durations, and the results were then once incorporated into the LEM calculations. The results of these variations are shown as new fragility curves, shifted to the left or right of the ones from the mean analyses.

2.2. Internal Erosion (Piping) Evaluation

The most prominent trigger for internal erosion is the high-water event, and as such this has been subject of many studies. To investigate the erosional behaviour of soil at the microscale (granular) and macroscale (levee), various methods have been used from physical models [40,41]—numerical simulations such as FEM, FDM, DEM [13–15,42,43], the material point method [44] and random lattice models [4]. Other tools such as neural networks were also used to help predict soil behaviour under seepage forces based on laboratory and field tests [12,45,46]. Despite many advantages of these advanced tools, the geotechnical community in many cases still relies on the simple empirical or semiempirical rules [47].

Within this study the closed-form analytical Sellmeijer 2-force rule [46], which resulted from a neural network based on field and laboratory tests and numerical analyses, was used for piping analyses. This approach is used in many state-of-the-art levee risk assessment methodologies, such as the VNK2 approach [47,48]. The closed-form solution predicts that for a given head difference a pipe of specific length will form. Once the critical head difference value is reached (ΔH_{crit}), the pipe will start to progress continuously until failure:

$$\frac{\Delta H_{crit}}{L} = F_R \cdot F_G \cdot F_S \tag{2}$$

In Equation (2), L (m) is a seepage path, equal to the width of the levee, and the factors F are the resistance, geometry, and scale term, respectively, and are functions of unit weight, drag force factor, angle of repose, relative density, effective grain size (d_{70}), kinematic viscosity, coefficient of uniformity and particle angularity. The critical head difference needs to be higher than the actual head difference reduced by the value of $0.3 \cdot D_{blanket}$ (thickness of the top clay layer that covers the aquifer), assuming the levee lays on a clay cover over the underlaying sandy aquifer. The parameter which was used as the random variable is the hydraulic conductivity of the aquifer, while all the other values were kept constant at their measured mean values or suggested mean values for the parameters for which measurements or correlations were not available. To assess the validity of the results obtained by the Sellmeijer's equation, the levee and the subsoil geometry and parameters should fall within certain limitations for which the rule was developed. The levee should lay on top of a homogeneous sandy aquifer of finite thickness, with horizontal ground surface in the cross-section direction [47]. Some guidelines [31] suggest applying the Sellmeijer method only if the thickness of the aquifer is less than the seepage length. Regarding the range of the parameters, Sellmeijer and Koenders [49] note that the routine is stable over the entire range of practically feasible parameters, while Sellmeijer et al. [42] give ranges for some new parameters introduced in the formula. The ratio of seepage length to hydraulic head difference for which the formula should be applied is $L/\Delta H > 10$.

3. Case Study Example: River Drava Levee

River Drava, with the overall length of 710 km, flows from Italy to eastern Croatia where it merges with Danube, and is historically known for major flood events [50], where prominent events have occurred in the last several years. The case study levee stretches across 6.8 km of the Drava old riverbed, from county Selnica to accumulation lake Dubrava. The levee is fragmented into three segments because of the presence of two smaller rivers, Bednja and Plitvica, flowing perpendicularly to the Drava (Figure 3).

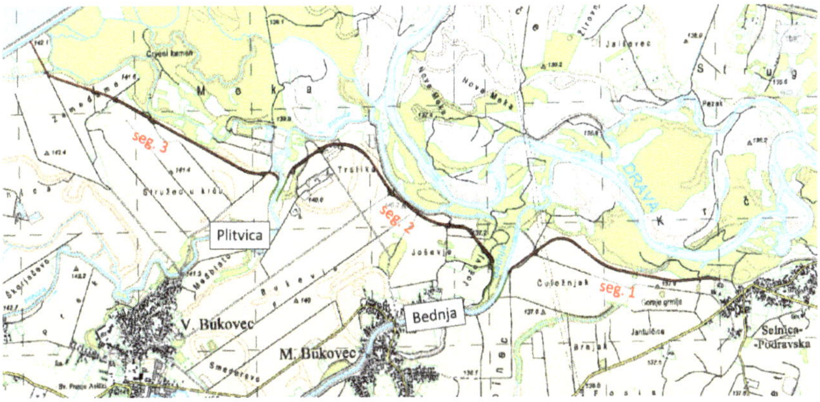

Figure 3. An overall layout of the Selnica–Dubovica levee with its distinctive segments.

The reach of interest for this study is defined by height and is the starting section of second segment, just after the Bednja river, where the levee's highest cross-section is present (Figure 4). By the request of the stakeholders, the designed crown level is 0.5 m above the 100-year high water, while the crown is 4.0 m wide. The levee slopes are at 1:3 and the service road is located on the levee's landside toe.

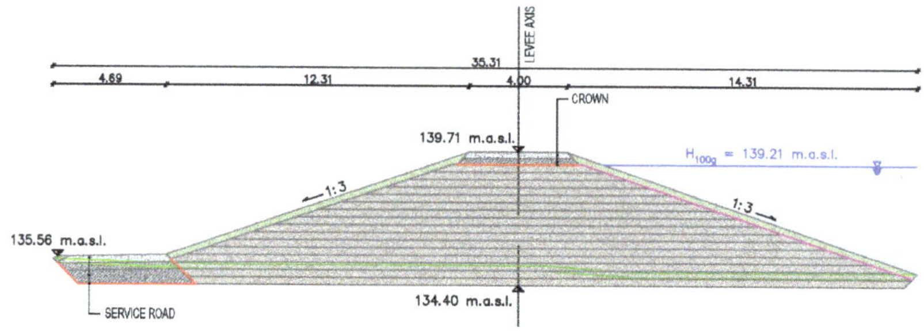

Figure 4. A cross-section of the case study levee.

3.1. Conducted Investigation Works

To obtain insight into the layering and physical-mechanical characteristics of the subsoil, an extensive geotechnical investigation campaign was conducted, consisting of 12 boreholes at equal spacings along with conduction of Standard Penetration Tests (SPTs), 12 Cone Penetration Tests with pore water measurements (CPTus) and 12 seismic refraction geophysical profiles. Both undisturbed and disturbed samples were taken during the geotechnical drilling, which were tested in a laboratory to determine their physical and mechanical characteristics. In addition to the in situ and laboratory direct test results, transformation models were implemented to relate the on-site and laboratory test results with the design parameters to infer geotechnical properties from indirect measurements. Based on these field and laboratory investigation works, a reliable geotechnical model of the subsoil was formed for the reach of interest regarding stratification and soil parameters to conduct calculations.

The subsoil was divided into a top layer of lower permeability underlined by the thick coarse-grained layer. At the location of the analysed cross-section, the upper low permeability layer was not detected ($D_{blanket} = 0$) and only coarse-grained soil was identified. Considering the investigation data scattering because of the mentioned inherent soil variability, measurement error and transformation uncertainty, to develop fragility curves by the means of probabilistic analyses some variation in the soil parameters, need to be considered.

3.2. Probabilistic Characterization of Soil Parameters

Since the soil parameter distributions can vary significantly, they should be limited to keep the values in the realm of possibility for a specific soil, thus avoiding illogical values. Phoon and Kulhawy [20] give a detailed literature review showing the ranges and number of samples for certain obtained statistics. These values for the internal friction angle are taken as guidelines for specifying the limits of their distributions. Since the slope stability of the levee is governed dominantly by the body and berm materials, these materials were probabilistically evaluated. Effective cohesion has seldom been reported in the literature on soil parameters variability, where it is considered as either normally or log-normally distributed, with CoV values similar to those for undrained shear strength reported in the literature [51–53]. This study assumes log-normal distribution and the commonly accepted CoVs, thus only the upper limit would be needed. However, as the mean values are already very low, the range for the distribution of $\pm 5\sigma$ is acceptable. Various authors reported that neglecting the correlation coefficient between cohesion and internal friction angle yields conservative results in slope stability calculations if their correlation is actually negative [54,55]. Results from tests conducted by Lumb [54] show a strong negative correlation for the compacted samples, and since the levee in the case study is compacted during construction, a negative correlation, thus a correlation of zero, can be assumed. Some sources suggest using some other value for the correlation coefficient [55,56]. The values and statistics (μ as mean value and CoV as coefficient of variation) of each random variable for stability analysis were assumed from the literature [20,27,57] and are shown in Table 1. As the levee was constructed from materials from an undefined borrow site, the mean values of the levee soil parameters were obtained during the deterministic design phase of the levee, such that the stability criteria were met (Table 3) and represent the minimum required values that the materials must have to deterministically ensure slope stability.

Table 1. Parameters and the statistics used for slope stability analyses.

Material	γ [kN/m^3]	φ [°]		c [kPa]		k_x [m/s] (Mean)		k_y/k_x [−]
		μ [kPa]	CoV [−]	μ [kPa]	CoV [−]	SDC1	SDC2	
Levee body	18	26	0.15	2	0.30	1×10^{-8}	1×10^{-8}	0.5
Crown and berm	20	30	0.12	1	0.30	1×10^{-4}	1×10^{-8}	0.5
Foundation soil	19	36	-	0	-	1×10^{-5}		0.5
Distribution	constant	normal		log-normal		constant		constant

Two stability design cases (SDCs) were analysed regarding hydraulic conductivity, one where the crown/berm is constructed from a more permeable material, and the other where the whole levee is homogeneous—i.e., the crown/berm and the body have the same conductivity.

Hydraulic conductivities for the subsoil were obtained from correlations with CPTu tests, while the hydraulic conductivities for levee materials were determined from deterministic steady seepage analyses, as minimally required to maintain hydraulic stability of levee in terms of the critical exit hydraulic gradient and free water surface position. To assess the sensitivity of the slope stability to the hydraulic conductivity of the levee body material, an arbitrary value of two orders of magnitude was selected since the range of reported hydraulic conductivity's CoV is too great to assume a value (30–750%) [26], while the foundation's and crown/berm's conductivities were kept the same as those defined for each SDC. For the piping analysis, the statistics of the hydraulic conductivity of the subsoil for case study location were estimated from the correlation with the CPTu test and assumed a log-normal distribution [58,59]. Since the CPTu showed a mixed subsoil profile with lenses of fine-grained soil, the profile was idealized to just one layer with one highly variable hydraulic conductivity. The distribution was described by the median (1×10^{-5} m/s) and the extremely high CoV due to the soil profile idealization by averaging over the whole CPTu profile. However, if it is assumed that the water flows around the fine-grained soil lenses, the conductivity distribution can be defined only for the sandy material, where the median is similar (3×10^{-5} m/s), but the variation is significantly decreased. Taking this into consideration, the piping calculations were conducted for two piping design cases (PDCs)—first (PDC1), where the subsoil is modelled as a homogeneous soil layer with highly variable hydraulic conductivity estimated from CPTu data for every material in the subsoil, and the second (PDC2), where the hydraulic conductivity of the layer was estimated only from the CPTu data for sand. Parameters chosen for piping analyses are shown in Table 2.

The hydraulic anisotropy ratio, defined as the ratio of the vertical to horizontal conductivity, can vary over a wide range of values [60]. As the levee materials are usually compacted dry of optimum [61], and soils dry of optimum have lower hydraulic anisotropy [62], the used anisotropy ratio values are reduced mean values from the literature [60]. The relative density (D_r) has been obtained via correlation with an SPT test.

The distribution of the hydraulic conductivity was defined through the median instead of the mean value since the parameter values vary over a few orders of magnitude, so the median was more intuitive and simpler to obtain as it is the geometric mean of the available data.

Table 2. Parameters and their statistics used for piping analyses.

Material	k_x [m/s]		D_r [%]	k_y/k_x [−]
	Median [m/s]	**CoV [−]**		
Foundation soil	PDC1		80	0.5
	1×10^{-5}	32.5		
	PDC2			
	3×10^{-5}	5.0		
Distribution	log-normal		constant	constant

4. The Probabilistic Analyses Background

The methodology for the development of fragility curves for levee stability implies that the proper stress state, as well the proper water pressure state, is established.

To obtain the appropriate stress state for the slope stability analyses, the methodology suggests conduction of the load deformation total stress analyses. Since soil plastification is not relevant to this study, the linear-elastic constitutive model was used, and this required input of soil stiffness and unit weights, as well Poisson's ratio, whose variation was not considered, even though it may have had some effect on the results [63]. Further, for the design situations, which include water level up to the top of levee crown, numerical analyses were carried out, including commonly used boundary conditions of properly defined hydraulic heads on the riverside and landside. However, if the water level is higher—that is, if surge overflow is considered—the stress analyses should be supplemented with the additional boundary shear stress along the crown and landside slope. Given that this aspect goes beyond standard analyses, a cautious evaluation of these shear stresses is required. The boundaries of both analyses were defined far from the levee region, enough to not affect the results. The constraints of load deformation analyses consisted of fixing movement of lateral soil elements of the model in the horizontal direction, and the bottom elements in two perpendicular directions. Seepage analyses require only hydraulic boundary conditions, which in this case consist of constant or varying hydraulic head values applied on lateral boundaries and on the top boundary of the model, up to the required height. During surge overflow, the water velocity increases down the slope until a terminal velocity is reached at equilibrium between water momentum and slope frictional resistance, after which the flow becomes steady and the velocity can be calculated by the following equation:

$$v_0 = \left[\frac{\sqrt{\sin\theta}}{n}\right]^{3/5} \cdot q_0^{2/5}\ [\mathrm{m/s}] \quad (3)$$

where v_0 (m/s) is the steady flow velocity, θ (°) is the landside slope angle, n (−) is Manning's coefficient, and q_0 (m^2/s) is the steady discharge [32]. For supercritical flow which develops on the landside slope, Figure 5, Hewlett et al. [64] proposed a value of Manning's coefficient of n = 0.02, relevant for slopes of 1:3.

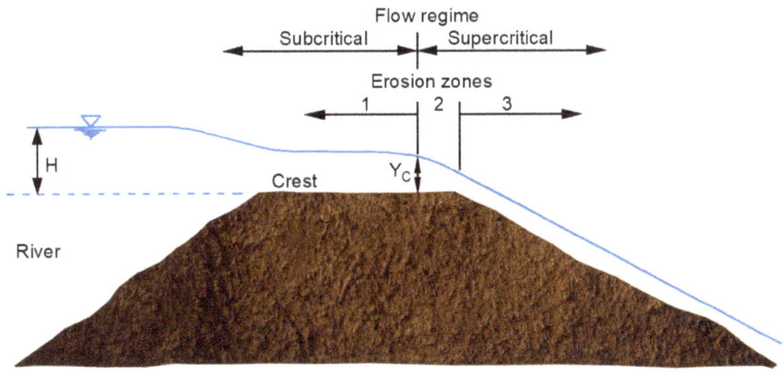

Figure 5. Flow regimes during overflow of a dam, redrawn from [65].

The discharge over the levee crown can be calculated using the equation for flow over a broad-crown weir, which gives slightly conservative results due to not taking into consideration frictional losses [66]:

$$q = \left(\frac{2}{3}\right)^{3/2} \cdot \sqrt{g} \cdot h_1^{3/2} \left[m^2/s\right] \quad (4)$$

where g $\left(m/s^2\right)$ is the gravitational acceleration and h_1 (m) is the upstream head (elevation over the levee crown). If steady flow is assumed, the discharge is constant along the slope. Therefore, the height of water is perpendicular to the slope in the steady, uniform flow area for unit length of the levee and can be calculated from Equations (3) and (4) as:

$$h = \frac{q}{v_0} \ [m] \quad (5)$$

Finally, when steady, uniform flow is reached, the shear stress from surge overflow is equal to:

$$\tau_0 = \gamma_w \cdot h \cdot \sin\theta \ [kPa] \quad (6)$$

where γ_w $\left(kN/m^3\right)$ is the unit weight of water. Equation (6) conservatively overestimates results since the resulting pressure is a little bit higher than the pressure in area above the steady flow [34]. Such calculated shear stress is applied along the crown and landside slope, as shown in Figure 6 for the case study numerical model.

With the full stress state properly defined, for all water levels including the surge overflow, stability analyses aim to find the critical slip surface out of the number of generated slip surfaces. To generate several slip surfaces, as well as to evaluate their safety margins, this study adopted a "Grid and Radius" method incorporated into the commercial software GeoStudio [67]. With this method, a grid of slip centres and a grid of slip tangents are created, and these define the number of analysed slip surfaces. However, a larger number of defined slip surfaces yields much longer calculation times, and conducting a single deterministic analysis prior to the probabilistic analyses is recommended. This analysis was conducted with a large grid covering a large area for potential slip surface centres, and a relatively large tangent grid to also cover a large range of possible surface depths. After the critical surface was found, the grid size and number of centre points and tangents were reduced around the critical surface point and tangent, to make smaller, denser grids and possibly find more critical surfaces around the original critical surface, while also minimizing the run time of the probabilistic calculations. This does not guarantee that the critical surface—defined in this case as the surface with maximum probability of failure—will be the same one as the deterministic critical surface, but it is a reasonable starting assumption that it will at least be close to the deterministic surface.

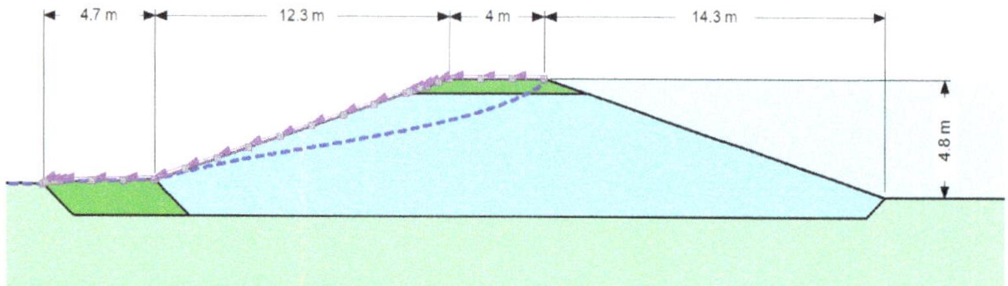

Figure 6. Numerical model for analyses of case study levee.

For the probabilistic slope stability analyses, four random variables were assigned, and these include cohesion and angle of internal friction for both levee body material as well the crown road base and berm material. The variability of the soil can generally be modelled by a random field described by the CoV and scale of fluctuation [20]. As opposed to the creation of random fields, the method used in this study sampled a random variable for each material only once and then applied it to all the slices found in the corresponding material. This kind of simulation usually gives conservative results. Within this study, the number of Monte Carlo trials was sufficient to obtain a relatively constant value p_f for each water level, which has been estimated at 15,000.

In regard to piping mechanism, for each hydraulic head difference, 25,000 Monte Carlo simulations were performed using an excel spreadsheet and its built-in random number generator function.

As the aquifer thickness is required for the Sellmeijer rule, two runs were carried out with minimum thickness from soil investigations to values after which further increasing of thickness no longer affected results. Therefore, aquifer thickness in numerical analyses ranged from 5 to 50 m.

5. Results and Discussion

5.1. Fragility Curves for Levee Slope Stability

To serve as a benchmark for the full probability analyses results, results of semi-probabilistic analyses adopted in Eurocode 7 are shown in Table 3. The analysed numerical model included idealized subsoil layering with mean values of both hydraulic and strength parameters, determined from the laboratory and transformation model data. Based on the Eurocode 7 design approach 3 (DA3), the design strength parameters were obtained from characteristic values by applying a prescribed PSFs. Several design situations were analysed by following relevant norms and guidelines for geotechnical design [16,68–70], with stability evaluation for both riverside and landslide slopes in drained and undrained conditions. The slope stability was assessed by the means of the LEM utilizing the pore pressure distributions from seepage analyses.

Deterministically obtained factors of safety are higher than unity ones, with a safety margin from 10 to 60%, indicating the stable levee slopes for all relevant design situations. However, these analyses neglect the variability of strength parameters as the ones governing the obtained safety values. Therefore, full probabilistic analyses were conducted to develop fragility curves for the levee's landside slope stability.

Figure 7 shows the fragility curves for the levee's landside slope stability, through the relation of the hydraulic head on the river side vs. probability of failure. As the case study levee crown also acts as a road base, the crown material, constructed up to 0.5 m above the 100-return period high water, consisted of coarser material mixed with fines. Therefore, the hydraulic conductivity of this layer is higher than levee body material conductivity, so when the water goes over the 100-year water level, the free water surface shifts towards the landside slope yielding a more unfavourable situation. The presented fragility curves were developed for steady seepage and include stability evaluation for levee material conductivity of 10^{-8} m/s and for increased conductivity of 10^{-6} m/s, while the crown material conductivity was kept constant for each SDC. The curves with higher p_f, marked in blue, refer to the levee constructed with the more permeable crown layer (SDC1), while the curves with lower p_f, marked in red, refer to the crown constructed of same permeability material as the levee body (SDC2). For the 100-year water level event, at 139.21 m a.s.l., there was an abrupt increase in p_f for the case with the more permeable layer on top, while the curve smoothly increased for the case of homogeneous levee. While the curves of SDC1 show an approximate linear trend, the SDC2 fragility curves show a bilinear trend, with the intersection at head value of around 143 m a.s.l. The point of slope change indicates the sudden shift from deeper (> 2 m) to shallower (< 0.7 m) slip surfaces, which do not exist for SDC1 as there is a slow transition from deeper to shallower surfaces. The increase in the levee body conductivity by two orders of magnitude (1×10^{-6}) at first had a slight positive effect for SDC1 because of the smaller difference in conductivities, but afterwards the negative effect was evident for both design cases.

Table 3. Deterministic safety factors for the case study levee, exposed to various design situations.

Design Situation				Safety Factor
Low water	Riverside	Static	Drained	1.4
		Seismic (475-year RP)	Undrained	1.1
	Landside	Static	Drained	1.4
		Seismic (475-year RP)	Undrained	1.1
High water	Riverside	Static	Drained	1.6
		Seismic (475-year RP)	Undrained	1.6
	Landside	Static	Drained	1.3
		Seismic (475-year RP)	Undrained	1.1
Rapid drawdown			Drained	1.2

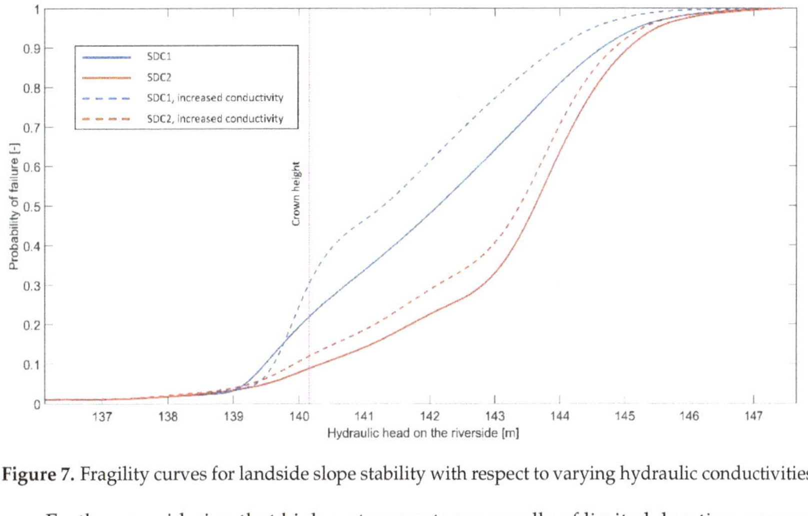

Figure 7. Fragility curves for landside slope stability with respect to varying hydraulic conductivities.

Further, considering that high-water events are usually of limited duration, preventing the development of a steady seepage, the fragility curves were further evaluated with consideration of transient seepage for a high-water event duration of 5 days. Failure probabilities for transient situations, up to the crown height, are shown in Figure 8. Even though the discrepancy in curves representing different levee conductivities is very low for 5-

day high water duration, it should be noted that the time required to numerically reach steady seepage with hydraulic conductivity 1×10^{-8} m/s is higher than 500 days, while for 1×10^{-6} m/s levee conductivity it is less than 50 days.

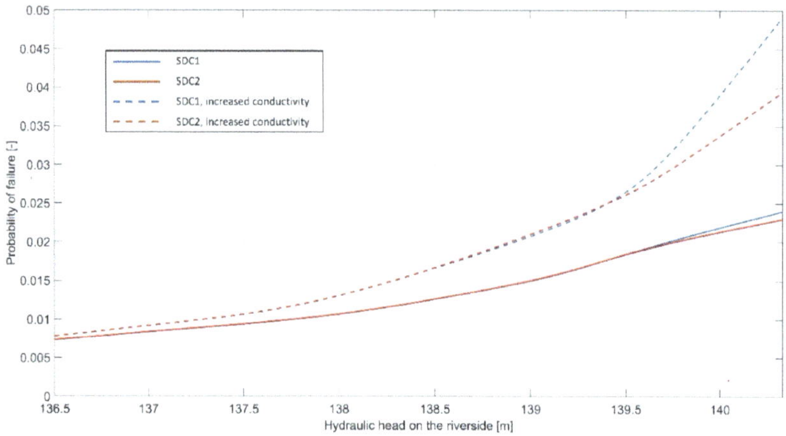

Figure 8. Fragility curves for transient seepage of 5-day duration.

The fragility curves for the varied statistics of strength parameters are given for various CoV values obtained by reducing the standard deviation. For both friction angle and cohesion, the CoVs are halved. Figure 9 shows that, by lowering the friction angle's standard deviations, the stability increased up to a certain hydraulic head value (marked with a point on the curves), after which the stability was reduced when compared to the original case with nonreduced variability. Such behaviour is expected since less variability means less probability of obtaining lower strength values, but also less probability of obtaining higher strength values which increase stability. Thus, for lower head values, when the slope is deterministically stable, less variability is favourable, while for higher head values, when the slope is deterministically unstable (or in equilibrium), less variability is unfavourable. It should be noted that for higher variability the distribution was truncated for a range of realistically possible values, while for lower variability the range was limited by the distribution itself and is lower than the truncated range for higher variability. For cohesion, the curves are practically unchanged, thus no point is marked on the corresponding curves in Figure 9.

Figure 10 shows the relation between reliability indices and probabilities of failure, for both calculated indices and their theoretical values for normally distributed safety factors. The figure shows results of the analysis with reduced friction angle variability, but all other calculations yielded similar curves. The reliability indices of critical slips were calculated as:

$$\beta = \frac{\mu - 1}{\sigma} \tag{7}$$

The numerically obtained curve shows very good concurrence with the theoretical value up to $\beta = 2.7$, which indicates that the safety factor follows a normal distribution. For lower p_fs (higher β), the numerically obtained curve deviates from the theoretical one, which might indicate that for lower probabilities of failure the safety factor no longer follows a normal distribution.

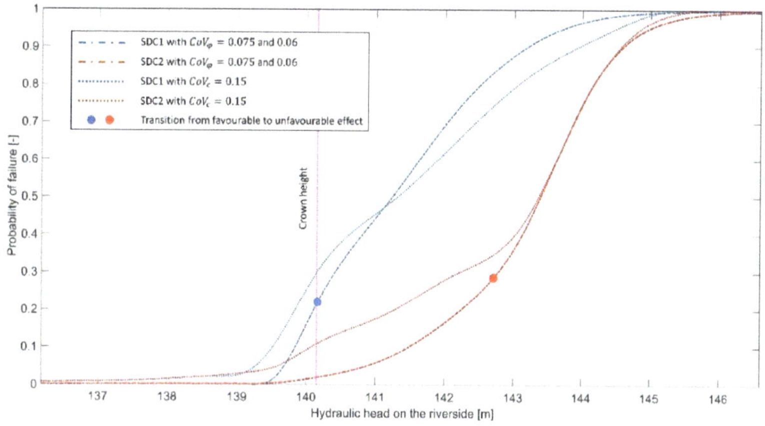

Figure 9. Fragility curves for variability of reduced strength parameters.

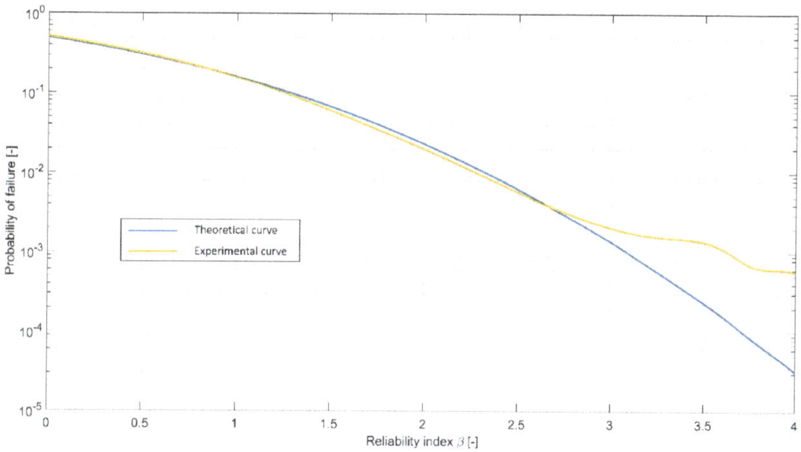

Figure 10. Probabilities of failure and respective reliability indices for levee slope stability.

5.2. Fragility Curves for Internal Erosion (Piping)

The piping analyses included hydraulic conductivity as a random variable, whereas simulations were conducted for two piping design cases (PDC1 and PDC2), depending on the procedure used to obtain the hydraulic conductivity distributions and statistics.

Additionally, to investigate the influence of aquifer thickness on the results, piping calculations included the deterministic variations of the thickness, starting from a 5 m value identified by the investigation works, to the value after which further increase does not affect the results (i.e., 50 m for the given analyses). Furthermore, the effective grain size d_{70} was varied between the minimum and maximum values (150 − 430 μm) which were used for the development of the Sellmeijer's model [46]. To assess the validity of the results obtained by Sellmeijer's equation, the levee and the subsoil geometry and parameters should fall within certain limitations for which the procedure is developed. Following the suggestion to apply the Sellmeijer's procedure only if the thickness of the aquifer is less than the seepage length [31], the maximum aquifer thickness of 35 m should actually be considered for the case study example. The curves in Figures 11 and 12 show the probability of failure for the water rising to the top of the crown for PDC1 and PDC2, respectively;

however, for the specified seepage length, the actual hydraulic head for which the formula still applies is around 1 m below the crown.

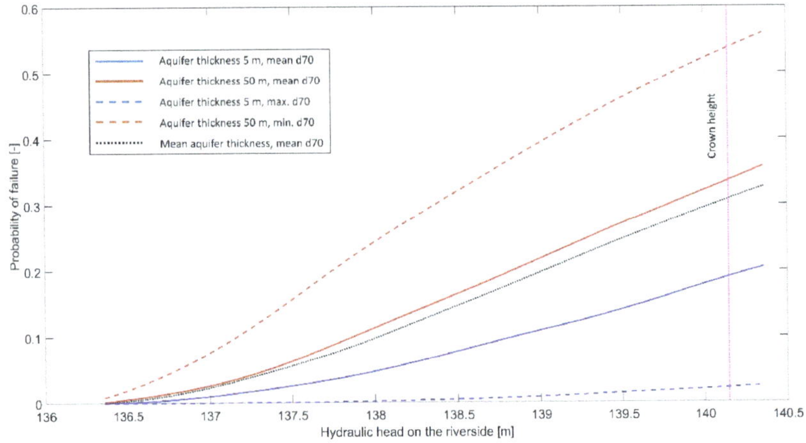

Figure 11. Probability of piping failure for piping design case 1 (PDC1).

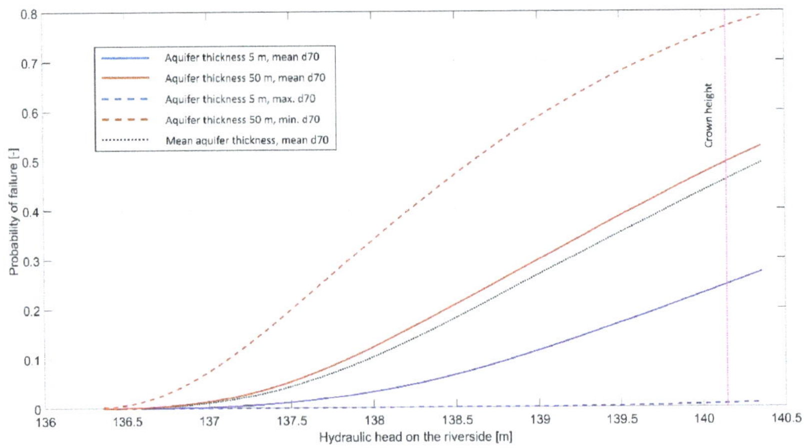

Figure 12. Probability of piping failure for PDC2.

The "real value" fragility curves for the analysed section are somewhere between the two extreme curves (dashed lines), which vary from p_f of only few percent up to the p_f of 50% for PDC1 and 75% for PDC2. This clearly demonstrates that quantities and quality of in situ and laboratory investigations, required to estimate the key parameters—i.e., aquifer thickness, d_{70}, and hydraulic conductivity are of paramount importance. Otherwise, the p_f for the backward erosion piping failure mechanism cannot be reliably estimated using the Sellmeijer 2-force rule. However, development of the shown curves provides a valuable insight into the effect that certain parameters have on the p_f. By analysing the mean curves for both design cases, as show in Figure 13, it can be expected that for smaller variations of the hydraulic conductivity, the probability of failure decreases in the lower range of hydraulic heads, but afterwards it drastically increases instead. The reason for this is the change in mean value which, even though is very subtle, significantly affects the results and seems to have much more impact on the p_f than the actual variability of the parameter.

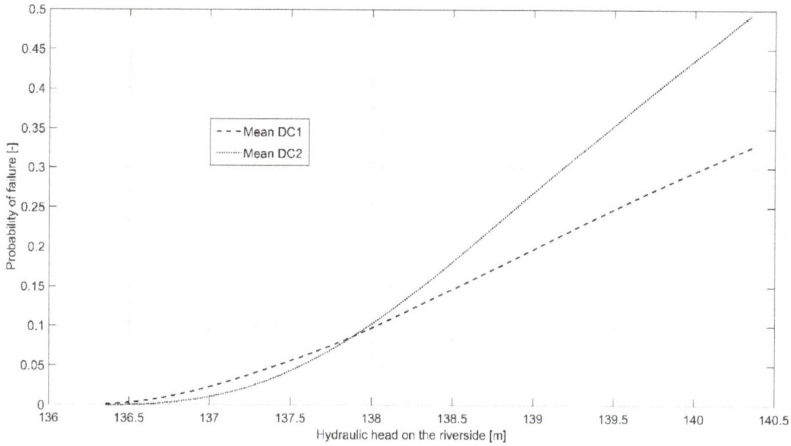

Figure 13. Mean curves for PDC1 and PDC2.

Overall, by utilizing USACE [28] classification and considering the 100-year flood event (139.21 m a.s.l.), the case study levee fits into the "hazardous" performance regarding piping mechanisms if the mean fragility curves are considered. Regarding the slope stability, for the worst-case scenario of steady seepage, the case study levee fits into the "poor" performance category. If the fragility curves for transient seepage of 5-day duration of the high-water event are analysed, then the probabilities of failure for slope stability indicate the levee have "below average" performances. With a lower variable friction angle, the situation significantly changes in favour of both SDCs, where the levee performance would be classified as "above average", while for the less variable cohesion the situation remains almost unchanged.

5.3. *Discussion on Calculation Assumptions and Recommendations for Future Work*

Several assumptions are considered for the sake of calculation simplification and/or because of lack of data. These assumptions, as well their effect on the calculation results, are discussed.

Considering that the levee will be constructed of material of an undefined borrow site, there are no soil investigations to compute its scale of fluctuation, which is therefore assumed as infinite, meaning that all points in the soil region have the same properties. This yields conservative reliability calculations of levee stability. Additionally, the hydraulic conductivity is assumed as constant, and only the effects of its mean value, without inherent variability, are investigated. To consider the variability of hydraulic conductivity, with extremely high range of CoV values as reported by Baecher and Christian [26], a random field seepage analysis should be implemented if Monte Carlo procedure is utilized.

Further, water table on the landside of the levee was fixed at the levee landside toe level and this raised the free water surface inside the levee body during the high-water event. Such a realistic assumption results in higher probabilities of failure. Further, this study considered water to affect slope stability only in terms of pore pressures which lower the shear strength of the material. However, rising and lowering water levels induce cumulative internal erosional effects, eventually leading to levee material degradation. Since this effect is more pronounced with an increasing number of flooding events, numerical models which consider the internal erosion propagation caused by water flow through soil [4,13,15,42–44] and its effects on mechanical and hydraulic properties [14] should be implemented in future probabilistic studies of levee stability.

Slope stability analyses were conducted with rising water levels until certain failure was reached. Van der Meer et al. [71] note that levees can endure an overflow of 1L/s (litre per second) if grass-cover is installed atop the crown and landside slope. However, it is unlikely that the landside slope with a clay and grass-cover will fail at discharges of less than 30 L/s [71]. If the latter is considered as representative for the case-study levee, and by utilizing Equation (4), such discharge occurs at a water height of around 7 cm above the crown, which means the slope would actually fail before reaching the hydraulic heads used for slope stability calculation during the overflow.

This probabilistic study calculates levee slope stability by utilizing finite element and limit equilibrium analyses, coupled with Monte Carlo simulations to determine the probability of failure. For each slip surface, a fixed number of calculations was used (in this case 15,000) with randomly sampled soil parameters according to the assigned distributions, providing the p_f of each slip surface. The slip surface with the highest p_f was then pinpointed as the critical slip surface. However, such a procedure might underestimate the probability of failure of the levee slope, as it considers only specific slip surfaces one by one, without considering the possibility of a different slip surface occurring for each different set of soil parameters. Running the multiple deterministic analysis with variations of soil parameter values will lead to different critical slip surfaces. A combination of shear strength parameters may generate deeper surfaces, while others may generate shallow ones. Thus, imposing a slip surface onto a set of parameters, instead of determining the slip surface based on the parameters, will yield a lower p_f. Combining the various deterministic critical slip surfaces from one Monte Carlo simulation would be a collection of the most critical slip surfaces for each set of random variables realizations and would result in the probability of failure of the levee, instead of a specific slip surface. The quantitative effect of this change in probability calculation procedure could be investigated in future studies.

6. Conclusions

To provide the probabilistic evaluation of stability and piping as a failure mechanism which lead to the larger levee breaches, this study proposes a methodology for the development of fragility curves which give an insight into the probability of failure for identified mechanisms with respect to the riverside water level, including the overflow surge. As the variability of soil parameters, resulting from inherent soil variability, measurement error and transformation uncertainty govern the shape of fragility curves, the necessity for proper selection of each parameter statistic is stresses out. The methodology for development of stability fragility curves is based on the fusion of different types of numerical analyses including the total stress load deformation analysis to obtain a reliable levee stress state and seepage analysis to obtain distribution of pore water pressures. The results of these two analyses feed into the probabilistic LEM analysis. Considering how computationally expensive Monte Carlo simulations are, the presented methodology minimises this disadvantage by combining the numerical analyses with LEM, which has the effect of decreasing the critical failure surface determination time. The necessity of separate stress analysis is additionally emphasized when overflow surge is considered, where equivalent shear stress, caused by water flow, should be applied. For the probabilistic evaluation of piping mechanism, the closed-form analytical Sellmeijer 2-force rule is the one being dominantly used in many state-of-the-art levee risk assessment methodologies.

The methodology was applied to a case study location of River Drava levee, a site which has shown a continuous trend of increased water levels in recent years. From the resulting fragility curves, it can be noted that the permeable crest layer affects stability substantially in cases where the water rises enough to start flowing through it. However, this effect becomes less notable as the ratio of the crown to body conductivities approaches 1. Moreover, if the duration of the high-water event is small enough so that it cannot achieve steady seepage through the levee, the effect also becomes less notable. Considering the soil variability, smaller variability offers increased stability up to a certain point, after which it has unfavourable effects. For the SDC1, this point was found at the crown height, but for SDC2 it was found only at hydraulic heads more than 2 m above the crown.

Regarding piping, even though values for parameters which were not available have been assumed based on correlations and recommendations, meaningful conclusions can still be obtained from the constructed curves. It has been shown how much the results can vary with changes to the investigated parameters—i.e., hydraulic conductivity statistics, effective grain size d_{70} and aquifer thickness, which emphasizes the importance of gathering relevant data for analyses. Additionally, the reduced variability of the hydraulic conductivity shows a favourable effect until a certain head height which depends on (not exclusively) d_{70} and aquifer thickness. After that point, the p_f increases. Since the mean value also changed together with the CoV, the effect shown in Figure 13 cannot be completely attributed to the change in variability, but by knowing the amount of change in each statistic their relative contribution is implied.

Since the proposed methodology includes several assumptions for the sake of calculation simplification and/or because of lack of data, this paper discusses them. However, with the lack of reliable data, conservative assumptions were usually made (e.g., higher soil parameter variability, longer flood duration, higher water levels, etc.). When each assumption introduces a small conservative effect, the effects stack and the probability of failure could be overestimated. With more data regarding variability of the levee and foundation soil's parameters, water levels and their durations, more reliable probabilistic analyses can be conducted.

Author Contributions: Conceptualization, N.R. and M.S.K.; methodology, N.R. and M.S.K.; software, N.R.; validation, N.R., M.B. and M.S.K.; formal analysis, N.R.; investigation, N.R., M.B., M.S.K. and L.L.; writing—original draft preparation, N.R. and M.B.; writing—review and editing, M.S.K. and L.L.; visualization, N.R.; supervision, M.S.K. All authors have read and agreed to the published version of the manuscript.

Funding: This research was funded by European Union Civil Protection Mechanism, under UCPM-2019-PP-AG call, Grant Agreement Number 874421, oVERFLOw project (Vulnerability assessment of embankments and bridges exposed to flooding hazard).

Conflicts of Interest: The authors declare no conflict of interest. The funders had no role in the design of the study; in the collection, analyses, or interpretation of data; in the writing of the manuscript, or in the decision to publish the results.

References

1. Wolff, T.F. Reliability of Levee Systems. In *Reliability-Based Design in Geotechnical Engineering*; Phoon, K.-K., Ed.; Taylor & Francis: Abingdon, UK, 2008; pp. 448–496.
2. Özer, I.E.; van Damme, M.; Jonkman, S.N. Towards an International Levee Performance Database (ILPD) and Its Use for Macro-Scale Analysis of Levee Breaches and Failures. *Water* **2020**, *12*, 119. [CrossRef]
3. USACE. *Investigation of Underseepage and Its Control, Lower Mississippi River Levees*; USACE: Vicksburg, MI, USA, 1956.
4. Fascetti, A.; Oskay, C. Dual Random Lattice Modeling of Backward Erosion Piping. *Comput. Geotech.* **2019**, *105*, 265–276. [CrossRef]
5. Martinović, K.; Reale, C.; Gavin, K. Fragility Curves for Rainfall-Induced Shallow Landslides on Transport Networks. *Can. Geotech. J.* **2017**, *55*, 852–861. [CrossRef]
6. Doumbouya, L.; Guan, C.S.; Bowa, V.M. Influence of Rainfall Patterns on the Slope Stability of the Lumwana (the Malundwe) Open Pit. *Geotech. Geol. Eng.* **2020**, *38*, 1337–1346. [CrossRef]
7. Far, M.S.; Huang, H. A Hybrid Monte Carlo-Simulated Annealing Approach for Reliability Analysis of Slope Stability Considering the Uncertainty in Water Table Level. *Procedia Struct. Integr.* **2019**, *22*, 345–352. [CrossRef]
8. Wang, L.; Wu, C.; Gu, X.; Liu, H.; Mei, G.; Zhang, W. Probabilistic Stability Analysis of Earth Dam Slope under Transient Seepage Using Multivariate Adaptive Regression Splines. *Bull. Eng. Geol. Environ.* **2020**, *79*, 2763–2775. [CrossRef]
9. Hu, H.; Huang, Y.; Chen, Z. Seismic Fragility Functions for Slope Stability Analysis with Multiple Vulnerability States. *Environ. Earth Sci.* **2019**, *78*, 690. [CrossRef]
10. Pan, Q.; Qu, X.; Wang, X. Probabilistic Seismic Stability of Three-Dimensional Slopes by Pseudo-Dynamic Approach. *J. Cent. South Univ.* **2019**, *26*, 1687–1695. [CrossRef]
11. Xiong, X.; Shi, Z.; Xiong, Y.; Peng, M.; Ma, X.; Zhang, F. Unsaturated Slope Stability around the Three Gorges Reservoir under Various Combinations of Rainfall and Water Level Fluctuation. *Eng. Geol.* **2019**, *261*, 105231. [CrossRef]
12. Kaunda, R.B. A Neural Network Assessment Tool for Estimating the Potential for Backward Erosion in Internal Erosion Studies. *Comput. Geotech.* **2015**, *69*, 1–6. [CrossRef]
13. Robbins, B.A. Numerical Modeling of Backward Erosion Piping. *Appl. Numer. Model. Geomech.* **2016**, *2016*, 551–558.
14. Aboul Hosn, R.; Sibille, L.; Benahmed, N.; Chareyre, B. A Discrete Numerical Model Involving Partial Fluid-Solid Coupling to Describe Suffusion Effects in Soils. *Comput. Geotech.* **2018**, *95*, 30–39. [CrossRef]

15. Yang, J.; Yin, Z.-Y.; Laouafa, F.; Hicher, P.-Y. Analysis of Suffusion in Cohesionless Soils with Randomly Distributed Porosity and Fines Content. *Comput. Geotech.* **2019**, *111*, 157–171. [CrossRef]
16. European Committee for Standardization (CEN). *Eurocode 7—Geotechnical Design-Part 1: General Rules*; CEN: Brussels, Belgium, 2004.
17. Thinh, K.D. Probabilistic Approach to the Design of Spread Footings: Application of Eurocode 7. In Proceedings of the 16th International Conference on Soil Mechanics and Geotechnical Engineering, Osaka, Japan, 12–16 September 2005; pp. 2797–2800.
18. European Committee for Standardization (CEN). *Eurocode—Basis of Structural Design*; CEN: Brussels, Belgium, 2005.
19. Canadian Standards Association (CSA). *Canadian Highway Bridge Design Code*; Canadian Standards Association: Mississauga, ON, Canada, 2014.
20. Phoon, K.-K.; Kulhawy, F.H. Characterization of Geotechnical Variability. *Can. Geotech. J.* **1999**, *36*, 612–624. [CrossRef]
21. Low, B.K.; Phoon, K.-K. Reliability-Based Design and Its Complementary Role to Eurocode 7 Design Approach. *Comput. Geotech.* **2015**, *65*, 30–44. [CrossRef]
22. Fenton, G.A.; Naghibi, F.; Dundas, D.; Bathurst, R.J.; Griffiths, D.V. Reliability-Based Geotechnical Design in the 2014 Canadian Highway Bridge Design Code. *Can. Geotech. J.* **2016**, *53*, 236–251. [CrossRef]
23. Hamedifar, H.; Bea, R.G.; Pestana-Nascimento, J.M.; Roe, E.M. Role of Probabilistic Methods in Sustainable Geotechnical Slope Stability Analysis. *Procedia Earth Planet. Sci.* **2014**, *9*, 132–142. [CrossRef]
24. Mazzoleni, M.; Bacchi, B.; Barontini, S.; Di Baldassarre, G.; Pilotti, M.; Ranzi, R. Flooding Hazard Mapping in Floodplain Areas Affected by Piping Breaches in the Po River, Italy. *J. Hydrol. Eng.* **2014**, *19*, 717–731. [CrossRef]
25. Schweckendiek, T.; Vrouwenvelder, A.C.W.M.; Calle, E.O.F. Updating Piping Reliability with Field Performance Observations. *Struct. Saf.* **2014**, *47*, 13–23. [CrossRef]
26. Baecher, G.B.; Christian, J.T. *Reliability and Statistics in Geotechnical Engineering*; Wiley: Hoboken, NJ, USA, 2003.
27. Nadim, F. Tools and Strategies for Dealing with Uncertainty in Geotechnics. In *Probabilistic Methods in Geotechnical Engineering*; Griffiths, D.V., Fenton, G.A., Eds.; CISM Courses and Lectures; Springer: Berlin/Heidelberg, Germany, 2007.
28. USACE. *Introduction to Probability and Reliability Methods for Use in Geotechnical Engineering*; Engineering and Design; USACE: Washington, DC, USA, 1997.
29. Santamarina, J.C.; Altschaeffl, A.G.; Chameau, J.L. Reliability of Slopes: Incorporating Qualitative Information. In *Rockfall Prediction and Control and Landslide Case Histories*; Transportation Research Record; Transportation Research Board: Washigton, DC, USA, 1992; pp. 1–5. ISBN 0-309-05206-8.
30. Simm, J.; Tarrant, O. Development of Fragility Curves to Describe the Performance of UK Levee Systems. In Proceedings of the Twenty-Sixth International Congress on Large Dams, Vienna, Austria, 1–7 July 2018.
31. TAW/ENW. *Technical Report on Sand Boils (Piping)*; Hydraulic Engineering Reports; Rijkswaterstaat: Amsterdam, The Netherlands, 1999.
32. Hughes, S.A.; Nadal, N.C. Laboratory Study of Combined Wave Overtopping and Storm Surge Overflow of a Levee. *Coast. Eng.* **2009**, *56*, 244–259. [CrossRef]
33. Hughes, S.A.; Shaw, J.M.; Howard, I.L. Earthen Levee Shear Stress Estimates for Combined Wave Overtopping and Surge Overflow. *J. Waterw. Port Coast. Ocean Eng.* **2012**, *138*, 267–273. [CrossRef]
34. Xu, Y.; Li, L.; Amini, F. Slope Stability of Earthen Levee Strengthened by Roller-Compacted Concrete under Hurricane Overtopping Flow Conditions. *Geomech. Geoengin.* **2013**, *8*, 76–85. [CrossRef]
35. Kaur, A.; Sharma, D.R.K. Slope stability analysis techniques: A review. *Int. J. Eng. Appl. Sci. Technol.* **2016**, *1*, 52–57.
36. Petterson, K.E. The Early History of Circular Sliding Surfaces. *Géotechnique* **1955**, *5*, 275–296. [CrossRef]
37. Morgenstern, N.R.; Price, V.E. The Analysis of the Stability of General Slip Surfaces. *Géotechnique* **1965**, *15*, 79–93. [CrossRef]
38. Spencer, E. A Method of Analysis of the Stability of Embankments Assuming Parallel Inter-Slice Forces. *Géotechnique* **1967**, *17*, 11–26. [CrossRef]
39. Akbas, B.; Huvaj, N. *Probabilistic Slope Stability Analyses Using Limit Equilibrium and Finite Element Methods*; IOS Press: Rotterdam, The Netherlands, 2015.
40. Planès, T.; Mooney, M.A.; Rittgers, J.B.R.; Parekh, M.L.; Behm, M.; Sneider, R. Time-Lapse Monitoring of Internal Erosion in Earthen Dams and Levees Using Ambient Seismic Noise. *Géotechnique* **2016**, *66*, 301–312. [CrossRef]
41. Hunter, R.P.; Bowman, E.T. Visualisation of Seepage-Induced Suffusion and Suffosion within Internally Erodible Granular Media. *Géotechnique* **2018**, *68*, 918–930. [CrossRef]
42. El Shamy, U.; Aydin, F. Multiscale Modeling of Flood-Induced Piping in River Levees. *J. Geotech. Geoenviron. Eng.* **2008**, *134*, 1385–1398. [CrossRef]
43. Liang, Y.; Jim Yeh, T.-C.; Wang, Y.-L.; Liu, M.; Wang, J.; Hao, Y. Numerical Simulation of Backward Erosion Piping in Heterogeneous Fields. *Water Resour. Res.* **2017**, *53*, 3246–3261. [CrossRef]
44. Yerro, A.; Rohe, A.; Soga, K. Modelling Internal Erosion with the Material Point Method. *Procedia Eng.* **2017**, *175*, 365–372. [CrossRef]
45. Sellmeijer, J.B. Numerical Computation of Seepage Erosion below Dams (Piping). In Proceedings of the 3rd International Conference on Scour and Erosion (ICSE-3), Amsterdam, The Netherlands, 1–3 November 2006.
46. Sellmeijer, H.; de la Cruz, J.L.; van Beek, V.M.; Knoeff, H. Fine-Tuning of the Backward Erosion Piping Model through Small-Scale, Medium-Scale and IJkdijk Experiments. *Eur. J. Environ. Civ. Eng.* **2011**, *15*, 1139–1154. [CrossRef]
47. Kramer, R. Piping Under Transient Conditions. Master's Thesis, University of Twente, Enschede, The Netherlands, 9 October 2014.
48. VNK2 Project Office Flood Risk in the Netherlands. *VNK2: The Method in Brief*; VNK2: Utrecht, The Netherlands, 2012.
49. Sellmeijer, J.B.; Koenders, M.A. A Mathematical Model for Piping. *Appl. Math. Model.* **1991**, *15*, 646–651. [CrossRef]

50. Krušelj, Ž. Katastrofalne Poplave u Koprivničkoj i Đurđevačkoj Podravini 1965., 1966. i 1972. godine. *Podrav. Časopis za Multidiscip. Istraživanja* **2017**, *16*, 5–35.
51. Schneider, H.R. *Definition and Determination of Characteristic Soil Properties*; Balkema: Hamburg, Germany, 1997.
52. Ahmed, A.; Soubra, A.-H. Extension of Subset Simulation Approach for Uncertainty Propagation and Global Sensitivity Analysis. *Georisk* **2012**, *6*, 162–176. [CrossRef]
53. Andreev, S.; Zhivkov Zhelyazkov, A. Probability of Failure of an Embankment Dam Due to Slope Instability and Overtopping: First Order Second Moment Method for Assessment of Uncertainty. In Proceedings of the 13th International Benchmark Workshop on Numerical Analysis of Dams, Lausanne, Switzerland, 9–11 September 2015; pp. 207–215.
54. Lumb, P. Safety Factors and the Probability Distribution of Soil Strength. *Can. Geotech. J.* **1970**, *7*, 225–242. [CrossRef]
55. Wolff, T.F. Analysis and Design of Embankment Dam Slopes: A Probabilistic Approach. Ph.D. Thesis, Purdue University, West Laffayete, IN, USA, 6 May 1985.
56. Correlation Coefficient (Cohesion and Friction Angle) in an SWedge Probabilistic. Available online: https://www.rocscience.com/help/swedge/swedge/Correlation_Coefficient.htm (accessed on 18 May 2020).
57. Phoon, K.-K.; Kulhawy, F.H. Evaluation of Geotechnical Property Variability. *Can. Geotech. J.* **1999**, *36*, 625–639. [CrossRef]
58. Fenton, G.A.; Griffiths, D.V. Statistics of Free Surface Flow through Stochastic Earth Dam. *J. Geotech. Eng.* **1996**, *122*, 427–436. [CrossRef]
59. Griffiths, D.V.; Fenton, G.A. The Random Finite Element Method (RFEM) in Stead Seepage Analysis. In *Probabilistic Methods in Geotechnical Engineering*; CISM Courses and Lectures; Springer: Berlin/Heidelberg, Germany, 2007; pp. 225–241.
60. Chapuis, R.P.; Gill, D.E. Hydraulic Anisotropy of Homogeneous Soils and Rocks: Influence of the Densification Process. *Bull. Int. Assoc. Eng. Geol.* **1989**, *39*, 75–86. [CrossRef]
61. Budhu, M. *Soil Mechanics and Foundations*, 3rd ed.; John Wiley & Sons, Inc.: Hoboken, NJ, USA, 2011.
62. Rahardjo, H.; Satyanaga, A.; Choon, L.E. Hydraulic Anisotropy Behavior of Compacted Soil. In Proceedings of the 19th International Conference on Soil Mechanics and Geotechnical Engineering, Seoul, Korea, 17–21 September 2017.
63. Fredlund, D.G.; Scoular, R.E.G. Using Limit Equilibrium Concepts in Finite Element Slope Stability Analysis. In Proceedings of the International Symposium on Slope Stability Engineering, IS-Shikoku '99, Matsuyama, Shikoku, Japan, 8–11 November 1999; pp. 31–47.
64. Hewlett, H.W.M.; Boorman, L.A.; Bramley, M.E. *Design of Reinforced Grass Waterways*; Construction and Industry Research and Information Association: London, UK, 1987; p. 116.
65. Powledge, G.R.; Ralston, D.C.; Miller, P.; Chen, Y.H.; Clopper, P.E.; Temple, D.M. Mechanics of Overflow Erosion on Embankments. II: Hydraulic and Design Considerations. *J. Hydraul. Eng.* **1989**, *115*, 1056–1075. [CrossRef]
66. Hughes, S.A. *Combined Wave and Surge Overtopping of Levees: Flow Hydrodynamics and Articulated Concrete Mat. Stability*; USACE, Engineer Research and Developement Center, Coastal and Hydraulics Laboratory: Washington, DC, USA, 2008.
67. GEO-SLOPE. *Stability Modeling with GeoStudio*; GEO-SLOPE International Ltd.: Calgary, AB, Canada, 2017.
68. European Committee for Standardization (CEN). *Eurocode 8—Design of Structures for Earthquake Resistance–Part. 1: General Rules, Seismic Actions and Rules for Buildings*; CEN: Brussels, Belgium, 2004.
69. European Committee for Standardization (CEN). *Eurocode 8–Design of Structures for Earthquake Resistance–Part. 5: Foundations, Retaining Structures and Geotechnical Aspects*; CEN: Brussels, Belgium, 2004.
70. Pickles, A.; Sandham, R. *Application of Eurocode 7 to the Design of Flood Embankments (C749)*; CIRIA: London, UK, 2014; ISBN 978-0-86017-754-8.
71. Van der Meer, J.W.; ter Horst, W.L.A.; van Valzen, E.H. Calculation of Fragility Curves for Flood Defence Assets. In *Flood Risk Management: Research and Practice*; Taylor & Francis Group: London, UK, 2009; pp. 567–573.

MDPI
St. Alban-Anlage 66
4052 Basel
Switzerland
Tel. +41 61 683 77 34
Fax +41 61 302 89 18
www.mdpi.com

Water Editorial Office
E-mail: water@mdpi.com
www.mdpi.com/journal/water

www.ingramcontent.com/pod-product-compliance
Ingram Content Group UK Ltd.
Pitfield, Milton Keynes, MK11 3LW, UK
UKHW060123010826
14089UKWH00006B/125

9 783036 556550